CHANGES IN SENSORY MOTOR BEHAVIOR IN AGING

ADVANCES IN PSYCHOLOGY

114

Editors:

G. E. STELMACH

P. A. VROON

ELSEVIER
Amsterdam – Lausanne – New York – Oxford – Shannon – Tokyo

CHANGES IN SENSORY MOTOR BEHAVIOR IN AGING

Edited by

Anne-Marie FERRANDEZ

CNRS URA 1166
Université de la Méditerranée
Marseille, France

Normand TEASDALE

Laboratoire de Performance Motrice Humaine
Université Laval
Quebec, Canada

1996

ELSEVIER
Amsterdam – Lausanne – New York – Oxford – Shannon – Tokyo

NORTH - HOLLAND
ELSEVIER SCIENCE B.V.
Sara Burgerhartstraat 25
P.O. Box 211, 1000 AE Amsterdam, The Netherlands

ISBN: 0 444 82101 5

This book is printed on acid-free paper.

Transferred to digital printing 2005

Preface

For the last two or three decades, studies on aging processes and age-related changes in behavior have been expanding considerably, probably due to the dramatic changes observed in the demographics. This increase in the overall age and proportion of elderly people has heightened the severity of problems related to the safety and well-being of elderly persons in everyday life. Many researchers working on motor control have thus focused more intensely on the effects of age on motor control. This new avenue of research has led to programs for alleviating or delaying the specific sensory-motor limitations encountered by the elderly (falls for example) in an attempt to make elderly people more autonomous.

The aggregation of studies from different perspectives is often fascinating, especially when the same field can serve as a common ground between researchers. Nearly all contributors to this book work on sensory-motor aging; they represent a large range of affiliations and backgrounds including psychology, neurobiology, cognitive sciences, kinesiology, neuropsychology, neuropharmacology, motor performance, physical therapy, exercise science, and human development.

Addressing age-related behavioral changes can also furnish some crucial reflections in the debate about motor coordination: aging is the product of both maturational and environmental processes, and studies on aging must determine how the intricate interrelationships between these processes evolve. The study of aging allows us to determine how compensatory mechanisms, operating on different subsystems and each aging at its own rate, compensate for biological degenerations and changing external demands. This book should contribute to demonstrating that the study of the aging process raises important theoretical questions.

In this book, some models of aging in motor control are presented. Greene and Williams, through a dynamic-system perspective, describe changes in coordination with aging. They focus mainly on how aging affects the coordination of movements with multiple degrees of freedom. They speculate on underlying neural mechanisms and non-neural control parameters which could account for contradictory evidence of both reduced and maintained coordination across the adult life span. According to this theory, aging may be viewed as a non-linear, thermodynamic process in which constraints are altered in ways that affect behavioral stability and the ability to cope with environmental demands. Jänicke and Coper discuss some areas of gerontological research on the basis of animal experiments: they endeavor to assess the possibilities, limits, and validity of animal tests for evaluating age-related changes in sensory-motor behavior. Studies on animals make it possible to systematically clarify the functional association of a sensory-motor behavior that diminishes with chronological age, and the delay in the reduction of performance due to physical training.

Many of the studies in this book are at least partially devoted to the control of balance and locomotion (Ferrandez, Durup, and Farioli; Greene and Williams; Hay; Hill and Vandervoort; Lajoie, Teasdale, Bard, and Fleury; Patla, Prentice, and Gobbi; Tang and Woollacott). This topic seems to have been a general trend for about fifteen years: researchers focus more and more on the coordination of multi-degree of freedom actions, rather than on unilateral and uniarticular movements. Moreover, this question is of particular interest in research on aging, insofar as inefficient control of balance and locomotion is often responsible for falls, so frequent in the elderly, and can have dramatic consequences on their autonomy.

Through various contributions, the book addresses the issue of behavioral plasticity. It is well known that one characteristic feature of aging is the loss of adaptability to environmental perturbations. Jänicke and Coper, and Greene and Williams discuss the reduced age-related ability to adapt. The general theme of adaptability is covered through the study of compensation strategies to counteract disturbances in the environment (Ferrandez, Durup, and Farioli; Hay; Patla, Prentice, and Gobbi) and of cognitive regulations in static balance and locomotion (Lajoie, Teasdale, Bard, and Fleury). The study of the effects of practice or training (Brown; Tang and Woollacott) and of adaptation to different levels of task complexity (Roy, Weir, and Leavitt) also shed some light on age-related adaptive behavior and plasticity. The question of how organisms (and especially humans) deal with the various degenerations that occur with increasing age is addressed by Brown, and by Hill and Vandervoort. These two studies consider how elderly people learn to cope with deficits in the motor system (cerebellar degeneration, or consequences of a stroke). One possible line of research consists of exploring how best to optimize neuromuscular function at all ages.

Slowness in cognitive and sensory-motor processes is a major characteristic of elderly people's behavior. This feature is highlighted in nearly all of the chapters in the book. Salthouse and Earles and Amrhein address the question of general or common factors contributing to age-related slowing. Salthouse and Earles examine the influence of health factors on the age-related slowing exhibited in simple measures of sensory-motor and perceptual speed. This study certainly contributes to discriminating between general and localized factors in the age-related slowing-down process. Amrhein supplies some new arguments to the debate over cognitive and sensory-motor slowing (general-slowing proponents versus localized-slowing proponents), by analyzing a wide range of data in tasks where reaction time and movement time have been measured.

The majority of the studies presented here were conducted on a healthy population. However, all researchers engaged in studies on aging are necessarily confronted with the problem of discriminating between pathological and physiological aging. Aging is accompanied by ever-increasing vulnerability which makes elderly subjects more likely to contract diseases and less able to resist. Because the probability of illness increases with age, how can we define "healthy elderly"? Does "normal aging" mean "free from disease" or "statistically normal"? These questions cannot be answered.

An increasingly large number of studies on aging involve a wide range of ages (from young adulthood or even childhood, to old elderly). This procedure is highly suited to improving our understanding of aging. Due to substantial interindividual differences, one needs both an extended scale of ages and a great number of subjects to investigate aging. Life-span studies are certainly destined to become more and more numerous. Considering the aging process as a part of the life-span development process is probably the most successful way to gain insight into the links between changes in age, vulnerability, and adaptation.

Acknowledgements

We are indebted to many individuals who helped us make this book a reality.

First, of course, we thank the contributing authors for their hard work and excellent chapters, and for their patience. They never complained when referees asked them to rewrite complete sections or do additional data processing. They always answered quickly when asked to provide better quality figures.

We gratefully acknowledge the assistance of Richard A. Abrams, Christine Assaiante, James E. Birren, Pierre. B. Boucher, John Cerella, John Dobbs, Sylvia Dobbs, Pertti Era, Michelle Fleury, Yves Girouard, Noreen Goggin, Jürgen Harting, Donald K. Ingram, Brian E. Maki, Jean Massion, Theo Mulder, Hajime Nakagawa, Jim G. Phillips, Jay Pratt, Ilari Pyykkö, Gregor Schöner, Albert B. Schultz, Deborah J. Serrien, Ann Shumway-Cook, Waneen W. Spirduso, Siegfried Stoll, Stephan Swinnen, Amy E. Tyler, and Carole P. Winstein, who reviewed the manuscripts.

We also thank Vivian E. Waltz for revising the preface and the chapters written by non-English speakers. She never failed to consider the emergency of the situation and gave this job priority each time.

Last, but not the least, we warmly thank Françoise Joubaud, managing editor of *Current Psychology of Cognition*, and *Revue de Neuropsychologie*. Since the contract called for delivery to the publisher of camera-ready copy, in a real sense, the printer of this book was Françoise Joubaud. She carried out many of the required tasks with constant diligence and professionalism. Her extended experience also proved highly fruitful in contacts with authors and referees. This book could definitely not have been achieved without her.

Anne-Marie Ferrandez and Normand Teasdale

Contributors

Paul C. Amrhein
Department of Psychology, University of New Mexico, Logan Hall, Terrace and Redondo Streets, NE, Albuquerque, NM 87131, U.S.A.

Chantal Bard
Université Laval, Laboratoire de Performance Motrice Humaine, PEPS, Québec, PQ G1K 7P4, Canada

Susan H. Brown
Center for Human Motor Research, Division of Kinesiology, University of Michigan, 401 Washtenaw Avenue, Ann Arbor, MI 48109-2214, U.S.A.

Helmut Coper
Free University of Berlin, Institute for Neuropsychopharmacology, Ulmenallee 30, 14050 Berlin, Germany

Madeleine Durup
Cognition et Mouvement, URA CNRS 1166, Université de la Méditerranée, IBHOP, Traverse Charles Susini, 13388 Marseille Cedex 13, France

Julie L. Earles
Department of Psychology, Furman University, Greenville, SC 29613, U.S.A.

Farioli Fernand
CREPCO, URA CNRS 182, Université de Provence, 13621 Aix-en-Provence Cedex 1, France

Ferrandez Anne-Marie
Cognition et Mouvement, URA CNRS 1166, Université de la Méditerranée, IBHOP, Traverse Charles Susini, 13388 Marseille Cedex 13, France

Michelle Fleury
Université Laval, Laboratoire de Performance Motrice Humaine, PEPS, Québec, PQ G1K 7P4, Canada

Lilian T. Gobbi
Neural Control Laboratory, Department of Kinesiology, University of Waterloo, Waterloo, Ontario N2L 3G1, Canada

Laurence S. Greene
University of Colorado at Boulder, Department of Kinesiology, Boulder, CO 80309-0354, U.S.A.

Laurette Hay
Laboratoire de Neurobiologie Humaine, URA CNRS 372, Université de Provence, Avenue Escadrille Normandie-Niemen, 13397 Marseille Cedex 20, France

Karen Hill
Group Health Centre, 240 McNabb Street, Sault Ste. Marie, Ontario P6B 1Y5, Canada

Bernhard Jänicke
Free University of Berlin, Institute for Neuropsychopharmacology, Ulmenallee 30, 14050 Berlin, Germany

Yves Lajoie
Université Laval, Laboratoire de Performance Motrice Humaine, PEPS, Québec, PQ G1K 7P4, Canada

Jack L. Leavitt
Department of Kinesiology, University of Windsor, Windsor, Ontario N9B 3P4, Canada

Aftab E. Patla
Neural Control Laboratory, Department of Kinesiology, University of Waterloo, Waterloo, Ontario N2L 3G1, Canada

Stephen D. Prentice
Département de Physiologie, Faculté de Médecine, Université de Montréal CP 6128, Succursale A, Montréal, Quebec H3C 3J7, Canada

Eric A. Roy
Department of Kinesiology, University of Waterloo, Waterloo, Ontario N2L 3G1, Canada

Timothy A. Salthouse
School of Psychology, Georgia Institute of Technology, Atlanta, GA 30332-0170, U.S.A.

Pei-Fang Tang
Department of Exercise and Movement Science, and Institute of Neuroscience, University of Oregon, Eugene, OR 97403-1240, U.S.A.

Normand Teasdale
Université Laval, Laboratoire de Performance Motrice Humaine, PEPS, Québec, PQ G1K 7P4, Canada

Anthony A. Vandervoort
University of Western Ontario, Department of Physical Therapy, London, Ontario N6G 1H1, Canada

Patricia L. Weir
Department of Kinesiology, University of Windsor, Windsor, Ontario N9B 3P4, Canada

Harriett G. Williams
University of South Carolina, Department of Exercise Science, Columbia, SC 29208, U.S.A.

Marjorie H. Woollacott
Department of Exercise and Movement Science, and Institute of Neuroscience, University of Oregon, Eugene, OR 97403-1240, U.S.A.

Contents

Changes in sensory motor behavior in aging
A.-M. Ferrandez and N. Teasdale (Editors)

AGE-RELATED SLOWING IN MOVEMENT PARAMETERIZATION STUDIES: NOT WHAT YOU MIGHT THINK

Paul C. AMRHEIN

University of New Mexico

Abstract

In this chapter, the nature of age-related slowing in speeded motor performance is explored. In particular, experiments assessing movement parameterization are reviewed. In these studies, specific movement parameters (e.g., arm, direction, extent) comprising a motor program are assessed concerning their preparation, maintenance, restructuring and execution within a movement plan. An advantage of movement parameterization studies is that they assess cognitive processing latency to assess a movement response (reaction time, RT) distinct from the latency to complete the movement response (movement time, MT). In general, most speeded tasks assess both of these latencies in aggregate (and refer to this aggregate latency as simply "RT"). As such, parameterization studies allow a test of prevailing response slowing theories of aging using components of task performance. Separate "Brinley plot" regressions of RT and Total Time (TT, TT = RT + MT) from these studies reveals additive slowing, but nominal (if any) multiplicative slowing. Moreover, the intercept difference between the best-fitting RT and TT lines validates the additive impact of MT in these studies. Even at a global level, these studies are inconsistent with claims of negligible additive slowing (i.e., small positive or negative intercept), but substan-

---

Correspondence should be sent to Paul C. Amrhein, Department of Psychology, University of New Mexico, Logan Hall, Terrace and Redondo Streets, NE, Albuquerque, NM 87131, U.S.A. (e-mail: amrhein@hydra.unm.edu).

tial multiplicative slowing (slope approximating 2.0) for "nonlexical" tasks espoused by General Slowing theorists (e.g., Lima, Myerson, & Hale, 1991). In addition, review of the individual studies indicates what the Brinley plot approach misses: Age Group × Condition interactions from some of these studies actually indicate speed increases in elderly relative to young subjects, due to apparent differences in parameter preparation maintenance and restructuring processes for the two age groups.

Key words: Aging, aimed movement, Brinley plot, movement time, reaction time, slowing.

INTRODUCTION

One of the staple, if not classic, methodologies used to study the effects of aging on human performance has been the reaction time (RT) task (see, e.g., Salthouse, 1985; Welford, 1959, 1977; Spirduso & MacRae, 1990). In particular, two reaction time tasks have been used extensively: Simple reaction time (SRT) and choice reaction time (CRT). Based, respectively, on Donders' (1869/1969) Type A and B tasks, they provide a means to separately assess age effects on sensory-motor and intervening cognitive processes (Dawson, 1988; Teichner & Krebs, 1974). As such, they provide a useful way to assess at a process level the pervasive response slowing seen in older persons (see Botwinick, 1984; Goggin & Stelmach, 1990; Welford, 1977).

In the SRT task, a pre-specified stimulus is presented and the subject responds with a pre-instructed response. (A variant of this task is where the stimulus is presented but subjects respond upon a latent "GO" signal; for example, see the delayed pronunciation task of Balota & Duchek, 1988.) By knowing the stimulus and the response to it, subjects are likely to prepare this response prior to actually receiving the stimulus (or "GO" signal). In a typical CRT task, subjects respond to one of a number of stimuli with a pre-instructed response unique to each potential stimulus. SRT and CRT tasks share perceptual and motor aspects in their task demands; that is, in both tasks (excluding the latent "GO" signal version), subjects must detect that a stimulus has been presented, and (including the latent "GO" signal version) the corresponding response must be prepared and executed. What distinguishes SRT and CRT tasks is the uncertainty concerning which stimulus is actually presented. Whereas there is no stimulus uncertainty for the SRT task, there is for the CRT task. As numerous studies have reported over the years, increases in this uncertainty yield corresponding increases in response latency across the adult lifespan (see, e.g., Kausler, 1991; Salthouse, 1985; Welford, 1959, 1977).

Many motor performance tasks are built upon SRT and CRT task methodologies. Indeed, SRT and CRT tasks typically require a manual (i.e., aimed movement) response. In the SRT task, response parameters (concerning which finger, hand, arm, foot or leg will be used) are prepared by the subject prior to target stimulus onset (see e.g., Amrhein, Stelmach, & Goggin, 1991). In the CRT task, by contrast, such preparation does not appear to occur (Amrhein et al., 1991; Klapp, Wyatt, & Lingo, 1974). Thus, SRT and CRT tasks actually represent two extremes on the scale of response preparation, and as such represent useful reference points when studying movement plan preparation, maintenance, restructuring and execution. While most SRT and CRT studies have assessed response initiation (reaction time, RT) and execution (movement time, MT) in aggregate (but still refer to the data as "RT" even though it might be better referred to as "Total Time", TT), there have been some studies which have used RT/MT assessment.

Methodologically, what distinguishes the larger set of "RT" from the smaller set of "RT/MT" studies is that subjects in the former set simply press a target button upon stimulus response, often with little experimental control over the initial resting location of their responding body part, whereas in the latter set, upon stimulus presentation, subjects release a button (often called a "Home button" or "Home key"), and then move to press a target button. In the aging literature, these RT/MT studies include: Amrhein et al. (1991), Amrhein, Von Dras, and Anderson (1993), Clarkson (1978), Goggin, Stelmach, and Amrhein (1989), Larish and Stelmach (1982), Spirduso (1975), Stelmach, Amrhein, and Goggin (1988), Stelmach, Goggin, and Amrhein (1988), Stelmach, Goggin, and Garcia-Colera (1987), Singleton (1954), Szafran (1951), and Welford (1959, 1977). Generally, these studies have revealed slower RTs and MTs for older (e.g., age range 50-87 years) relative to younger (e.g., age range 18-31 years) individuals. However, many of these studies failed to separate the role of visual guidance from motor performance. That is, subjects in the other studies were allowed to use vision to guide their movement responses. This is not a trivial problem; there is a sizable literature which documents perceptual-motor interaction (e.g., see Rosenbaum, 1991; Szafran, 1951). Accordingly, I will constrain the scope of this chapter to cover only those studies where the role of visual perception is limited to stimulus processing. Moreover, each of the studies reviewed assessed healthy, community dwelling elderly (age range 63-80 years) and young (age range 18-31 years) individuals. Also, subjects in these studies received sufficiently numerous trials to allow an assumption that both subjects attained their respective asymptotic levels of practice on the various tasks (see, e.g., Spirduso & MacRae, 1990, concerning differential practice effects preceding asymptotic performance).

Two popular aimed movement tasks in the aging literature are the movement plan specification and restructuring tasks. Common to both tasks is the manipulation of movement parameters such as arm, direction, extent, velocity, force, etc. Such parameters take on values which are specific to a generalized motor program that defines a particular pattern of physical activity (Schmidt, 1988). Latency to initiate (RT) and execute (MT) a planned movement is assessed, as well as errors which may occur for movement initiation and execution. Overall, both tasks have exhibited age-related slowing like that seen for the SRT/CRT tasks; this is not surprising, because these movement tasks also manipulate stimulus uncertainty in a fashion similar to SRT and CRT tasks (Amrhein et al., 1991). However, the age-related slowing observed in these movement tasks has rarely received statistical analysis beyond experiment-specific determination of elderly/young latency ratios (see Goggin & Stelmach, 1990). As will be detailed later, regression analysis across-experiments may offer a more detailed assessment of the loci of this slowing, specifically concerning "sensory-motor" and intervening "computational" processes (Cerella, 1990). Details of each task are presented below.

Movement plan specification task

The *movement plan specification task* involves a precuing paradigm in which the subject is given partial or complete parameter information concerning the impending movement response (e.g., use of the left arm in a movement away from oneself). RTs indicating additional latency to respond to a stimulus, relative to that for a completely specified response, are related to the parameters left to be specified for that stimulus after preceding partial specification of the other parameters by the precue stimulus (e.g., left arm is specified but direction – away or toward oneself – is not). However, MTs are expected to be immune from such parameter specification effects because execution of the movement plan assessed by MT is assumed to occur only after full parameter specification has taken place. Two studies which have investigated aging and movement plan specification are Stelmach et al. (1987) and Stelmach, Amrhein, and Goggin (1988).

Stelmach et al. (1987) presented elderly (67-74 years), middle age (40-49 years), and young (18-22 years) subjects with precue-target stimuli pairs. Using an eight-light (four row × two column) display, the position of the precue stimulus specified in varying levels of completeness, the values of three movement parameters, arm (left or right), direction (away or toward oneself), and extent (short or long), to be implemented in response to a subsequent target stimulus. This response was carried out by releasing either a left or right Home button and then moving to and pressing a target button which was compatible in position with the target stimulus.

Stelmach et al. (1987) found slower RTs and MTs for elderly relative to middle age subjects, which were in turn slower than those for young subjects. Elderly subjects were also slower in preparing non-precue specified parameter values relative to middle age and young subjects, who were equivalent. Interestingly, direction required more time to prepare relative to arm, which in turn required more time to prepare than extent.

Stelmach, Amrhein, and Goggin (1988) investigated bimanual and unimanual movement preparation and execution in elderly (67-75 years) and young (21-25 years) subjects. Subjects were presented with precue stimuli which indicated whether the impending target stimulus response would require the left or right arm or both arms, and whether lateral movement of the arm(s) would be a shorter or longer distance from the Home button(s). This meant that bimanual responses were either symmetric (same movement extent for left and right arms) or asymmetric (different movement extent for left and right arms).

These researchers found, beyond slower RTs and MTs for elderly subjects, that elderly subjects initiated long movements faster than short movements, unlike their younger counterparts who (weakly) exhibited the opposite pattern. Also, elderly subjects showed a greater MT increase from unimanual to bimanual movements relative to young subjects, although this pattern was not seen in the RT data. Arguably the most interesting finding, though, was that elderly subjects had poorer bimanual coordination than young subjects; they exhibited less symmetry both in bimanual movement initiation and subsequent completion, indicating that age impacts on the coordination of movement plan execution for two limbs.

Movement plan restructuring task

Based on the movement specification task, the *movement plan restructuring task* (or motor reprogramming task, see Rosenbaum & Kornblum, 1982) has the added characteristic of variable precue validity. Here, the precue specifies the target stimulus response with a biasing probability (75%-80%). Such a probability induces subjects to prepare the movement parameters, thus enabling a quick response on "valid precue" trials where precue and target stimuli match. On the "invalid precue" trials, however, the precue and target stimuli do not match, requiring subjects to "restructure" their planned movement responses. This restructuring can involve one or more parameters. Additional RT for these invalid precue trials over the valid precue trials has been shown to be due to attentional processing and restructuring cost for individual movement parameters (Amrhein et al., 1991, 1993). As was the case for the movement plan specification task, MTs are expected to be immune from such restructuring effects, because execution of the movement plan assessed by MT is assumed to occur only after full and final movement plan preparation has taken place. Several studies have employed this paradigm (e.g., Amrhein et al., 1991, 1993; Goggin et al., 1989; Larish & Stelmach, 1982; Stelmach, Goggin, & Amrhein, 1988). These studies will be discussed chronologically.

Larish and Stelmach (1982) presented elderly (M = 69.1 years) and young (M = 21.9 years) subjects with precue stimuli that matched subsequently presented target stimuli with varying probability (20%, 50% or 80%). Movement direction was also manipulated: the precue stimuli specified the direction of a possible movement response, left or right from a single Home button. As expected, elderly subjects exhibited slower RTs and MTs compared to the younger subjects. Finally, both groups exhibited equivalent increases in RT with corresponding decreases in precue validity (i.e., from 80% to 50% to 20%).

Using the apparatus of the movement plan specification task of Stelmach et al. (1987), Stelmach, Goggin, and Amrhein (1988) presented elderly (65-75 years) and young (21-30 years) subjects with precue stimuli which specified the target stimulus with 75% probability. Their precue stimuli indicated the values of three movement parameters: arm (left or right), direction (away or toward oneself) and extent (short or long). On the remaining invalid precue trials, the precue stimuli incorrectly indicated the target stimulus response with regard to one or more of these parameters. Here, the precue was displayed for 1000 ms followed by preparation interval (PI) between precue offset and target onset of 1000 ms. Beyond elderly slowing for RT and MT, the results of this study indicated age group similarity concerning movement plan restructuring for all three parameters. However, like the bimanual study of Stelmach, Amrhein and Goggin (1988), elderly subjects were slower in initiating short relative to long movements, whereas young subjects (weakly) showed the opposite pattern. In addition, extent also impacted differentially on age group concerning MT: elderly subjects exhibited a smaller increase for executing long over short movements relative to young subjects. These RT and MT findings concerning extent suggest that short movements are more difficult to plan and execute for older persons compared to their younger counterparts.

Goggin et al. (1989) presented elderly (63-76 years) and young (21-26 years) subjects with precue stimuli which specified the target stimulus (and response) with 75% probability. Specifically, the precue stimuli indicated the values of two movement parameters: arm (left or right) and direction (away or toward oneself) using a four-light display. On the remaining invalid precue trials, the precue stimuli incorrectly indicated the target stimulus response with regard to arm, direction or both, thus requiring restructuring of these parameters upon target stimulus onset. Unlike Larish and Stelmach (1982) and Stelmach, Goggin, and Amrhein (1988), the precue display interval was limited to 250 ms and the PI was varied (500, 1000, 1500 or 2000 ms) to provide a measure of the time course of movement plan preparation, maintenance and restructuring (see Amrhein et al., 1991, for a detailed discussion).

Overall, elderly subjects had slower RTs and MTs than the young subjects, though both groups had slower RTs for the invalid precue relative to the valid precue trials. Importantly, age groups differed among the individual invalid precue trials: relative to young subjects, RT for elderly subjects to change direction was faster than RTs to change arm or both parameters, which were equivalent. This finding suggested that elderly subjects failed to prepare or lost direction preparation by a PI of 500 ms, to such a degree that restructuring it no longer incurred additional latency, unlike the young subjects. Moreover, because this direction preparation had been lost, there was no difference when restructuring the remaining arm preparation alone or in combination with direction.

Using a more comprehensive methodology, Amrhein et al. (1991) conducted two experiments. In Experiment 1, elderly (65-78 years) and young (21-28 years) subjects performed a movement plan restructuring task like that used by Goggin et al. (1989); in addition they also performed SRT and CRT tasks. These tasks were included to provide baselines with which to compare performance on the restructuring task. In the SRT task, the precue always correctly indicated the target stimulus response; in the CRT task, the precue never indicated the target stimulus response. Finally, the precue display interval was 250 ms, while the PI was varied (250, 500, 750 or 1000 ms) using a more sensitive range of values than Goggin et al. (1989). In Experiment 2, elderly (70-77 years) and young (20-24 years) subjects performed the same set of tasks but the precue display interval was subject-determined (denoted by a foot-pedal release) and the PI was fixed at 250 ms.

Across both experiments, elderly subjects exhibited slower RTs and MTs. Moreover, both groups exhibited additional latency for invalid precue trials relative to CRT trials, indicating a temporal cost to restructuring a pre-existing movement plan relative to responding with no such plan (Klapp et al., 1974). Also, in Experiment 2, both age groups exhibited a linear increase in latency to view the precue stimulus (PT, precue viewing time) with increases in its validity in predicting the target stimulus; PT increased from the CRT task (precue validity = 0%) to the restructuring task (precue validity = 75%) to the SRT task (precue validity = 100%).

Most interesting and consistent with Goggin et al. (1989), for both experiments, RT for elderly subjects to change direction was faster than RTs to change arm or both parameters, which were again equivalent. In Experiment 1, this occurred at a PI of 1000 ms (although it occurred at an earlier PI of 500 ms for some of the elderly subjects). These findings indicated that direction preparation loss was occurring in the Goggin et al. (1989) study rather than a failure to initially prepare direction. Moreover, results from Experiment 2 indicated that this loss could not be modulated by subject-control of the duration of the precue stimulus.

The final study to be reviewed was conducted by Amrhein et al. (1993). Elderly (65-80 years) and young (18-31 years) subjects performed a movement plan restructuring task like that used by Amrhein et al. (1991) and Goggin et al. (1989); but, in addition, these subjects also performed a matched spatial orienting task. Procedural events were identical for the two tasks with the exception that in the spatial orienting task, subjects simply released the Home button upon target stimulus onset. A post-trial spatial memory test was administered to ensure that subjects were equally compelled to attend to the precue and target stimuli in the two tasks. For both tasks, the precue display interval was 250 ms, whereas the PI was varied (250, 500, 1000 or 2000 ms).

Again, elderly subjects had slower RTs and MTs relative to the young subjects. Importantly, the RT pattern for the parameter change trials for the restructuring task replicated that of Amrhein et al. (1991) and Goggin et al. (1989). Moreover, this pattern was not found for the spatial orienting task, indicating that an attentional resource allocation account of these parameter change results was not tenable. The Amrhein et al. (1991, 1993) and Goggin et al. (1989) studies thus provide corroborative evidence of specific loss in elderly movement plan preparation concerning the parameter of direction. I will return to these studies in particular at the end of the next section of this chapter.

PARAMETERIZATION STUDIES AND THE AGE-RELATED SLOWING DEBATE

A pervasive debate in the cognitive aging literature concerns the nature of slowing of elderly individuals in speeded information processing tasks. The parties in this debate consist of proponents and opponents of a general slowing account of the increased response latency reliably observed in elderly individuals. The primary mode of empirical support for this "General Slowing" theory has been from meta-analyses based on statistical regressions of elderly on young group condition latencies (Hale, Lima, & Myerson, 1991; Hale, Myerson, & Wagstaff, 1987; Lima, Hale, & Myerson, 1991) and more recently, condition latency differences (e.g., Myerson, Ferraro, Hale, & Lima, 1992). Based on comments by Cerella (e.g., Cerella & Hale, 1994), Perfect (1994) has argued recently that General Slowing theory is "anti-Cognitive Psychology" because it reduces all age-based performance differences to a mathematical description of changes in neurological function efficiency, thus removing the need to reference stimulus or task characteristics (beyond a dimension of "complexity" – but see below) to predict and explain age-related slowing. Indeed, Cerella and Hale (1994) argue that General Slowing theory is a one parameter theory which can account for the inverted U shape of processing speed from childhood to late adulthood.

There are two criticisms typically levelled at this theory: One criticism concerns the mode of analysis typically employed by General Slowing proponents: Meta-analysis using nonlinear or more often linear regression. For example, Perfect (1994) states that the results of the "Brinley" plot regression approach can misrepresent the underlying task parameters that determine an age group's overall performance. The other criticism comes directly from studies using a range of tasks, the data from which either fail to exhibit Age Group × Condition interactions or exhibit Age Group × Condition interactions that indicate nonlinear or non-monotonic slowing in the elderly subjects (e.g., Amrhein et al., 1991, 1993; Goggin et al., 1989; Stelmach, Amrhein, & Goggin, 1988; Stelmach, Goggin, & Amrhein, 1988).

Another kind of evidence against General Slowing theory are cases where the age-related slowing observed from a meta-analysis does not indicate the type of proportional slowing typically reported in the meta-analyses of General Slowing proponents (i.e., that the slope of the best-fitting line falls near 1.5 for lexical tasks or near 2.0 for non-lexical tasks, the intercept is negligible (positive or negative), and that line accounts for at least 80% of the elderly condition mean variance). Finally, evidence from meta-analyses indicating domain or task specificity concerning age-related slowing (or lack of such slowing) also argues against at least a simple single parameter value account (e.g., see Amrhein, 1995). For example, as already indicated, General Slowing proponents have themselves reported that elderly slowing for lexical and nonlexical tasks differs (e.g., Lima et al., 1991). (However, this conclusion concerning domain specificity is qualified by the evidence for task specificity revealed by Amrhein, 1995, for a number of studies Lima et al., 1991, included in their meta-analysis.)

A critical assumption made by those researchers using the "Brinley plot" regression approach is that task complexity can be readily defined. But the definition of "task complexity" itself appears to be circular (e.g., see Myerson & Hale, 1993). To elaborate, in the *a priori* application of this approach, increases in the number of specifiable processes underlying task performance should correspondingly increase overall response time. However, it is often difficult to specify exactly what these additional processes would be, so the *ad hoc* application is then used. In the *ad hoc* application, greater response latency for a condition (which is not compromised by a speed-accuracy tradeoff) is taken as *prima facie* evidence that that condition is more "complex" in an information processing sense. Regardless of how complexity is defined, General Slowing theory predicts that elderly subjects will exhibit proportionally longer response latencies for more "complex" experimental conditions relative to young subjects.

For the present set of motor control studies, greater complexity would be expected for experimental conditions where the precue either incompletely (movement plan specification task) or incorrectly (movement plan restructuring task) specifies the values of the movement parameters for the impending response to the target stimulus. Beyond this, particular parameter differences may reflect differential complexity inherent within them. For example, as I have suggested elsewhere (Amrhein et al., 1991, 1993), movement direction seems inherently more complex than arm of movement, because the former is potentially continuous (0°-360°), while the latter is simply binary (left or right). If so, then movement direction should take longer to prepare prior to response execution; and this finding should be more pronounced for the elderly than young subjects because such complexity as defined here would be expected to increase response latency. However, because direction is a more complex movement parameter, elderly subjects may have greater difficulty in maintaining their preparation for it. If so, its greater complexity may result in *faster* direction change latency relative to the other change conditions (which concern parameters with more easily maintained preparation) in the movement plan restructuring task. Such a finding would seem to compromise the "greater task complexity-greater response latency" assumption that underlies the Brinley plot regression approach to revealing the nature of age-related slowing.

To date, meta-analyses have been conducted separately for speeded lexical tasks (e.g., Lima et al., 1991; Myerson et al., 1992) and non-lexical tasks (including SRT and RT tasks; e.g., Hale et al. 1987, 1991; Lima et al., 1991). A common finding of these meta-analyses is that age-related slowing for nonlexical tasks appears to differ from that of lexical tasks. Specifically, slowing for nonlexical tasks has been shown to be nonlinear, and best accounted for by a power law (Hale et al., 1987, 1991), whereas slowing for lexical tasks is linear and best accounted for by a regression line with a slope of around 1.5 with a negligible positive or negative intercept. However, according to Lima et al. (1991), if the response latencies fall within the modal range of 0-3000 ms for both age groups, a straight line provides a good approximation of the relationship between elderly and young lexical and nonlexical latencies. For lexical tasks, this line is expected to have a slope around 1.5, with a negligible positive or negative intercept, whereas for non-lexical tasks the line is expected to have a slope around 2.0 (i.e., at least a slope significantly greater than 1.5), again with a negligible positive or negative intercept. Unfortunately, the nonlexical tasks analyzed to date have represented a mixed bag of stimuli and task types – including diagrams used in image rotation tasks as well as simple light displays used in CRT tasks. This is not a trivial problem: An aggregate analysis of dissimilar tasks can fortuitously produce a slowing function with a theoretically consistent slope but task types when analyzed separately can reveal different slowing functions (see Amrhein, 1995, concerning reanalysis of studies given in Table 4 of Lima et al., 1991).

I should also point out that "General Slowing" theory actually covers a family of slowing models, only one of which actually represents strict generalized slowing. Of relevance here, Cerella (1990) has distinguished two linear slowing models, *generalized* and *multilayered*. In the generalized slowing model, the relationship between elderly and young response latencies is wholly multiplicative (where the slope of the best-fitting line approximates 1.5 or 2.0), thus exhibiting no additive slowing (i.e., the line intercept is negligibly positive or negative), whereas in the multilayered model, this relationship is both multiplicative (as just defined) and additive (i.e., line intercept is positive and not negligible). In both models, multiplicative slowing is interpreted as due to age-based neurological changes impacting on "computational processes", whereas additive slowing is interpreted as being due to age-based changes in neurophysiological functioning of "sensory-motor" processes. A third possible model is an additive model in which elderly slowing is wholly additive (slope of best-fitting line approximates 1.0 but line intercept is positive and not negligible), indicating that the elderly slowing is due strictly to these sensory-motor changes. To date, the regression analyses concerning lexical and nonlexical tasks reported by General Slowing theory proponents (Hale et al., 1987, 1991; Lima et al., 1991; Myerson et al., 1992) have supported the generalized slowing model with the exception of a more recent lexical analysis by Laver and Burke (1993) which supported a simple additive slowing model.

Curiously, none of the motor control studies reviewed here have been included in the extant non-lexical meta-analyses cited earlier. For this reason, I conducted a meta-analysis of these studies to determine their contribution to the debate on age-related slowing. Mean latencies from these studies were obtained either from tables or appendices, or were estimated from figures presented in the published articles. Table 1 presents the details concerning number of conditions contributed from each study and their source in each article.

Scatter plots of the condition RTs and TTs (where TT = RT + MT) plotted according to age group are given in Figure 1. If, as assumed, MT simply measures latency to execute a movement plan and is thus immune to experimental manipulations which impact movement plan preparation, maintenance or restructuring (see e.g., Singleton, 1954), then a line fitting TT should parallel that fitting RT. This is essentially what is found. The slope of the RT line is 1.17 with an additive intercept of 65.6 ms; the line accounts for 89.3% of the elderly condition mean RT variance. The slope of the TT line has a slope of 1.02 with an additive intercept of 271.5 ms; however, this line accounts for only 70.9% of the elderly condition mean TT variance, indicating that the MTs for these studies are quite variable. Before addressing this issue however, I want to point out that the slopes of the RT and TT lines are both significantly less than the slope of 2.0 [RT: $t(94) = -20.0$, $p < .001$; TT: $t(94) = -14.6$, $p < .001$] predicted by the meta-analyses of non-lexical tasks presented by Lima et al. (1991); indeed, both slopes are significantly less than the apparent modal slope of 1.5 espoused for other tasks (e.g., lexical tasks; see Lima et al., 1991; Myerson et al., 1992) [RT: $t(94) = -8.00, p < .001$; TT: $t(94) = -7.19, p < .001$]. The point here is that these lines are indicating nominal slowing (17% for RT), if not negligible slowing (2% for TT) in elderly computational processes.

TABLE 1.
Meta-analysis results.

Study	Conditions	Source
Aging and movement parameterization studies		
Larish & Stelmach (1982, Experiment 1)	6	Figures 2 & 3
Stelmach, Goggin, & Garcia-Colera (1987)	8	Figure 3 & Appendix B
Stelmach, Amrhein, & Goggin (1988)	3	Tables 2 & 3
Stelmach, Goggin, & Amrhein (1988)	8	Table 4
Goggin, Stelmach, & Amrhein (1989)	16	Figure 1[a]
Amrhein, Stelmach, & Goggin (1991)		
Experiment 1	24	Figure 2 & Appendix A
Experiment 2	6	Figure 3 & Appendix B
Amrhein, Von Dras, & Anderson (1993)	16	Figure 2
Best-fitting lines: $RT_{ELDERLY} = 1.17RT_{YOUNG} + 65.6$ ms ($r^2 = .893$)		
$TT_{ELDERLY} = 1.02TT_{YOUNG} + 271.5$ ms ($r^2 = .709$)		
Studies not assessing movement extent		
Larish & Stelmach (1982, Experiment 2)	6	Figures 2 & 3
Goggin, Stelmach, & Amrhein (1989)	16	Figure 1[a]
Amrhein, Stelmach, & Goggin (1991)		
Experiment 1	24	Figure 2
Experiment 2	6	Figure 3
Amrhein, Von Dras, & Anderson (1993)	16	Figure 2
Best-fitting lines: $RT_{ELDERLY} = 1.17RT_{YOUNG} + 60.8$ ms ($r^2 = .937$)		
$TT_{ELDERLY} = 1.17TT_{YOUNG} + 170.9$ ms ($r^2 = .849$)		
Studies assessing movement extent		
Stelmach, Goggin, & Garcia-Colera (1987)	8	Figure 2
Stelmach, Amrhein, & Goggin (1988)	3	Tables 2 & 3
Stelmach, Goggin, & Amrhein (1988)	8	Table 4
Best-fitting lines: $RT_{ELDERLY} = 1.28RT_{YOUNG} + 37.8$ ms ($r^2 = .751$)		
$TT_{ELDERLY} = .89TT_{YOUNG} + 370.7$ ms ($r^2 = .422$)		

TABLE 1. Following

Study	Conditions	Source
Movement plan restructuring task studies reporting direction preparation loss		
Goggin, Stelmach, & Amrhein (1989)	16 $(12^b)(8^c)$	Figure 1[a]
Amrhein, Stelmach, & Goggin (1991)		
Experiment 1	24 $(12^b)(8^c)$	Figure 2 & Appendix A
Experiment 2	6 $(3^b)(2^c)$	Figure 3 & Appendix B
Amrhein, Von Dras, & Anderson (1993)	16 $(12^b)(8^c)$	Figure 2

Best-fitting lines:

- *Parameter change conditions:*

$RT_{ELDERLY} = 1.07RT_{YOUNG} + 107.1$ ms($r^2 = .859$)

$TT_{ELDERLY} = 1.00TT_{YOUNG} + 287.5$ ms($r^2 = .876$)

- *Direction change condition excluded:*

$RT_{ELDERLY} = .96RT_{YOUNG} + 164.3$ ms($r^2 = .916$)

$TT_{ELDERLY} = .93TT_{YOUNG} + 341.1$ ms($r^2 = .908$)

Note: RT = reaction time, MT = movement time, TT = total time (RT + MT). Studies are listed in the order of discussion in the text.

[a] MT data were drawn from original data files and are available by request from the author.

[b] Number of parameter change conditions.

[c] Number of parameter change conditions with direction change condition excluded.

Moreover, the difference in the intercepts for the two lines indicates that after response initiation, an additional 205.9 ms is needed, on average, for elderly subjects to move to and press a target button in these studies; indeed, the additive intercept for the TT line indicates the only substantial age-related slowing (subsuming the additive intercept of the RT line) occurring in these motor control tasks. One might argue that "of course" this should be the case, because these studies represent "sensory-motor" tasks, and sensory-motor slowing should exhibit only an additive constant in the slowing function (see Botwinick, 1984;

Cerella, 1985). However, the movement plan specification and restructuring tasks used in these studies contain a stimulus uncertainty component like that found in CRT tasks (see Amrhein et al., 1991), and CRT tasks purportedly exhibit proportional slowing due to inferred age-related changes in "computational" processes which intervene between sensory and motor processes (Cerella, 1990; Goggin & Stelmach, 1990; Hale et al., 1987, 1991; Lima et al., 1991).

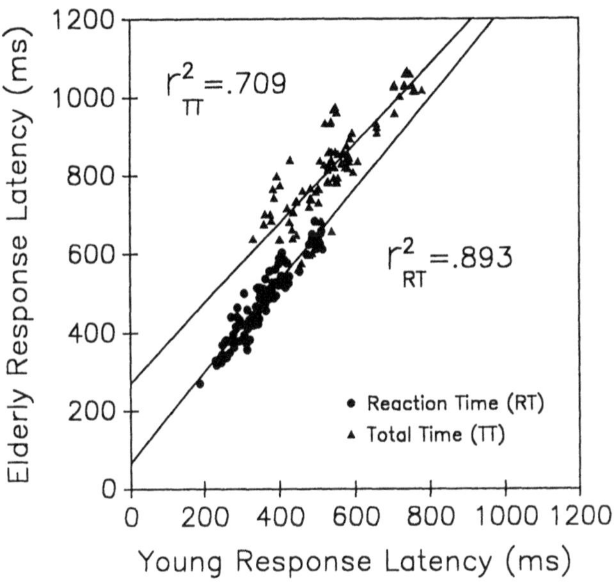

FIGURE 1. Scatter plot and best-fitting lines for Reaction Time (RT) and Total Time (TT, TT = RT + MT) for parameterization studies in the aging literature.

As noted above, the TT line in Figure 1 does not strikingly account for the corresponding elderly condition mean variance. One possibility for this is that the TT latencies reflect the summation of MTs for which the age-relation does not remain constant across the studies. This situation is particularly relevant for those studies which manipulated the parameter of movement extent (Stelmach, Amrhein, & Goggin, 1988; Stelmach, Goggin, & Amrhein, 1988; Stelmach et al., 1987). In all of these experiments, MT varied with movement extent (short or long), an effect which sometimes occurred for RT and/or interacted significantly with age group (e.g., Stelmach, Amrhein, & Goggin, 1988; Stelmach, Goggin, & Amrhein, 1988). Accordingly, RT and TT regression lines with the latencies of these studies removed should show an increase in r^2. Indeed, as can be seen in Figure 2, this is the case. Now, the RT and TT lines have identical slopes of 1.17; the RT line has an additive intercept of 60.8 ms whereas the TT line has an additive intercept of 170.9 ms. Importantly, the RT and TT lines now account for 93.7% and 84.9% of their respective elderly condition mean variance. The slopes of the RT and TT lines are both significantly less than 2.0 [RT: $t(66) = -22.3$, $p<.001$; TT: $t(66) = -13.8$, $p<.001$] as well as less than 1.5 [RT: $t(66) = -8.89$, $p<.001$; TT: $t(66) = -5.52$, $p<.001$]. These lines are again indicating nominal slowing (17% for RT and TT) in elderly computational processes.

Moreover, the difference in the intercepts for the two lines indicates that after response initiation, an additional 110.1 ms is needed, on average, for elderly subjects to move to and press a target button in these studies, again suggesting that the additive intercept for the TT line (subsuming the additive intercept of the RT line) indicates the only substantial age-related slowing occurring in these motor performance tasks.

Finally, comparing Figure 1 and 2, it can be seen that it was the MT age-relationship that was primarily impacted by the studies manipulating movement extent. Removal of the RT and TT latencies of those studies from the regression analysis altered only the TT line (which contains the additional latency for MT), and now both RT and TT lines have the same slope, thus validating across experiments that MT assesses motor control processes (i.e., movement plan execution) that differ from those assessed by RT (i.e., stimulus perception, movement plan preparation, maintenance and for some experimental conditions, restructuring). The experimental manipulations of the studies plotted in Figure 2 were designed to impact RT but not MT; this is borne out by the simple additive shift for the TT line from the RT line. That is, aging simply increases MT in a constant manner across these studies.

Conversely, the RT line and especially the TT line for the studies manipulating the parameter of extent should show poorer fits when their RT and TT latencies are analyzed. As can be seen in Figure 3 this is also the case. Respectively, the RT and TT lines have more disparate slopes (1.28 and .89) and intercepts (37.8 ms and 370.7 ms). Lastly, the RT and TT lines account for 75.1% and 42.4% of their respective elderly condition mean variance. The markedly poorer fit for the TT line in Figure 3 occurs because the latencies plotted reflect the age-differential MT effects for movement extent. The slopes of the RT and TT lines are both significantly less than 2.0 [RT: $t(26) = -4.96$, $p<.01$; TT: $t(26) = -5.39$, $p<.01$], but while the slope of the TT line is significantly less than 1.5 [TT: $t(26) = -8.05$, $p<.001$], the RT line is not [$t(26) =$

-1.51, $p > .05$]. These lines indicate some slowing (28%) for RT but a slight speed *increase* (11%) for TT, concerning elderly computational processes.

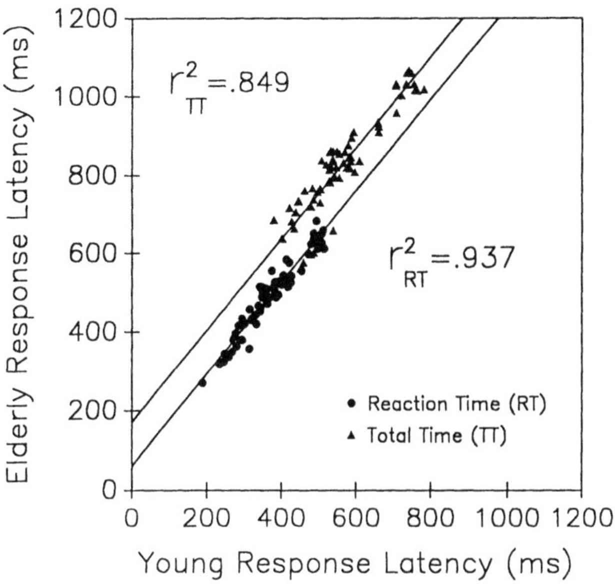

FIGURE 2. Scatter plot and best-fitting lines for Reaction Time (RT) and Total Time (TT, TT = RT + MT) for parameterization studies not assessing movement extent.

The difference in the intercepts for the two lines indicates that after response initiation, an additional 332.9 ms is needed, on average, for elderly subjects to move to and press a target button in these studies. However, given the poorer fits of these lines (especially the TT line) compared to those found in the other studies plotted in Figure 2, the magnitudes of the slopes and intercepts of the best-fitting lines in Figure 3 are somewhat suspect. Importantly, what Figures 1, 2, and 3 do indicate is that without knowledge of the impact of specific condition manipulations of the individual experiments, neither can a good linear fit of the data be obtained nor can the poor regression line fit be adequately explained. More specifically, these Figures demonstrate that elderly slowing is not always well-accounted for by a simple linear function when the response latencies fall within the modal range (0-3000 ms) stipulated by Lima et al. (1991).

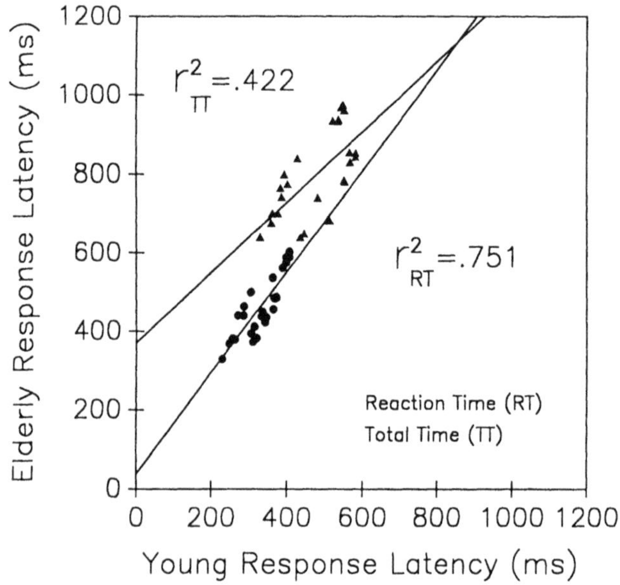

FIGURE 3. Scatter plot and best-fitting lines for Reaction Time (RT) and Total Time (TT, TT = RT + MT) for parameterization studies assessing movement extent.

My final set of regression analyses concern those movement plan restructuring studies which have reliably demonstrated faster direction change RT for elderly subjects relative to arm, and arm and direction change RTs (Amrhein et al., 1991, 1993; Goggin et al., 1989). This pattern provides a critical test of General Slowing theory, and more generally, the utility of the Brinley plot regression approach in revealing the nature of age-related slowing in speeded cognitive and cognitive-motor tasks. That is, finding an example where elderly subjects are faster in a condition that young subjects are not, qualifies as an instance stipulated by Cerella (1990) which is problematic for a generalized or multilayered slowing function account of elderly task performance, and suggests, rather, a qualitative age difference in cognitive-motor processes.

Of interest here, firstly, is the slowing pattern exhibited by the elderly RTs and TTs for these studies; an extremely high degree of linear fit would seem unlikely because of the direction change effect. Secondly, by removing direction change RTs for both age groups, thus effectively removing this Age Group × Condition interaction, the fit should actually improve. As can be seen by comparing Figures 4 and 5, however, this only nominally occurs.

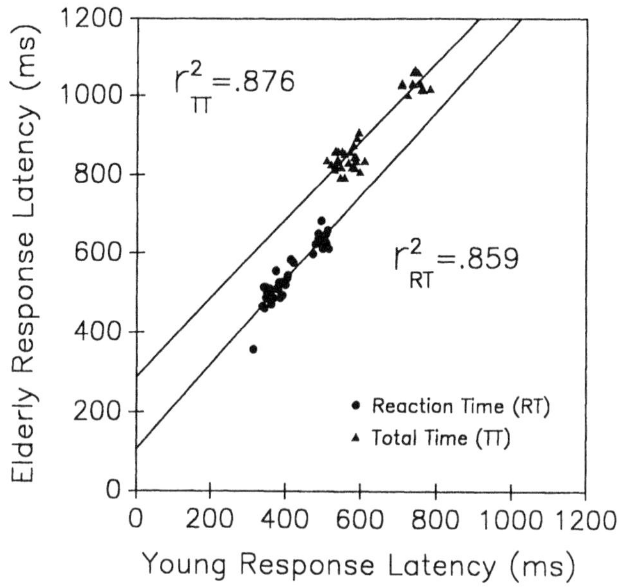

Figure 4. Scatter plot and best-fitting lines for Reaction Time (RT) and Total Time (TT, TT = RT + MT) for movement plan restructuring studies reporting direction preparation loss for elderly subjects.

In Figure 4, RT and TT latencies for the parameter change conditions of the invalid precue trials for these three studies (see Table 1) are plotted. The RT line has a slope of 1.07, an additive intercept of 107.1 ms, and accounts for 85.9% of the elderly condition mean RT variance. The TT line has a slope of 1.00, an additive intercept of 287.5 ms and accounts for 87.6% of the elderly condition mean TT variance. The slopes of these RT and TT lines are both significantly less than 2.0 [RT: $t(37) = -13.1$, $p<.001$; TT: $t(37) = -16.1$, $p<.001$] and 1.5 [RT: $t(37) = -6.06$, $p<.001$; TT: $t(37) = -8.05$, $p<.001$]. These lines indicate nominal slowing (7%) for RT, but no slowing for TT, concerning elderly computational processes. Moreover, the difference in the intercepts for the two lines indicates that after response initiation, an additional 180.4 ms is needed, on average, for elderly subjects to move to and press a target button for the parameter change conditions in these restructuring studies. This difference clearly indicates that the additive intercept for the TT line (subsuming the additive intercept of the RT line) reflects the only substantial age-related slowing occurring in these motor performance task conditions.

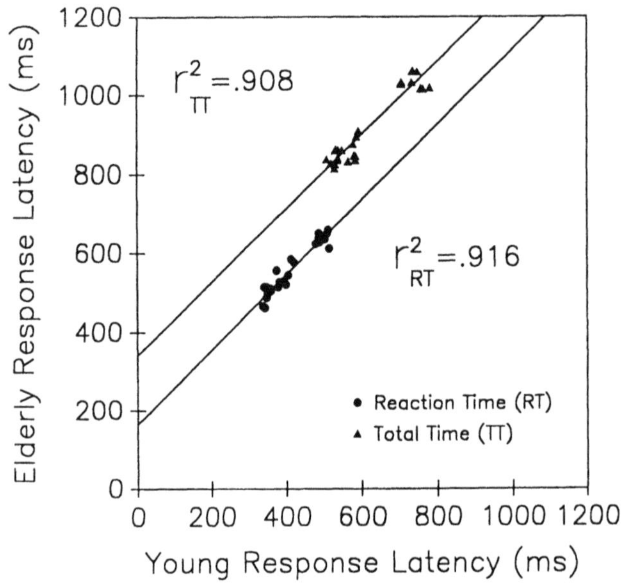

Figure 5. Scatter plot and best-fitting lines for Reaction Time (RT) and Total Time (TT, TT = RT + MT) for movement plan restructuring studies reporting direction preparation loss for elderly subjects (direction change trials excluded).

In Figure 5, latencies for the parameter change conditions are again plotted with the exclusion of direction change RT and TT latencies. Now, the RT line has a slope of .96, an additive intercept of 164.3 ms, and accounts for 91.6% of the elderly condition mean RT variance. The TT line has a slope of .93, an additive intercept of 341.1 ms, and accounts for 90.8% of the elderly condition mean TT variance. The slopes of the RT and TT lines are both significantly less than 2.0 [RT: $t(24) = -17.8$, $p<.001$; TT: $t(24) = -17.7$, $p<.001$] and 1.5 [RT: $t(24) = -9.29$, $p<.001$; TT: $t(24) = -9.40$, $p<.001$]. These lines are again indicating no slowing [actually a negligible speed *increase* for RT (4%) and TT (7%)] in elderly computational processes.

Additionally, the difference in the intercepts for the two lines changes little from that in Figure 4: After response initiation, an additional 176.8 ms is needed, on average, for elderly subjects to move to and press a target button in the arm and arm and direction change conditions of these restructuring studies. This difference again clearly indicates that the additive intercept for the TT line (subsuming the additive intercept of the RT line) reflects the only substantial age-related slowing occurring in these motor performance task conditions.

Thus, by removing the differential latency pattern not explainable by appeal to changes in either sensory-motor slowing (i.e., the additive intercept) or computational slowing (i.e., that the slope remains near 1.00 indicates that there is none), the regression fit is minimally improved. In other words, the age-differential effect for direction change revealed in these studies would likely go unnoticed if the response latencies were simply submitted to a Brinley plot regression analysis. In short, as I (Amrhein, 1995; Amrhein & Theios, 1993) and others (Fisk & Fisher, 1994; Fisk, Fisher, & Rogers, 1992; Perfect, 1994) have argued elsewhere, Brinley plots can provide incomplete and sometimes misleading information about the nature of task- and underlying process-specific slowing when contrasting elderly and young subjects' speeded performance.

CONCLUSIONS

Regression analysis of movement parameterization studies assessing aging effects reveal sensory-motor slowing without any substantial, intervening computational slowing, contrary to the predictions for non-lexical tasks derived from extant meta-analyses reported by proponents of the General Slowing theory (e.g., Hale et al., 1987, 1991; Lima et al., 1991). For RT, the range of slopes was .96-1.28 with an average of 1.13 (1.09 with the RT line with $r^2 < .80$ removed). For the TT line, the range of slopes was .89-1.17 with an average of 1.00 (1.03 with the two TT lines with $r^2 < .80$ removed). These slope values are much less than the 2.0 slope reported by General Slowing proponents for non-lexical (e.g., SRT, CRT) tasks (and even less than the 1.5 slope reported for lexical tasks; see e.g., Lima et al., 1991).

Rather, the locus of slowing across the RT and TT lines is seen in their sizable, positive intercepts, owing to slowed sensory-motor processes (see e.g., Botwinick, 1984; Cerella, 1990). For RT, the range of intercepts was 60.8-164.3 ms with an average of 87.2 ms (99.5 ms when the RT line with $r^2 < .80$ is removed). For TT, the range of intercepts was 170.9-370.7 ms with an average of 288.3 ms (266.5 ms when the two TT lines with $r^2 < .80$ are removed). Thus, the contribution of elderly MT to their TT is generally additive across a number of movement plan specification and restructuring tasks. However, when movement extent is manipulated, non-proportional age differences in MT (and to a lesser degree, RT) increase response variability. Given that all speeded cognitive tasks (even those "RT" tasks which assess RT and MT in aggregate) contain a motor response component (excluding passive EEG, ERP, PET or MRI studies), these findings are not trivial.

Finally, Brinley plot regression analysis obscures non-proportional age differences in RT concerning apparent loss of preparation for movement direction; a loss which indicates a qualitative age difference in motor control. As such, the analyses presented in this chapter provide corroborative support for those researchers critical of the utility of the Brinley plot approach in uncovering the nature of age-related slowing in speeded cognitive and cognitive-motor tasks (e.g., Amrhein, 1995; Amrhein & Theios, 1993; Fisk & Fisher, 1994; Fisk et al., 1992; Perfect, 1994).

REFERENCES

Amrhein, P. C. (1995). Evidence for task specificity in age-related slowing: A review of speeded picture-word processing studies. In P. Allen & T. Bashore (Eds.), *Age differences in word and language processing* (pp. 144-171). Amsterdam: Elsevier Science Publishers, B.V.

Amrhein, P. C., & Theios, J. (1993). The time it takes elderly and young individuals to draw pictures and write words. *Psychology and Aging, 8,* 197-206.

Amrhein, P. C., Stelmach, G. E., & Goggin, N. L. (1991). Age differences in the maintenance and restructuring of movement preparation. *Psychology and Aging, 6,* 451-466.

Amrhein, P. C., Von Dras, D., & Anderson, M. (1993). Evidence for direction loss in elderly movement preparation is not due to spatial orienting effects. *Experimental Aging Research, 19,* 71-95.

Balota, D. A., & Duchek, J. M. (1988). Age-related differences in lexical access, spreading activation, and simple pronunciation. *Psychology and Aging, 3,* 84-93.

Botwinick, J. (1984). *Aging and behavior.* New York: Springer.

Cerella, J. (1985). Information processing rates in the elderly. *Psychological Bulletin, 98,* 67-83.

Cerella, J. (1990). Aging and information-processing rate. In J. Birren & K. W. Schaie (Eds.), *Handbook of the psychology of aging* (pp. 201-221). San Diego, CA: Academic Press.

Cerella, J., & Hale, S. (1994). The rise and fall of information processing rates over the life span. *Acta Psychologica, 86,* 109-198.

Clarkson, P. M. (1978). The effect of age and activity level on simple and choice fractionated response time. *European Journal of Applied Physiology, 40,* 17-25.

Dawson, M. R. W. (1988). Fitting the ex-Gaussian equation to reaction time distributions. *Behavior Research Methods, Instruments, & Computers, 20,* 54-57.

Donders, F. C. (1969). On the speed of mental processes. *Acta Psychol-ogica, 30,* 412-431. (Originally published in 1869).

Fisk, A. D., & Fisher, D. L. (1994). Brinley plots and theories of aging: The explicit, muddled, and implicit debates. *Journal of Gerontology, 49,* P81-89.

Fisk, A. D., Fisher, D. L., & Rogers, W. A. (1992). General slowing alone cannot explain age-related search effects: Reply to Cerella (1991). *Journal of Experimental Psychology: General, 121,* 73-78.

Goggin, N. L., & Stelmach, G. E. (1990). Age-related deficits in cognitive-motor skills. In E. A. Lovelace (Ed.), *Aging and cognition: Mental processes, self-awareness and interventions* (pp. 135-155). New York: Elsevier Science Publishers B.V.

Goggin, N. L., Stelmach, G. E., & Amrhein, P. C. (1989). Effects of age on motor preparation and restructuring. *Bulletin of the Psychonomic Society, 27,* 199-202.

Hale, S., Lima, S. D., & Myerson, J. (1991). General cognitive slowing in the nonlexical domain: An experimental validation. *Psychology and Aging, 6,* 512-521.

Hale, S., Myerson, J., & Wagstaff, D. (1987). General slowing of nonverbal information processing: Evidence for a power law. *Journal of Gerontology, 42,* 131-136.

Kausler, D. H. (1991). *Experimental psychology, cognition, and human aging*. New York: Springer-Verlag.

Klapp, S. T., Wyatt, E. P., & Lingo, W. M. (1974). Response programming in simple and choice reactions. *Journal of Motor Behavior, 6,* 263-271.

Larish, D., & Stelmach, G. E. (1982). Preprogramming, programming, and reprogramming of aimed hand movements as a function of age. *Journal of Motor Behavior, 14,* 322-340.

Laver, G. D., & Burke, D. M. (1993). Why do semantic priming effects increase in old age? A meta-analysis. *Psychology and Aging, 8,* 34-43.

Light, K. E., & Spirduso, W. (1990). Effects of the movement complexity factor of response programming. *Journal of Gerontology: Psychological Sciences, 45,* P107-109.

Lima, S. D., Hale, S., & Myerson, J. (1991). How general is general slowing? Evidence from the lexical domain. *Psychology and Aging, 6,* 416-425.

Myerson, J., & Hale, S. (1993). General slowing and age invariance in cognitive processing: The other side of the coin. In J. Cerella, J. Rybash, W. Hoyer, & M. L. Commons (Eds.), *Adult information processing: Limits on loss* (pp. 115-141). San Diego, CA: Academic Press.

Myerson, J., Ferraro, F. R., Hale, S., & Lima, S. D. (1992). General slowing in semantic priming and word recognition. *Psychology and Aging, 7,* 257-270.

Perfect, T. J. (1994). What can Brinley plots tell us about cognitive aging? *Journal of Gerontology, 49,* P60-64.

Rosenbaum, D. A. (1991). *Human motor control.* San Diego, CA: Academic Press.

Rosenbaum, D. A., & Kornblum, S. (1982). A priming method for investigating the selection of motor responses. *Acta Psychologica, 51,* 223-243.

Salthouse, T. A. (1985). Speed of behavior and its implications for cognition. In J. E. Birren & K. W. Schaie (Eds.), *Handbook of the psychology of aging* (pp. 400-426). New York: Van Nostrand Reinhold.

Schmidt, R. A. (1988). *Motor control and learning: A behavioral emphasis.* Champaign, IL: Human Kinetics Press.

Singleton, W. T. (1954). The change of movement timing with age. *British Journal of Psychology, 45,* 166-172.

Spirduso, W. W. (1975). Reaction and movement time as a function of age and physical activity level. *Journal of Gerontology, 30,* 435-440.

Spirduso, W. W., & MacRae, P. G. (1990). Motor performance and aging. In J. E. Birren & K. W. Schaie (Eds.), *Handbook of the psychology of aging* (pp. 184-200). San Diego, CA: Academic Press.

Stelmach, G. E., Amrhein, P. C., & Goggin, N. L. (1988). Age differences in bimanual coordination. *Journals of Gerontology: Psychological Sciences, 43,* P18-23.

Stelmach, G. E., Goggin, N. L., & Amrhein, P. C. (1988). Aging and the restructuring of precued movements. *Psychology and Aging, 3,* 151-157.

Stelmach, G. E., Goggin, N. L., & Garcia-Colera, A. (1987). Movement specification time with age. *Experimental Aging Research, 13,* 39-46.

Szafran, J. (1951). Changes with age and with exclusion of vision in performance at an aiming task. *Quarterly Journal of Experimental Psychology, 3,* 111-118.

Teichner, W. H., & Krebs, M. J. (1974). Laws of visual reaction time. *Psychological Review, 81,* 75-98.

Welford, A. T. (1959). Psychomotor performance. In J. E. Birren (Ed.), *Handbook of aging and the individual* (pp. 562-613). Chicago, IL: University of Chicago Press.

Welford, A. T. (1977). Motor performance. In J. E. Birren & K. W. Schaie (Eds.), *Handbook of the psychology of aging* (pp. 450-496). New York: Van Nostrand.

Changes in sensory motor behavior in aging
A.-M. Ferrandez and N. Teasdale (Editors)

CONTROL OF SIMPLE ARM MOVEMENTS IN THE ELDERLY

Susan H. Brown

University of Michigan

Abstract

In this chapter, the characteristics of single-joint arm movements in the elderly are described in terms of kinematic relations, movement variability and muscle activation patterns. In studies involving visually-guided, step-tracking elbow movements of different amplitudes, no evidence of movement slowing was found. Elderly subjects were able to scale peak velocity suggesting no apparent impairment in movement initiation. However, analysis of the time course of movements revealed age-related changes in the ability to produce time symmetric velocity profiles where deceleration duration was consistently longer than the duration of the acceleratory phase. Compared to young subjects, trajectory variability was greater for both acceleratory and deceleratory phases regardless of movement amplitude. Changes in movement dynamics appeared related to altered control of antagonist muscle drive in that a normal pattern of phasic activation was replaced by either continuous or inappropriately timed phasic antagonist activity. The similarity of these findings to those observed in patients with mild cerebellar dysfunction is discussed. In other studies, elderly subjects were able to improve motor performance with practice as evidenced by a reduction in trajectory variability and improved control of antagonist muscle drive.

Correspondence should be sent to S. H. Brown, Ph.D., Center for Human Motor Research, Division of Kinesiology, University of Michigan, 401 Washtenaw Avenue, Ann Arbor, MI 48109-2214, U.S.A. (e-mail: shcb@ginger.kines.umich.edu).

Elderly subjects were also able to adapt to unexpected changes in visual display gain although the time course of adaptation, particularly in scaling of peak velocity, was often prolonged compared to young subjects. Changes in the performance of upper limb motor tasks are discussed in terms of adaptive strategies in the face of possible age-related cerebellar degeneration.

Key words: Elderly, movement kinematics, muscle activity, practice.

INTRODUCTION

The stereotyped image of the elderly individual draws heavily upon the decline in motor performance associated with the aging process. Stooped posture, slowed and hesitant movement and a shuffling gait pattern are particularly noticeable to varying degrees in the aged. In spite of these characteristic motor deficits which collectively can have devastating social and economic consequences, the exact nature and pathogenesis of age-related motor dysfunction remains to be fully understood.

Morphological changes in the central and peripheral nervous systems as well as alterations in muscle composition are well documented in the aging human. Although the relative importance of neuronal degeneration in the development of motor deficits is still in question, several studies have shown a significant reduction in neuronal density in brain areas thought to be involved in the planning and execution of voluntary movement. Age-related degeneration affects not only pyramidal cells in the motor cortex (Allen, Benton, Goodhardt, Haan, Sims, Smith, Spillane, Bowen, & Davidson, 1983; Brody, 1970; Scheibel, Tomiyasu, & Scheibel, 1977), but also cerebellar Purkinje cells (Hall, Miller, & Corsellis, 1975; Rogers, Silver, Shoemaker, & Bloom, 1980) and the substantia nigra of the basal ganglia (Bugiani, Salvarini, Perdelli, Mancardi, & Leonardi, 1978; McGeer, McGeer, & Suzuki, 1977). A reduction in the number of spinal motor neurons has been also described (Scheibel, 1979) as well as widespread dendritic degeneration (Sheibel et al., 1977). Motor-related neurotransmitter systems are also affected by aging. For instance, a fifty percent loss of dopaminergic neurons in the substantia nigra has been reported to occur between the twentieth and eightieth decades (McGeer et al., 1977). This latter finding has supported the view that the motor deficits seen in normal elderly humans reflect basal ganglia degeneration which, in accelerated cases, develops into the well recognized parkinsonian syndrome of rigidity, akinesia and tremor.

Age-related functional and structural changes in skeletal muscle also contribute to a reduction in motor performance in the elderly individual. There is significant widespread loss in voluntary muscle strength (Vandervoort, 1992) with up to 80 percent loss reported for upper extremity muscles (McDonagh, White, & Davies, 1983). Muscle contractile properties also change with age including an increase in the time to reach peak tension (Campbell, McComas, & Petito, 1973; Davies & White, 1983; Newton & Yemm, 1986). These changes have been linked to a reduction in motor unit number and size, particularly of fast twitch fibers (Aniannson, Hedberg, Henning, & Graimby, 1986, Grimby, 1988; Larsson, 1978; Lexell, Taylor, & Sjostrom, 1988). Fiber type grouping due to age-related degeneration and reinnervation has also been described (Lexell et al., 1988). As a result of these alterations in muscle fiber number and composition, total muscle cross sectional area may be reduced by up to one third over an eighty year lifespan (Tzankoff & Norris, 1977).

Apart from central neuronal and musculoskeletal degeneration, age-related deterioration in various sensory systems are well documented. Visual abnormalities in the elderly involve not only ocular changes (Cohen & Lessell, 1984; Sekuler, Klein, & Dismukes, 1982) but are also apparent in the execution of voluntary eye movements (Jenkyn et al., 1985; Sharpe & Sylvester, 1978; Warabi, Noda, & Kato, 1986). Loss of vestibular function (Baloh, 1984; Mulch & Peterman, 1979) along with proprioceptive and cutaneous sensibility (Kenshalo, 1979; Sabin & Venna, 1984; Shaumburg, Spencer, & Ochoa, 1983) also contribute to multi-system deterioration underlying impaired motor performance with age.

It is thus clear that the integrity of various sensorimotor systems can be significantly compromised by the aging process. Although it is not well understood how these cellular degenerative processes contribute to the decline in motor performance, our understanding of the behavioural consequences of such age-related changes has grown considerably over the past several years. For instance, there now exist considerable data on the control of posture, balance and gait with a view towards developing effective rehabilitative strategies aimed at maintaining functional independence in the elderly. Early studies on upper limb function focussed primarily on psychomotor variables such as reaction times and movement times. Advances in experimental technology, however, have permitted a more sophisticated analysis of normal motor function and have contributed to a more thorough understanding of the mechanisms underlying the generation of goal-directed movements. As a result, certain movement characteristics or properties have emerged which appear to reflect organizing strategies used by the central nervous system in generating limb movements. These developments have, in turn, led to renewed interest in upper limb motor performance in elderly populations.

This chapter will describe the results of studies which focus on the dynamics of single-joint arm movements in elderly individuals. As an experimental paradigm, the use of a single degree of freedom movement minimizes the number of dependent variables associated with a particular motor task which, otherwise, may confound the identification of those movement parameters sensitive to the aging process. While caution must be exercised when extrapolating findings from the single joint case to more complex motor tasks, there is increasing evidence suggesting that the central nervous system may utilize similar control strategies for both single and multi joint movements (see next section). A visually guided, step-tracking task involving horizontal flexion and extension movements about the elbow was employed (cf. Brown & Cooke, 1981, Brown, Hefter, Mertens, & Freund, 1990). The general experimental procedure required subjects to be seated and to grasp a handle which pivoted beneath the elbow. Handle (arm) and target position were displayed as vertical cursors on a monitor which was placed 1 m in front of the subject. In response to a step change in target location, subjects were instructed to place the handle cursor within a 5 deg target zone.

CHARACTERISTICS OF SIMPLE ARM MOVEMENTS IN YOUNG SUBJECTS

It is reasonable to assume that much of the decline seen in aging arises from widespread changes in neuromuscular and sensorimotor function as described above. It is unclear, however, as to what aspects of movement generation are particularly vulnerable to the aging process. For example, it might be expected that, compared to younger individuals, central planning and programming of voluntary movements is organized differently in the elderly, possibly reflecting adaptive strategies to compensate for gradual deterioration in a variety of motor-related systems. In many studies, age-related motor impairment has been typically described in terms of slowed reaction and movement times (Birren & Botwinick, 1961; Salthouse, 1979; 1985; Singleton, 1954; Stelmach, Goggin, & Amrhein, 1988; Welford, 1977; Welford, Norris, & Shock, 1969). In terms of movement dynamics, however, it has only been the last ten to fifteen years that an understanding of the control processes involved in movement programming and execution has begun to emerge.

There now exists a large body of literature describing common characteristics of simple, skilled movements which are thought to represent organizational strategies or "rules" used by the central nervous system in generating many motor tasks. For example, a wide variety of well-learned movements are characterized by bell-shaped, temporally symmetric velocity profiles in which the time required to accelerate and decelerate the limb is approximately equal. Examples of symmetric profiles are shown in Figure 1 for a 26 year old subject performing "fast and accurate" elbow flexion movements of different amplitudes. Temporally symmetric velocity profiles occur not only in simple, single-joint limb movements regardless of direction or gravitational load (Virji-Babul, Cooke & Brown, 1994) but also in speech movements (Ostry, 1986) and movements of the vocal folds (Munhall, Ostry, & Parush, 1985). In more complex, multi-joint reaching tasks, hand speed profiles are also time symmetric as are, in many cases, individual joint angular velocity profiles (Atkeson & Hollerbach, 1985; Flash, 1987; Morasso, 1981; Soechting, 1984; Virji-Babul & Cooke, 1995). One of most widely accepted view is that symmetric profiles represent the most energy efficient means of generating movement by minimizing jerk, that is, the rate of change of acceleration (Hogan, 1984, Nelson, 1983; Flash, 1987).

Another organizing feature underlying movement generation in younger populations is the relationship between maximum movement speed and the distance the limb moves. As shown in Figure 1, peak movement velocity increases with increasing movement amplitude. This relationship is highly linear and is maintained over a broad range of movement speeds. Scaling of velocity with distance serves to minimize movement duration and is accomplished by appropriate modulation of phasic agonist muscle activity (Brown & Cooke, 1984).

In addition to what appear to be invariant kinematic relations, many movements share similar muscle activation patterns. A reciprocally organized pattern of phasic agonist – antagonist – agonist muscle activity (cf. Fig 1D) occurs in a broad range of movements varying in both speed and amplitude (Brown & Cooke, 1981; Hallett & Marsden, 1979; Karst & Hasan, 1987; Wacholder & Altenburger, 1926). In single joint movements, the initial agonist burst produces the muscle force necessary

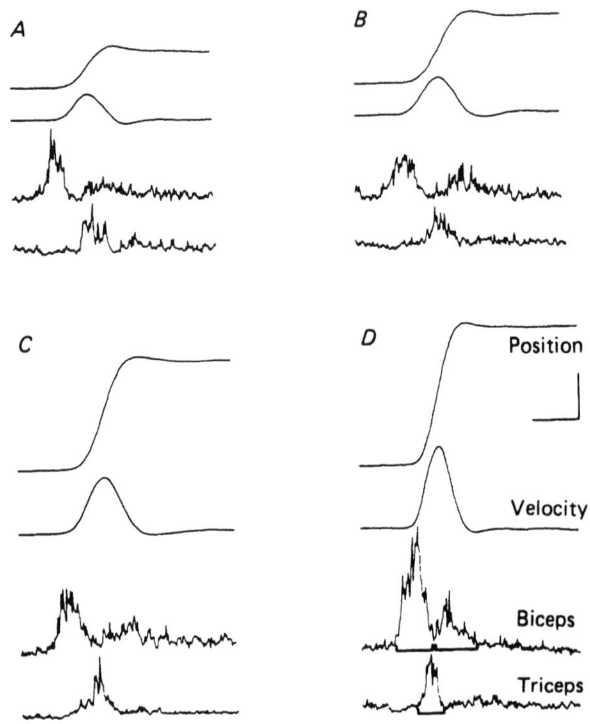

FIGURE 1. Kinematics and phasic muscle activity associated with horizontal elbow movements. The traces in A-D show averaged records of position, velocity, and biceps and triceps electromyograms from flexion movements of different amplitudes (A, 16 deg; B, 32 deg; C, 48 deg; D, 64 deg). All traces represent the average of fifteen "fast and accurate" movements. Horizontal bars below phasic muscle bursts in D demarcate the initial agonist burst, antagonist burst and second agonist burst. The vertical calibration in D represents 25 deg for position and 300 deg/s for velocity. The horizontal calibration represents 200 ms. (From Brown and Cooke, 1981).

to overcome limb inertia and initiate movement. The antagonist burst which occurs near or at the time of peak velocity assists in decelerating the movement. In rapid movements, an early phase of antagonist activity may be coactive with the initial agonist burst and is thought to contribute to termination of the acceleratory phase (Cooke & Brown, 1990). The second agonist burst, commonly associated with rapid movements, acts, in concert with the antagonist burst, to actively control limb deceleration (Brown, 1986; Cooke & Brown, 1990). The muscle pattern described above and shown in Figure 1 is characteristic of time symmetric movements. By altering duration, magnitude and relative timing of the burst components, it is possible to produce temporally asymmetric profiles (Brown & Cooke, 1990). Thus, the desired temporal structure of a movement is dependent upon the precise timing of phasic muscle activity and, in the case of the initial agonist burst, a direct relation between burst duration and acceleration duration has been recently described (Cooke & Brown, 1994).

CHARACTERISTICS OF SIMPLE ARM MOVEMENTS IN THE ELDERLY

Are the invariant features which appear to be characteristic of movements made by younger subjects preserved in the elderly? This was investigated in eleven elderly subjects ranging in age from 70 to 95 years (mean age = 81 +/-9 yrs). They were all independent community dwellers many of whom were recruited from a local lawn bowling club. Eight young subjects (mean age 22 +/- 0.5 yrs) served as controls. Subjects were asked to make movements of different amplitudes (10 to 80 deg) under "own speed" and "fast and accurate" instructions (Cooke, Brown, & Cunningham, 1989). One to two minutes were allowed for practice at each amplitude. Typical position and velocity records obtained from elbow flexion movements are shown in Figure 2 for a 22 year old (A) and an 81 year old (B) subject. In contrast to the temporally symmetric and highly reproducible movements seen in younger subjects, movements made by elderly subjects were found to be more variable in their time course, particularly during the deceleratory phase. Occasionally, elderly subjects were unable to stop smoothly but made corrective movements as they approached the target.

Although movements made by the elderly subject shown in Figure 2 were noticeably slower than the control subject, mean peak velocities did not differ significantly between control and elderly groups. As shown in Figure 3, the ability to scale peak velocity with movement amplitude was preserved in the elderly under both slower, "own speed" as well as "fast and accurate" instructions. Indeed, in the case of "own speed" movements, elderly subjects appeared more proficient at scaling peak velocity with movement amplitude compared to control subjects as evidenced by a more linear relationship between these two kinematic variables. When asked to move as fast and as accurately as possible, group mean values were remarkably similar at all movement amplitudes.

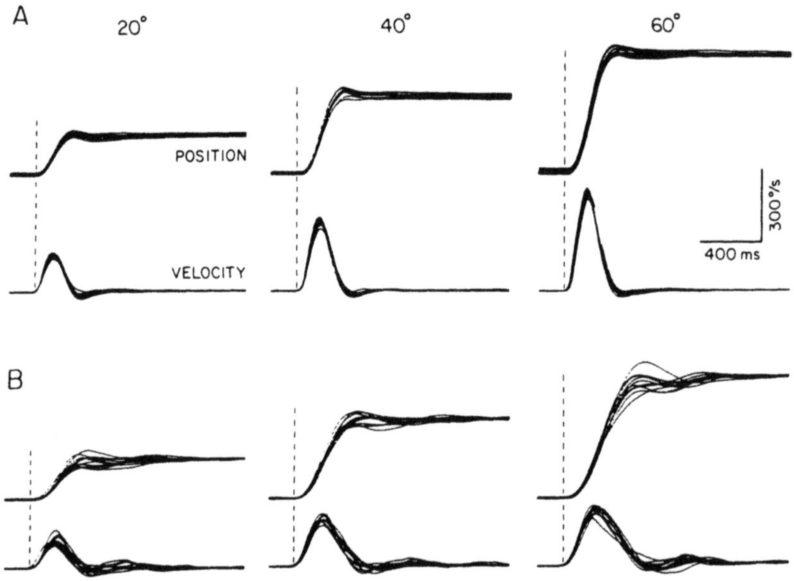

FIGURE 2. Elbow flexion movements made by young and elderly subjects. Superimposed records of position and velocity are shown from "fast and accurate" movements from a young (A) and elderly (B) subject. Records from movements of three amplitudes are shown (20, 40 and 60 deg). Movements were aligned to the defined start of movement (vertical dashed lines) for plotting (from Cooke et al., 1989).

Although peak movement speeds were comparable in the two groups, amplitude and instruction-dependent differences in total movement duration were observed. Under own speed conditions, movement durations in both groups were approximately similar for amplitudes ranging from 10 to 60 deg. For larger amplitude movements (70 and 80 deg), control subjects, on average, did not show a linear scaling of peak velocity with amplitude, resulting in prolonged movement durations at these larger distances. Under "fast and accurate" conditions, the elderly took longer to complete small amplitude (10 deg) movements (mean duration = 566 ms elderly, 450 ms young) but for amplitudes greater than 20 deg, no significant differences were observed (mean duration averaged across 30 - 80 deg movements = 733 ms young, 715 ms elderly).

As mentioned earlier, movements made by young subjects are relatively symmetric in their time course. One means of quantifying the shape of a movement is to calculate the ratio of acceleration duration to

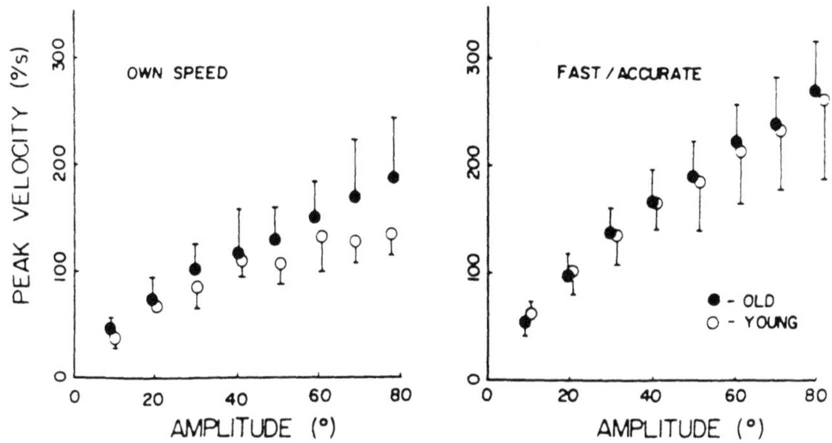

FIGURE 3. Peak velocity – amplitude relations in young and elderly subjects. Peak velocity is plotted as a function of movement amplitude for "own speed" (left hand graph) and "fast and accurate" (right hand graph) flexion movements. Each data point is the mean (+/- 1 S.D.) from all subjects. Open symbols – young subjects; closed symbols – elderly subjects (from Cooke et al., 1989).

deceleration duration. A perfectly symmetrical movement profile would have a ratio of 1.0. Typically, however, well learned movements are slightly asymmetric with acceleration durations being slightly shorter than deceleration durations (ratios = .85 - .95). Analysis of movement symmetry in the elderly group revealed clear age-related differences in the performance of these relatively simple, step-tracking movements. Mean symmetry ratios are shown in Figure 4. In young subjects, movements were relatively time symmetric for small to mid-range movement amplitudes. At larger distances, movements tended to become more asymmetric, particularly for amplitudes 60 deg and greater. However, across the full range of movement amplitudes, symmetry ratios in the control group varied only from .78 to .96. In the elderly subjects, however, temporal asymmetries were observed at all amplitudes and under both instructions. Deceleration duration was consistently longer than acceleration duration with symmetry ratios ranging from .61 to .78. The most noticeable changes in temporal symmetry occurred in small amplitude movements, particularly when subjects were moving fast and accurately.

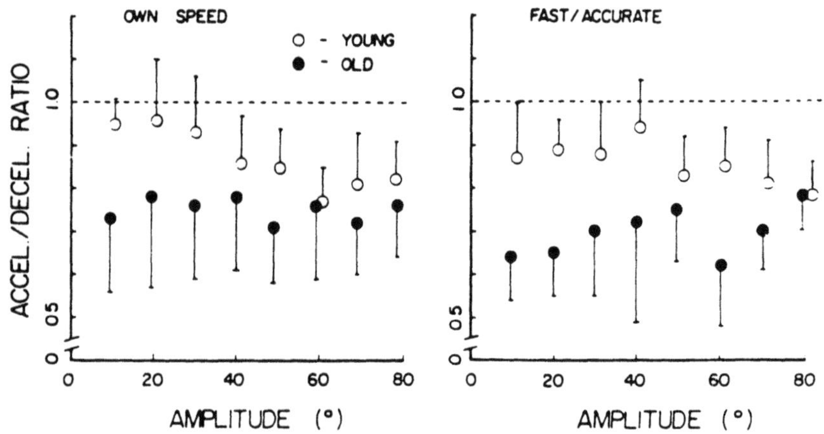

FIGURE 4. Flexion movement symmetry ratios in young and elderly subjects. The ratio of acceleration to deceleration durations is shown as a function of movement amplitude. The horizontal dashed line indicates a ratio of 1 (acceleration duration equal to deceleration duration). Each point is the mean (+/- 1 S.D.) from all subjects. Open symbols – young subjects; closed symbols – elderly subjects (from Cooke et al., 1989).

Thus, in this simple motor task having relatively low end-point accuracy requirements (+/- 10 percent of target distance), elderly subjects were able to move as quickly as their younger counterparts. The ability of elderly subjects to produce peak velocities comparable to those seen in control subjects would suggest that, on average, they are able to generate the appropriate muscular forces responsible for movement acceleration. This contrasts with other studies which describe age-related movement slowing (Goggin & Stelmach, 1990; Stelmach et al., 1988). These differences may be explained, in part, by differences in task requirements and the level of physical fitness of the subjects. As mentioned earlier, most of the subjects in this study actively participated in lawn bowling, a somewhat genteel sport but one which, for this age group, could be considered to represent a form of regular exercise and which also provided a valuable means of social interaction.

Despite normal movement speeds and durations, however, the presence of altered movement dynamics as reflected by temporally asymmetric movement profiles, would suggest age-related impairment in the organization of descending motor commands. Asymmetric profiles have since been confirmed in the elderly for arm movements made on a digitizing tablet (Goggin & Meeuwsen, 1992), wrist rotations (Pratt, Chasteen, & Abrams, 1994) and the transport phase of reaching movements (Bennett & Castiello, 1994). The next section presents data supporting the view that altered movement dynamics in the elderly may arise from impaired control of antagonistic muscle activity, possibly arising from age-dependent changes in cerebellar structures.

MOVEMENT VARIABILITY IN THE ELDERLY

One of the most striking findings in the production of voluntary movements in the elderly is the degree of variability in kinematic parameters. As shown in the overplotted records in Figure 2 and by the standard deviations for peak velocity and movement symmetry in Figures 3 and 4, there is an apparent loss in the ability to consistently produce the same movement profile over repetitive trials.

Another approach which has been used to examine movement variability has focussed on the relationship between position and velocity (phase plane trajectory) throughout the course of the movement (Darling & Cooke, 1987). Analysis of the moment to moment variability of both single and multi-joint arm movements has shown that, in young subjects, trajectory variability increases during the acceleratory phase with little change during movement deceleration (Darling & Cooke, 1987, Darling & Stephenson, 1993). When variability was examined in elderly subjects performing step-tracking movements, trajectory variability was greater in elderly subjects compared to younger controls for both the acceleratory and deceleratory phases of movements regardless of amplitude (Darling, Cooke, & Brown, 1989). This increase in movement variability is shown in the phase plane trajectories in Figure 5 where radii of variability ellipses are equivalent to 2 S.D. in position (horizontal axis) and velocity (vertical axis) calculated at 10 ms intervals throughout averaged elbow extension movements. Variability during the acceleratory phase increased more rapidly and to a greater extent in the elderly, contributing to the overall increase in trajectory variability.

In addition to increased trajectory variability, movement-related muscle activity was also more variable in elderly subjects. This was true not only for phasic activity (cf. initial agonist bursts in Fig. 5B, 30 - 70 deg) but also for relative levels of tonic activation preceding and following the movement. Although elderly subjects showed qualitatively normal control of the initial agonist burst in that both magnitude and duration increased with movement amplitude (cf. Brown & Cooke, 1981,

1984), clear phasic activation of the antagonist was not a consistent finding. Instead, elderly subjects often showed a gradual or step-like increase in the level of antagonist activity. This typically resulted in coactivation of agonist and antagonist muscles during the deceleratory phase of the movement. When a distinct burst of activity in the antagonist muscle was occasionally seen, it often occurred too early in the movement.

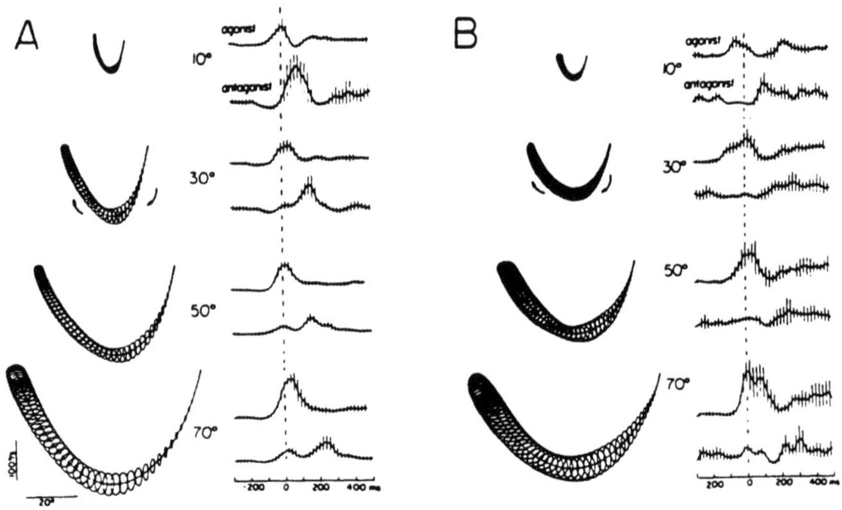

FIGURE 5. Movement trajectories and electromyographic activity (EMGs) for movements of different amplitudes. Shown are averaged phase plane trajectories (left side) and the associated EMGs (right side) for extension movements made by a control subject (A) and an elderly subject (B). Superimposed on the trajectories are ellipses which represent variability in the trajectories at 10 ms intervals. The vertical lines superimposed on the EMG records represent 2 S.D. and indicate variability in EMG amplitude at 20 ms intervals during the movements. EMGs were lowpass filtered (10 Hz, 0 phase lag) and aligned to acceleration onset prior to averaging. Dashed line indicates acceleration onset (from Darling et al., 1989).

Since appropriately timed, phasic antagonist activity plays a major role in the smooth deceleration of goal-directed movements (Brown & Cooke, 1990; Marsden, Obeso, & Rothwell, 1983), one may conclude that temporally asymmetric and variable movement trajectories in the elderly result, in part, from impaired control of phasic antagonist drive. In this regard, it is of interest to compare the above findings with those obtained from studies in patients with cerebellar dysfunction. In mild cerebellar ataxia, visually-guided elbow movements are characterized by prolonged decelerations which may contain two or more submovements (Brown et al., 1990; Brown, Kessler, Hefter, Cooke, & Freund, 1993). These movements, characterized by minimal dysmetria, are generally associated with either increased tonic antagonist activity which persists throughout the movement or phasic antagonist activity which occurs inappropriately early in the movement (Brown, Hefter, Cooke, & Freund, 1989). These and other studies support the view that the cerebellum plays a critical role in motor coordination (see Thach, Goodkin, & Keating, 1992, for a comprehensive review) which, at the single joint level, translates into temporal programming of reciprocally organized muscle groups (Brown, Hefter, Cooke & Freund, 1989; Hore, Wild, & Diener, 1991). Taken together with known age-related, neuronal loss in cerebellar structures, it is reasonable to conclude that impaired ability to consistently reproduce time symmetric movement profiles in the elderly may reflect subclinical cerebellar degeneration. Such a notion is in contrast to the view that, given the similarities between age-related and Parkinsonian motor impairment, deterioration of motor function in the elderly reflects degeneration of basal ganglia structures (Dobbs, Lubel, Charlett, Bowes, O'Neill, Weller, & Dobbs, 1992; Nagasaki, Itoh, Maruyama, & Hashizume, 1988; Stelmach et al., 1988). However, a recent comparative study of upper limb kinematics in Parkinson's disease patients and patients with mild cerebellar ataxia supports the view that age-related cerebellar degeneration may play a significant role in the generation of motor deficits in the elderly. In contrast to cerebellar patients who produced temporally asymmetric elbow flexion movements, movement symmetry was preserved in the Parkinson group despite reduced peak velocities and prolonged movement durations (Hefter, Brown, Cooke, & Freund, in press).

EFFECTS OF PRACTICE ON MOVEMENT KINEMATICS AND MUSCLE ACTIVATION PATTERNS

It is well established that practice leads to a significant improvement in motor performance most generally seen as a reduction in both reaction and movement times. In the elderly, practice effects appear to be more pronounced in tasks which form part of the individual's normal repertoire of daily activities. For example, interkey reaction times of older, experienced typists have been found to be comparable to reaction times of young typists (Salthouse, 1984). Studies involving experimental paradigms also exist which indicate that practice can enhance the performance of novel motor tasks in elderly subjects, particularly for reaction-time tasks (Clark, Lanphear, & Riddick, 1987; Murrell, 1970).

In addition to changes in point kinematic variables such as peak speed and movement times, it has also been shown that variability throughout the movement can be improved with practice (Darling & Cooke, 1987). Typically, young subjects show a greater reduction in variability during the deceleratory compared to the acceleratory phase of the movement. However, in cases where there is a noticeable increase in movement speed with practice, trajectory variability increases.

In order to determine if practice had similar effects on trajectory variability in elderly subjects and thus could improve consistency of performance, similar studies were performed on 5 subjects aged 68 - 95 years. Each subject performed a series of 180 flexion and 180 extension step tracking elbow movements. Movement amplitude was kept constant at 30 deg. Movement irregularities, particularly during the deceleratory phase, persisted in as many as 30 percent of the first 30-40 movements. With practice, however, moment to moment trajectory variability decreased across trials as shown in Figure 6 for an 89 year old subject. A comparison of the size of variability ellipses for the first 30 flexion movements versus movements performed near the end of the session (movements 121-150) indicates that prolonged practice of even a simple, single-joint motor task can lead to improved performance in the elderly. Practice appeared to have the greatest effect in reducing trajectory variability during movement deceleration, resulting in an approximate 50 percent reduction in the area of position-velocity ellipses. Small but significant decreases were also observed for the acceleratory phase of the movement. Variability in the relative durations of acceleration and deceleration (symmetry ratio) also decreased with practice. Thus, the kinematic profile of the movements became more stereotyped as a result of practice.

In addition to reducing variability of the movement trajectory, extended practice modified the pattern of movement-related muscle activity, particularly in the antagonist muscle. During the first set of averaged movements, a gradual increase in antagonist activity was observed in place of a distinct phasic burst characteristic of movements made by younger subjects (cf. Figure 1). With practice, variability in the level of agonist and antagonist premovement activity decreased and a more burst-like antagonist pattern began to emerge. Despite an overall reduction in magnitude variability however, timing of phasic antagonist activity often occurred early in the movement leading to considerable phasic agonist-antagonist coactivation.

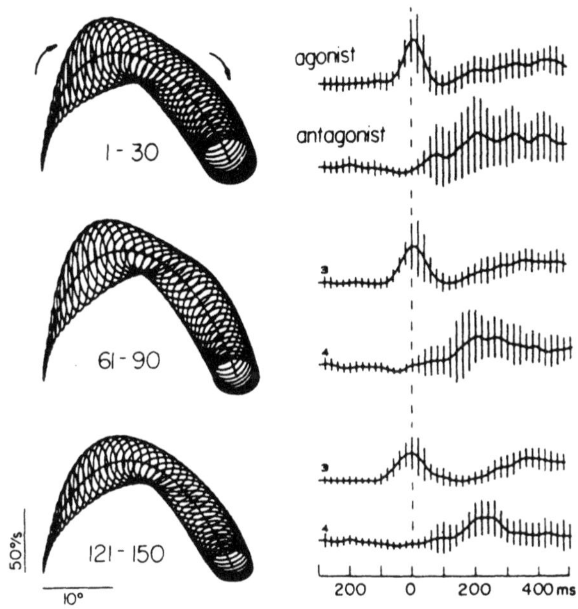

FIGURE 6. Effects of practice on variability of movement trajectories and related EMGs. On the left side are shown averaged movement trajectories with superimposed ellipses to show trajectory variability at 10 ms intervals. On the right side are the associated agonist and antagonist muscle EMGs with vertical lines superimposed to show variability in EMG magnitude at 20 ms intervals. Prior to averaging, the rectified EMGs and kinematic records were aligned to acceleration onset. The rectified EMGs were lowpass filtered (10 Hz, 0 phase lag) prior to averaging. The dashed line indicates acceleration onset. (From Darling et al., 1989.)

Practice-related decreases in movement variability support the view that motor performance can be enhanced in the elderly through repetition. Indeed, it has been suggested that "overpractice that occurs in activities of daily living appears to retard age-related deterioration of physical performance" (Spirduso & MacRae, 1990, p. 193). The observation that practice appears to modulate the pattern of movement-related antagonist muscle drive suggests that practice may lead to "fine tuning" of central motor commands underlying movement generation. This, in turn, would lead to a reduction in overall trajectory variability and thereby extend the limits of performance in the elderly. It should be noted, however, that practice may also enhance motor ability at the information processing stage (Clark et al., 1987) and it would be reasonable to assume that the facilitatory effects of practice in aged populations involve both aspects of the motor control process.

It should be also noted that conditions do exist where practice does not appear to facilitate motor performance in the elderly. In a recent study examining 37 deg rapid pronation and supination movements about the wrist, practice had no effect on modifying the time course of the movement (Pratt et al., 1994). Secondary, corrective submovements persisted even after 200 wrist rotations. However, spatial (end point) accuracy did show a significant improvement with practice. To what extent differences in task conditions and/or methods of data analysis may contribute to contradictory findings with respect to practice effects on motor performance in the elderly clearly remains to be determined.

ADAPTATION TO NOVEL TRACKING TASKS

The ability of the central nervous system to rapidly adapt to novel changes in visually guided motor tasks has been, perhaps, best illustrated by reversing prism studies (Baizer and Glickstein, 1974; Thach et al., 1992). In these experimental paradigms, limb position must be recalibrated with respect to target location in order to maintain accuracy. Such remapping of limb spatial coordinates requires processing of error information and appropriate modification of descending motor commands responsible for movement. We have recently examined the ability of elderly subjects to adapt to unexpected changes in a visually guided tracking paradigm where, instead of target location, the spatial relation between displayed and actual arm position was altered.

Six elderly subjects (mean age = 71 +/- 5 yrs) first performed a series of "fast and accurate" 10 deg elbow movements (20 flexion, 20 extension). Without warning, the handle display gain was suddenly changed so that subjects now had to make 40 deg elbow movements in order to accurately place the handle cursor within the target zone. Target display gain remained unchanged. Experiments were performed in a darkened room to prevent visual feedback from the moving arm.

The first movement following the change in handle display gain was characterized by marked irregularities in both young and elderly subjects. Maximum velocity of the initial movement segment reflected the speed at which the previous set of 10 deg movements had been made.

This resulted in pronounced undershooting of the target immediately following the gain change with one or more corrective movements occurring in order to reach the target zone. This was most clearly seen in the overplotted velocity records shown in Figure 7 for a young and two elderly subjects. Young subjects adapted rapidly to the change in gain so that after 4-5 consecutive flexion-extension movements, peak velocity was appropriately scaled to reflect an approximate four fold increase in movement amplitude (from 10 to 40 deg). Elderly subjects were also able to adapt to an altered display gain, although the number of movements required before smooth, unimodel velocity trajectories were achieved varied considerably between subjects. In 4 of the 6 subjects, up to 10 consecutive movements were required before peak velocity was adequately scaled. In contrast, 2 subjects were able to scale peak velocity as rapidly as younger subjects. No correlations between speed of adaptation age or gender were observed in this small sample.

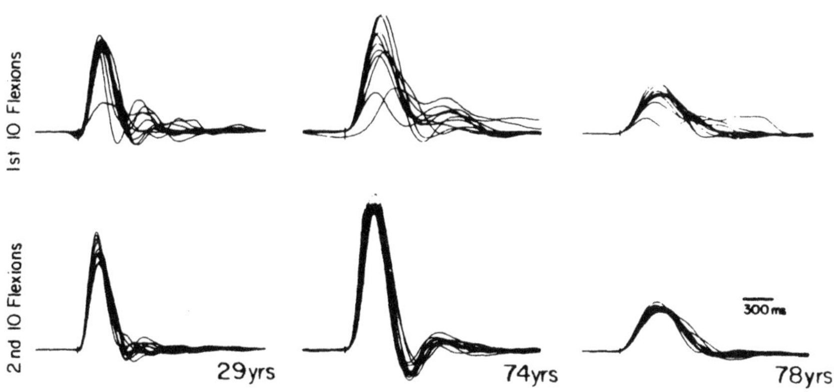

FIGURE 7. Adaptation to an unexpected change in handle display gain. Overplots of velocity records from the first ten (upper panel) and second ten (lower panel) elbow flexion movements following a four-fold change in display gain are shown for one young and two elderly subjects. In each plot, records have been aligned to start of movement.

The time course of peak velocity scaling is shown quantitatively for both young and elderly subjects in Figure 8 (upper graph). By the second flexion movement, young subjects had, on average, appropriately adjusted peak velocity to match the increase in movement amplitude. Over the course of subsequent movements, a small increase in velocity occurred. When averaged across subjects, the elderly showed a more gradual adaptation. Adjusted peak velocities (that is, peak velocity following adaptation), were more variable across young compared to elderly subjects which may simply reflect increased variability associated with higher movement speeds (Darling & Cooke, 1987; Schmidt, Zelaznik, Hawkins, Franks, & Quinn, 1979).

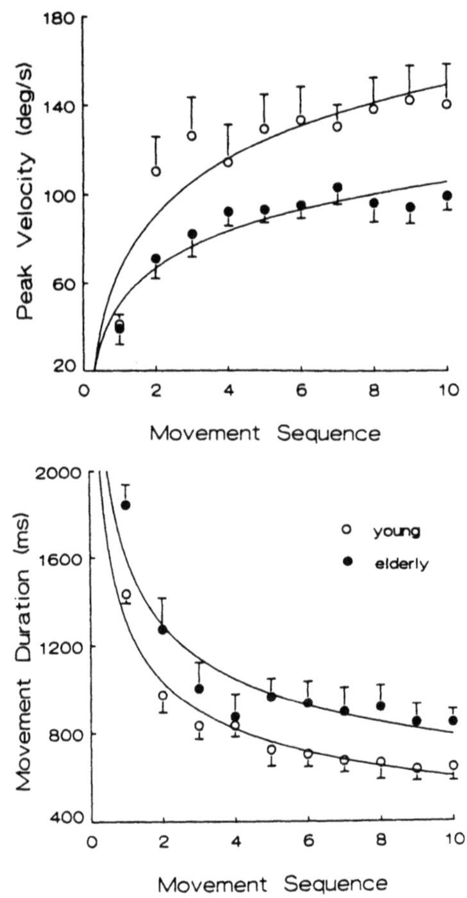

FIGURE 8. Scaling of peak velocity and movement duration. Data were obtained during adaptation to a four-fold change in handle display gain. Each data point represents group means (+/- 1 S.E.) for sequential flexion movements. Data points for movement sequence 1 represent the first flexion movement following the change in display gain. A logarithmic regression analysis provided the best fit for peak velocity data (young, r = .90; elderly, r = .93). Movement duration data were best described by a power function (young, r = .98; r = .92).

In both young and elderly subjects, scaling of movement duration was more gradual and, in the elderly group, highly variable. This is shown in the right hand graph in Figure 8. After approximately 10 consecutive flexion-extension movements, relatively stable movement durations were achieved with the elderly group taking 15 to 30 percent longer than young subjects to complete the movement.

It is important to note that, in this task, elderly subjects moved considerably more slowly than their younger counterparts. This contrasts with findings reported in the first series of experiments where peak speed was unaffected by age. While such differences may simply reflect widespread variability across small groups of elderly individuals, it is another example where direct comparisons of motor performance using different experimental conditions may give rise to conflicting observations.

The results of this study indicate that the ability to scale movements in response to unexpected changes in task requirements is not lost with age. What does appear to be age-dependent is the time required for modification of movement parameters to occur and even that may show considerable intersubject variability. It is certainly possible that an analysis of a larger and older population might show an even longer time course and more pronounced changes in movement kinematics. It is also unclear whether the delay in adaptation observed here is primarily perceptual-motor in origin or reflects a delay in the actual scaling of descending motor commands. However, no significant group differences in reaction time were observed, suggesting that, in these individuals, delay was related to an impaired ability to rapidly modulate muscle drive, in particular, the magnitude and duration of the initial agonist burst.

CONCLUSIONS

What do the findings presented here tell us about changes in motor performance with age? For certain movement parameters such as peak speed and movement duration, it might be concluded that aging has little apparent effect on the performance of visually- guided, single-joint movements. However, an analysis of the movement dynamics and, particularly movement-related muscle activation patterns would suggest otherwise. Thus, in terms of motor programming demands, even the simplest of limb movements show age-related changes compared to younger individuals. Most notably are prolonged decelerations, increased trajectory variability and impaired control of antagonist muscle activity.

It is unclear whether such movements made by the elderly can be considered "abnormal" or simply reflect adaptive strategies in order to accomplish a particular task. For instance, in those studies where both speed and endpoint accuracy were emphasized, movement deceleration was often prolonged leading to temporally asymmetric movement profiles. Prolonged decelerations have been also reported in younger subjects when accuracy demands are high (Carlton, 1981). Thus, asymmetric movements in the elderly may not necessarily be interpreted as impaired movement production per se but, in the face of age-related sensory, neuromuscular, and central neuronal loss, represent modification of a preferred movement profile in order to maximize task success. Similar strategy-related interpretations have been made in explaining single joint movement asymmetries seen in the early stages of cerebellar degeneration (Brown et al., 1990). Recently, Bennett, and Castiello (1994) have suggested that subtle changes in movement kinematics associated with reach and grasp movements may also reflect strategies developed to compensate for age-related degeneration in a variety of physiological systems.

The question remains as to the locus responsible for age-related changes in movement dynamics. In terms of motor-related brain structures, the findings reported here provide evidence that motor impairment in the elderly may be cerebellar in origin. This view contrasts with the widely held belief that age-related motor changes reflect loss of basal ganglia function. It would be naive to assume, however, that the decline in motor performance with aging can be attributable to a single brain structure or, for that matter, a specific aspect of the motor planning or programming process.

In summary, caution must be exercised in making sweeping conclusions from aging studies varying widely in task complexity, subject profiles, and specific movement parameters under investigation. This is particularly true for many laboratory tasks, where movement paradigms may not have a direct parallel in terms of activities of daily living. Thus, absolute measures of, for example, point kinematics such as peak velocity and movement duration may lead to inaccurate and misleading deductions regarding the limits of motor performance in the elderly individual. It is also clear that task familiarity, movement complexity in terms of interjoint and interlimb coordination as well as limb-posture interactions may also affect the level of motor performance. Despite these caveats, however, it is clear that, at least for relatively simple movements, the elderly can improve consistency of performance with practice and, given adequate time, can adapt to new motor tasks.

ACKNOWLEDGEMENTS

The author is grateful for the collaborative efforts of J. D. Cooke, Ph.D., Faculty of Applied Health Sciences, University of Western Ontario, London, Canada; W. G. Darling, Ph.D., Dept. of Exercise Science, University of Iowa, Iowa City, IA, USA; and H. Karbe, M.D., Dept. of Neurology, University of Cologne, Germany. These studies were supported, in part, by NSERC Canada and an Alexander von Humboldt award to SHB and HK.

REFERENCES

Allen, S. J., Benton, J. S., Goodhardt, M. J., Haan, E. A., Sims, N. R., Smith, C. C. T., Spillane, J. A., Bowen, D. M., & Davison, A. M. (1983). Biochemical evidence of selective nerve cell changes in the normal aging human and rat brain. *Journal of Neurochemistry, 41,* 256-265.

Aniansson, A., Hedberg, M., Henning, G., & Grimby, G. (1986). Muscle morphology, enzymatic activity and muscle strength in elderly men: A follow-up study. *Muscle and Nerve, 9,* 585-591.

Atkeson, C. G., & Hollerbach, J. M. (1985). Kinematic features of unrestrained vertical arm movements. *Journal of Neuroscience, 5,* 2318-2330.

Baizer, J. S., & Glickstein, M. (1974). Role of cerebellum in prism adaptation. *Journal of Physiology, 236,* 34-35.

Baloh, R. W. (1984). Neuro-otology of aging. In M. L. Albert (Ed.), *Clinical neurology of aging* (pp. 345-361). Oxford: Oxford University Press.

Bennett, K. M. B., & Castiello, U. (1994). Reach to grasp: changes with age. *Journal of Gerontology, 49* (1), 1-7.

Birren, J. E., & Botwinick, J. (1951). The relation of writing speed to age and the senile psychoses. *Journal of Consulting and Clinical Psychology, 15,* 243-249.

Brody, H. (1970). Structural changes in the aging nervous system. *Interdisciplinary Topics Gerontology, 7,* 9-21.

Brown, S. H. (1986). Control of movement initiation in humans. Unpublished doctoral thesis, University of Western Ontario, London.

Brown, S. H., & Cooke, J.D. (1981). Amplitude- and instruction-dependent modulation of movement-related electromyogram activity in humans. *Journal of Physiology, 316,* 97-107.

Brown, S. H., & Cooke, J. D. (1984). Initial agonist burst duration depends on movement amplitude. *Experimental Brain Research, 55,* 523-527.

Brown, S. H., & Cooke, J. D. (1990). Movement-related phasic muscle activation. I. Relations with temporal profile of movement. *Journal of Neurophysiology, 63* (3), 455-464.

Brown, S. H., Hefter, H., Cooke, J. D., & Freund, H.-J. (1989). Duration of movement-related EMG activity in patients with mild cerebellar dysfunction. *Proceedings of the 15th Annual Meeting of the Society for Neuroscience, 15,* 473.2.

Brown, S. H., Hefter, H., Mertens, M., & Freund, H.-J. (1990). Disturbances in movement trajectory due to cerebellar dysfunction. *Journal of Neurology, Neurosurgery, and Psychiatry, 53,* 306-313.

Brown, S. H., Kessler, K. R., Hefter, H., Cooke, J. D., & Freund, H.-J. (1993). Role of the cerebellum in visuomotor coordination. I. Delayed eye and arm initiation in patients with mild cerebellar ataxia. *Experimental Brain Research, 94,* 478-488.

Bugiani, O., Salvarini, S., Perdelli, G.L., Mancardi, G. L., & Leonardi, A. (1978). Nerve cell loss with aging in the putamen. *European Neurology, 17,* 286-291.

Campbell, M. J., McComas, A. J., & Petito, F. (1973). Physiological changes in aging muscles. *Journal of Neurology, Neurosurgery, and Psychiatry, 36,* 174-182.

Carleton, L. G. (1981). Processing visual feedback information for movement control. *Journal of Experimental Psychology: Human Perception and Performance, 7,* 1019-1030.

Clark, J. E., Lanphear, A. K., & Riddick, C. C. (1987). The effects of videogame playing on the response selection processing of elderly adults. *Journal of Gerontology, 42* (1), 82-85.

Cohen, M. M., & Lessell, S. (1984). Neuro-ophthalmology of aging. In M. L. Albert (Ed.), *Clinical neurology of aging* (pp. 313-344). Oxford: Oxford University Press.

Cooke, J. D., Brown, S. H., & Cunningham, D. A. (1989). Kinematics of arm movements in the elderly. *Neurobiology of Aging, 10,* 159-165.

Cooke, J. D., & Brown, S. H. (1990). Movement-related phasic muscle activation. II. Generation and functional role of the triphasic pattern. *Journal of Neurophysiology, 63,* 465-472.

Cooke, J. D., & Brown, S. H. (1994). Movement-related phasic muscle activation. III. The duration of phasic agonist activity initiating movement. *Experimental Brain Research, 99,* 473-482.

Darling, W. G., & Cooke, J. D. (1987). Changes in the variability of movement trajectories with practice. *Journal of Motor Behavior, 19,* 291-309.

Darling, W. G., Cooke, J. D., & Brown, S. H. (1989). Control of simple arm movements in elderly humans. *Neurobiology of Aging, 10,* 149-157.

Darling, W. G., & Stephenson, M. (1993). Directional effects on variability of upper limb movements. In K. M. Newell & D. M. Corcos (Eds.), *Variability and motor control* (pp. 65-88). Champaign, IL: Human Kinetics Publishers, Inc.

Davies, C. T. M., & White, M. J. (1983). The contractile properties of elderly human triceps surae. *Gerontology, 29,* 19-23.

Dobbs, R. J., Lubel, D. D., Charlett, A., Bowes, S. G., O'Neill, J., Weller, C., Dobbs, S. M. (1992). Hypothesis: Age-associated changes in gait represent, in part, a tendency towards Parkinsonism. *Age and Ageing, 21,* 221-225.

Flash, T. (1987). The control of hand equilibrium trajectories in multi-joint arm movements. *Biological Cybernetics, 57,* 257-274.

Goggin, N. L., & Meeuwsen, H. J. (1992). Age-related differences in the control of spatial aiming movements. *Research Quarterly for Exercise and Sport, 63,* 366-372.

Goggin, N. L., & Stelmach, G. E. (1990). Age-related differences in a kinematic analysis of precued movements. *Canadian Journal on Aging, 9,* 371-385.

Grimby, G. (1988). Physical activity and effects of muscle training in the elderly. *Annals of Clinical Research, 20,* 62-66.

Hall, T. C., Miller, A. K. H., & Corsellis, J. A. N. (1975). Variations in the human Purkinje cell population according to age and sex. *Neuropathology and Applied Neurobiology, 1,* 267-292.

Hallett, M., & Marsden, C. D. (1979). Ballistic flexion movements of the human thumb. *Journal of Physiology (London), 294,* 33-50.

Hefter, H., Brown, S. H., Cooke, J. D., & Freund, H.-J. (in press). Impairment of timing versus scaling: A comparison of forearm trajectories in cerebellar and Parkinson's patients. *Electromyography and Clinical Neurophysiology.*

Hore, J., Wild, B., & Diener, H.-C. (1991). Cerebellar dysmetria at the elbow, wrist and fingers. *Journal of Neurophysiology, 65,* 563-571.

Hogan, N. (1984). An organizing principle for a class of voluntary movements. *Journal of Neuroscience, 4,* 2745-2754.

Jenkyn, L. R., Reeves, A. G., Warren, T., Whiting, R. K., Clayton, R. J., Moore, W. W., Rizzo, A., Tuzun, I. M., Bonnett, J. C., &

Culpepper, B. W. (1985). Neurologic signs in senescence. *Archives of Neurology, 42,* 1154-1157.
Karst, G. M., & Hasan, Z. (1987). Antagonist muscle activity during human forearm movements under varying kinematic and loading conditions. *Experimental Brain Research, 67,* 391-401.
Kenshalo, D. R., Sr. (1979). Changes in the vestibular and somesthetic systems as a function of age. In J. M. Ordy & K. Brizzee (Eds.), *Sensory systems and communication in the elderly* (Aging, Vol. 10). New York: Raven Press.
Larsson, L. (1978). Morphological and functional characteristics of the ageing skeletal muscle in man. *Acta Physiologica Scandinavica (Suppl.), 457,* 1-29.
Lexell, J., Taylor, C. C., & Sjostrom, M. (1988). What is the cause of the ageing atrophy? Total number, size and proportion of different fibre types studied in whole vastus lateralis muscle from 15- to 83-year-old men. *Journal of the Neurological Sciences, 84,* 275-294.
McDonagh, M. J. N., White, M. J., & Davies, C. T. M. (1983). Different effects of ageing on the mechanical properties of human arm and leg muscles. *Gerontology, 30,* 49-54.
McGeer, P. L., McGeer, E. G., & Suzuki, J. S. (1977). Aging and extrapyramidal function. *Archives of Neurology (Chicago), 34,* 33-35.
Marsden, C. D., Obeso, J. A., & Rothwell, J. C. (1983). The function of the antagonist muscle during fast limb movements in man. *Journal of Physiology, 335,* 1-13.
Morasso, P. (1981). Spatial control of arm movements. *Experimental Brain Research, 42,* 223-227.
Mulch, G., & Peterman, W. (1979). Influence of age on results of vestibular function tests: Review of literature and presentation of caloric test results. *Annals of Otology, Rhinology, and Laryngology, 88,* 1-17.
Munhall, K. G., Ostry, D. J., & Parush, A. (1985). Characteristics of velocity profiles of speech movements. *Journal of Experimental Psychology: Human Perception and Performance, 11,* 457-474.
Murrell, F. H. (1970). The effect of extensive practice on age differences in reaction time. *Journal of Gerontology, 25,* 268-274.
Nagasaki, H., Itoh, H., Maruyama, H., & Hashizume, K. (1988). Characterstic difficulty in rhythmic movement with aging and its relation to Parkinson's disease. *Experimental Aging Research, 14,* 171-176.
Nelson, W. L. (1983). Physical principles for economies of skilled movements. *Biological Cybernetics, 46,* 135-147.

Newton, J. P. & Yemm, R. (1986). Changes in the contractile properties of human first dorsal interosseous muscle with age. *Gerontology, 32,* 98-104.
Ostry, D. J. (1986). Characteristics of human jaw movement in mastica-tion and speech. *Neuroscience Letters [Suppl.], 26,* S87.
Pratt, J., Chasteen, A. L., & Abrams, R. A. (1994). Rapid aimed limb movements: Age differences and practice effects in component sub-movements. *Psychology and Aging, 9,* 325-334.
Rogers, J., Silver, M. A., Shoemaker, W. J., & Bloom, F. E. (1980). Senescent changes in a neurobiological model system: Cerebellar Purkinje cell electrophysiology and correlative anatomy. *Neuro-biology and Aging, 1,* 3-11.
Sabin, T. D., & Venna, N. (1984). Peripheral nerve disorders in the elderly. In M. L. Albert (Ed.), *Clinical neurology of aging* (pp. 425-444). Oxford: Oxford University Press.
Salthouse, T. A. (1979). Adult age and the speed-accuracy trade-off. *Ergonomics, 22,* 811-821.
Salthouse, T. A. (1984). Effects of age and skill in typing. *Journal of Gerontology, 113,* 345-371.
Salthouse, T. A. (1985). Speed of behavior and its implications for cognition. In J. E. Birren & K. W. Schaie (Eds.), *Handbook of the psychology of aging* (pp. 400-426). New York: Van Nostrand Reinhold.
Schaumburg, H. H., Spencer, P. S., & Ochoa, J. (1983). The aging human peripheral nervous system. In R. Katzman & R. D. Terry (Eds.), *The neurology of aging* (pp. 111-122). Philadelphia: F.A. Davis Co.
Scheibel, M. E., Tomiyasu, U., & Scheibel, A. B. (1977). The aging human Betz cell. *Experimental Neurology, 56,* 598-609.
Scheibel, A. B. (1979). Aging in human motor control systems. In J. M. Ordy & K. Brizzee (Eds.), *Sensory systems and communication in the elderly* (Aging, Vol. 10, pp. 297-310). New York: Raven Press.
Schmidt, R. A., Zelaznik, H. N., Hawkins, B., Frank, J. S., & Quinn, J. T. (1979). Motor-output variability: A theory for the accuracy of rapid motor acts. *Psychological Review, 86,* 415-451.
Sekuler, R., Klein, D., & Dismukes, K. (1982). *Aging in human visual functions.* New York: Alan R. Liss.
Sharpe, J. A., & Sylvester, T. O. (1978). Effects of aging on horizontal smooth pursuit. *Investigative Ophtalmology and Visual Science, 17,* 465-468.

Singleton, W. T. (1954). The change of movement timing with age. *British Journal of Psychology, 45,* 166-172.

Soechting, J. F. (1984). Effect of target size on spatial and temporal characteristics of pointing movement in man. *Experimental Brain Research, 54,* 121-132.

Spirduso, W. W., & MacRae, P. G. (1990). Motor performance and aging. In J. E. Birren & K. W. Schaie (Eds.), *Handbook of the psychology of aging* (3rd ed., pp. 183-200). San Diego: Academic Press, Inc.

Stelmach, G. E., Goggin, N. L., & Amrhein, P. C. (1988). Aging and preprogramming: The restructuring of planned movements. *Psychology and Aging, 3,* 151-157.

Thach, W. T., Goodkin, H. P., & Keating, J. G. (1992). The cerebellum and the adaptive coordination of movement. *Annual Review of Neuroscience, 15,* 403-442.

Tzankoff, S. P., & Norris, A. H. (1977). Effect of muscle mass decrease on age-related BMR changes. *Journal of Applied Physiology, 43,* 1001-1006.

Vandervoort, A. A. (1992). Effects of ageing on human neuromuscular function: Implications for exercise. *Canadian Journal of Sports Sciences, 17* (3), 178-184.

Virji-Babul, N., Brown, S. H., & Cooke, J. D. (1994). A common movement profile is preserved by EMG changes under different gravitational loads. *Experimental Brain Research, 99,* 38-46.

Virji-Babul, N., & Cooke, J. D. (1995). Influence of joint interac-tional effects on the coordination of planar two-joint arm movements. *Experimental Brain Research, 103,* 451-459.

Wacholder, K., & Altenburger, H. (1926). Beitrage zur Physiologie der wurhlichen Bewegung. X. Einzelbewegungen. Pflugers Archiv-*European Journal of Physiology, 214,* 642-661.

Warabi, T., Noda, H., & Kato, T. (1986). Effect of aging on sensorimotor function of eye and hand movements. *Experimental Neurology, 92,* 686-697.

Welford, A. T., Norris, A. H., & Shock, N. W. (1969). Speed and accuracy of movement and their changes with age. *Acta Physiologica, 30,* 3-15.

Welford, A. T. (1977). Motor performance. In J. E. Birren & K. W. Schaie (Eds.), *Handbook for the psychology of aging*. New York: Van Nostrand Reinhold.

Changes in sensory motor behavior in aging
A.-M. Ferrandez and N. Teasdale (Editors)

SLOWNESS, VARIABILITY, AND MODULATIONS OF GAIT IN HEALTHY ELDERLY

Anne-Marie FERRANDEZ,[1] Madeleine DURUP,[1] and Fernand FARIOLI[2]

CNRS, Université de la Méditerranée et Université de Provence

Abstract

Slowness and intra-individual variability increase as age increases. The present study was designed to address the issues of slowness and variability in elderly gait. The effects of slowness were investigated. In two experiments, a young adult group matched on walking speed (walking at the same speed as elderly people) was used to determine whether the observed effects of aging can be attributed (at least in part) to slowness. Age-related changes in spatial modulations of gait were tested, and the implications of slowness and variability on these modulations were investigated. The experiments reported here showed that the gait of the elderly is normal if we take their speed into account: in particular, the ability to intentionally increase walking speed is still intact in elderly subjects. In steady-state walking, young adult subjects (even when walking slowly) exhibited less intra-individual variability than elderly subjects. However high steady-state variability did not alter these gait modulations in the elderly, since groups matched on speed were found to apply the same modulation strategies.

Key words: Aging, intra-individual variability, locomotion, slowness.

---

1. Cognition et Mouvement, URA CNRS 1166, Université de la Méditerranée, Faculté de Médecine, IBHOP, Traverse Charles Susini, 13388 Marseille Cedex 13, France (e-mail: ibhop@romarin.univ-aix.fr).
2. CREPCO, URA CNRS 182, Université de Provence, 13621 Aix-en-Provence Cedex 1, France (e-mail: farioli@romarin.univ-aix.fr).

INTRODUCTION

Motor variability in the elderly

Increasing variability with age can be demonstrated at several levels. As organisms age, inter- and intra-individual variability increases, not only in performance but at the morphological, biochemical, physiological, and behavioral levels, especially in humans.

This long-known fact has become a new issue. Two recent studies (Nelson & Dannefer, 1992, and Morse, 1993) have shed some light on this question. Nelson and Dannefer (1992) noted that 65% of the 185 gerontological studies they examined reported increases in inter-individual variability with age. Morse (1993) directly questioned the validity of the assertion that older people are more variable than younger people by calculating coefficients of inter-individual variability in measures of reaction time, memory, and intelligence for a large number of studies published between 1986 and 1990 in *Psychology and Aging* and in the *Journal of Gerontology*. In Lupien and Lecours' (1993) recent review of the literature two approaches to the aging heterogeneity phenomenon were distinguished: the experimental approach, which aims at describing the methodological factors that might create artifactual performance heterogeneity within the aged population, and the developmental approach, which explains this phenomenon by suggesting that senescence processes are not homogeneous, i.e., they do not occur at the same pace in all individuals.

Intra-individual variability in elderly subjects has not been investigated as much as inter-individual variability. Intra-individual variability has been studied in animals (Jänicke, Coper, & Schulze, 1988). A tentative explanation (Curcio, Buell, & Coleman, 1982) for increasing intra-subject variability with age can be given by referring to a heterogeneous panorama, at a given age, of the interactions between degenerative and regenerative phenomena in various regions of the central nervous system: "The static picture at any instant is a representation of the balance between degenerative and regenerative phenomena. At any specified age this balance will probably vary from one region to another. The factors that operate to differentiate CNS regions and species showing varying rates of degeneration and regeneration remain one of the important unknowns in gerontological research" (Curcio et al., p. 24).

Regarding motor actions, the question of intra-subject variability has been a crucial issue for many years. In 1967, Fetz (quoted by Hatze, 1986) used the reciprocal of the coefficient of variation as a measure of the accuracy of the outcomes of repeated motor acts. Bernstein (1967) considered decreased variability in a given complex movement as an indicator of progress in the acquisition of the new motor skill. The question of the link between variability and accuracy in movement was the subject of considerable debate about 15 years ago (reported by Worringham, 1991). In his elegant model proposed as early as 1986, Hatze (1986) combined the stochastic notion of motor variability (bandwidth, neuromotor noise, ...) with an approach which considers the evolution of a given variability over a given time period (here, several minutes). He described this evolution using a model of entropy. This approach is consistent with a large corpus of theoretical and empirical studies conducted at the end of the last decade, which focused on the importance of steady-states and their fluctuations (Kay, Saltzman, & Kelso, 1991; Scholtz & Kelso, 1989, 1990; Schöner & Kelso, 1988). Generally these studies were aimed at determining which parameters are controlled in order to return to a stable state after a perturbation, or to switch from one stable motor pattern to another.

Low intra-individual gait variability in young healthy humans has been demonstrated several times over the past few years (Inman, Ralston, & Todd, 1981; Patla, 1985; Winter, 1984), no matter what level is being observed: kinetics, kinematics, or EMG activity. In their tables of the main features of gait patterns in healthy young adults and healthy elderly to be used for diagnosis purposes, some authors have given indications about variability in EMG activity (Winter, 1987) and kinematic parameters (Winter, 1987; Winter, Patla, Frank, & Walt, 1990; Dobbs, Charlett, Bowes, Weller, Hughes, & Dobbs, 1993). They have also tried to find correlations between gait variability and age or pathology. The results in this matter are confusing. Winter's studies (Winter, 1987; Winter et al., 1990) reported lower gait variability in healthy elderly people than in young adults. This "more consistent motor pattern" in the elderly was interpreted by the authors as more "robotic" walking explained by the partial loss of neural plasticity. Less intra-individual variability could also have its origin in the more cautious gait pattern of this population. However note that these authors systematically normalized gait cycles to 100% before analyzing their data; this process contaminates the variability measures.

However, even though increasing intra-individual variability with age and pathology was not found in some cases, this is what the majority of researchers usually expected to find in their studies of gait patterns or upper limb movements. One of the most disappointing studies on this question was certainly Gabell and Nayak's (1984). These authors distinguished two kinds of locomotor parameters: (1) stride length and cycle duration, which are related to the automatic bases of locomotor pattern, and (2) stride width and double support duration, related to control of balance during walking. Although the variability (coefficient of variation on any of the chosen parameters) obtained on the parameters related to balance was higher than that obtained on those related to automatism, this study was unable to show any difference in intra-individual variability between the elderly and the young adult groups. As the elderly population was a carefully selected group (only 32 subjects were selected, on the basis of neurological examinations and the absence of recent falls, from an original population of more than 1100), these authors concluded that any high variability reported on gait parameters in the elderly could be interpreted as an indicator of pathology.

Higher intra-individual variability in parkinsonian subjects compared to healthy elderly subjects has been observed for certain gait parameters by comparing coefficients of variation of subjects in the two populations (Blin, Ferrandez, & Serratrice, 1990, on stride length but not cycle duration). However, Dobbs et al. (1993) could not find greater stride length variability (using standard deviations and not coefficients of variation) in parkinsonian subjects compared to healthy elderly. They explained this lack of a difference by the fact that "none of them was house bound or exhibited clinical dementia, factors which may be associated with irregularity of stride length (Imms & Edholm, 1981; Visser, 1983)" (Dobbs et al., 1993, p. 29).

Intra-individual variability in healthy or pathological aging is not any clearer for upper limb movements than it is for locomotion. Phillips, Stelmach, and Teasdale (1991) studied handwriting movements in young adults, healthy elderly adults, and parkinsonian elderly. They did not observe a difference between populations in the coefficients of variation for stroke length or stroke duration. By contrast, using a more sophisticated index (signal-to-noise ratio), Teulings and Stelmach (1993) reported that parkinsonians exhibited greater variability on the same kind of movements than did the elderly controls. However, the latter group was not more variable than the young adult group. Using another kind of arm movement (step-tracking task), Cooke, Brown, and Cunningham (1989) showed that movements made by elderly subjects were more variable than those of young subjects, particularly at smaller amplitudes and velocities. Their results also indicated that intra-individual variability decreased as amplitude increased.

To our knowledge, the study by Cooke, Brown, and Cunningham (1989) is one of the rare studies on elderly movement intra-individual variability in which velocity components were varied. Movement amplitude ranged from 10 to 80 degrees, allowing comparisons within a wide range of values. Velocity also was varied as an intentional instruction: subjects had to perform at their own speed on one series of trials, and as fast and accurately as possible on the other series. Given the importance of slowness as a general feature of motor behavior in the elderly, this experimental feature is of particular interest.

Behavioral slowing in the elderly

The age-related slowing of behavior is a widely investigated topic and has been studied through both cognitive and sensorimotor activities. Slowing seems to affect almost every function in the elderly (Birren, Woods, & Williams, 1980). Only a few sensorimotor actions do not exhibit a slower execution rate with age; this is the case, for example, for the patellar reflex, as Clarkson showed (1978). However, as Salthouse noted, "There is a strong tendency for the age differences in speed to increase with the cognitive complexity of the task, and thus it is not unreasonable to expect very slight differences on simple reflex activities" (Salthouse, 1985, p. 253).

The planning and controlling of a given movement during its execution involves several levels of functioning, including interactions between central processes and peripheral inputs and outputs, sensory modalities, and effector conditions (joints, muscles, etc.). All of these levels of functioning are gradually altered in the elderly. Muscle strength decreases, especially because of a reduction in the number and diameter of muscle fibers (fast-twitch fibers are damaged first; see Larsson, 1983; Lexell, 1993), and the span of joint openings also drops. Nearly every sensory modality is affected: visual information processing takes longer in older subjects, who need more contrasted stimuli (Sekuler & Hutman, 1980, Kline, Schieber, & Coyne, 1983); deteriorating changes occur in the vibration sense; and sensitivity to passive joint opening declines (Kenshalo, 1977, 1979; Kokmen, Bossemeyer, & Williams, 1978). However, Stemach and Sirica (1986) suggested for active joint sensation that the elderly rely more heavily on the active corollary of efferent discharges foı maintaining proprioceptive awareness.

For complex behaviors, Diggles-Buckles (1993) cites several causes of age-related slowing, such as health (depression, schizophrenia, fitness), attitude (cautiousness), disuse, arousal levels (older people may be over- or under-aroused), strategy differences (serial versus parallel processing), and differences in attentional capacity (capacity changes, distractibility, diminished inhibitory processes). She insists on the fact that a combination of these factors may contribute to central nervous system decline.

Several studies have shown that the characteristics of sensorimotor actions are the same in older and younger adults, but are slower in the former. Moreover, despite deficits in several sensory modalities, older adults can maintain good performance when sensory information is redundant and all types are available. For example, Teasdale and collaborators (Teasdale, Stelmach, & Breuning, 1991), who studied postural sway while varying the kind of information available to the subject (altered visual and/or support surface), reported that the "exclusion or disruption of one of the sensory inputs, alone, was not consistently sufficient to differentiate between elderly and young adults, because of compensation by the remaining sensory sources" (Teasdale et al., 1991, p. 239). These compensation mechanisms ensure elderly humans relatively good sensorimotor performance, given the magnitude of their deficits. As Williams (1990) remarked, "If one looks at the myriad changes that occur in muscular and neural functions with age it seems almost miraculous that eye-hand coordination behaviors are maintained to the extent that they are in the aging individual" (Williams, 1990, p. 351).

In kinematic research on human movement, an easy way to take slowness into account in the study of movement control in the elderly consists of varying or controlling velocity or related movement parameters (amplitude, duration). By matching walking speed (requesting young adult subjects to walk at very slow speeds), or speed and stride length in certain situations, one can better determine whether features known to be specific to the elderly can in fact be attributed to slowness. This is particularly helpful for a movement like locomotion, which, being what one might call a dynamic equilibrium, is particularly sensitive to the speed at which it is performed.

The present study was designed to address the issues of slowness and variability in elderly gait. Some of the questions raised were: Do phenomena such as the shortening of strides and the lengthening of the double support phase simply result from the low walking speed adopted by these persons? Despite their slowness, are elderly people able to modulate their speed efficiently? Does the ability to modulate speed evolve during aging? Is the gait of older adults more variable than that of younger adults? If so, can between-subject variability be explained by speed alone, or is it age related? Does increasing variability have any repercussions on the ability to make accurate modulations of stride length in the elderly? The following experiments were conducted in an attempt to answer these questions.

EXPERIMENT 1: EFFECTS OF AGING AND SLOWING ON GAIT

Subjects and instructions

Elderly subjects. The experiment was conducted in a hospital. Subjects were recruited from all wards, where they were spending one or two days for a checkup or preventive examinations. A wide variety of socioeconomic classes and educational levels were represented. All subjects were healthy and could perform outdoor activities normally. Sixty-seven elderly adults (31 males and 36 females) between the ages of 60 and 92 (mean: 72; median: 72; SD: 7.84) were tested. All subjects were examined by a neurologist who excluded those suffering from disorders causing pain in walking (bone or joint disorders in the lower limbs or spinal cord, vascular or neuromuscular disorders in the lower limbs); deficits of motor, sensorial, cerebellar, or vestibular origin; major visual defects; severe heart or breathing malfunction; asthenia or depressive tendencies (or patients under sedative medication); or severe cognitive disorders (poor comprehension or execution of instructions). Each subject was led into the experiment room (8 meters long and 6 meters wide), and after being fitted with an apparatus, was asked to walk to the assistant standing at the other end of the room. Subjects walked barefoot. Each subject signed an informed consent form, in compliance with university rules.

Young adult subjects. The young adult group was composed of 4 males and 4 females who were members of the laboratory staff. Their age range was 22 to 38 (mean: 31, median: 32, SD: 4.8). For this group, the experiment was conducted in a 20-meter corridor in the laboratory. Their instructions were: "You will be asked to walk six times, and each time you must walk a little faster than the time before. So the first walk should be very slow, as slow as possible, and the sixth walk should be very fast, as fast as possible". These subjects walked with their usual comfortable shoes. Each subject signed an informed consent form, in compliance with university rules.

Apparatus and materials

Locomotor parameters were automatically recorded using an apparatus designed by Bessou, Dupui, Montoya, and Pages (1989) which measures the longitudinal displacement of both feet during locomotion by means of potentiometers. This apparatus can be used to determine the characteristics of locomotor displacement (stride length, cycle, stance, swing and double support durations, stride and swing velocities) over a long distance (more than 10 meters). It does not necessitate any special walkway or specific lighting, and can be used to record natural locomotion without any discomfort to the subjects, which is very important for elderly persons. In the present experiment, the data were recorded at 50 Hz, then filtered using a Finite Impulsive Response filter (McClellan & Parks, 1973) with a 33-point window and a 10-Hz cutoff (-3 dB). The calibration of the apparatus defined the volt/meter coefficient to be used in computing spatial and velocity data.

Data analysis

The relationships between variables were analyzed using linear and second-order polynomial regressions. Forward stepwise multiple regressions served to define the best models describing the relationships between the independent variables and a given dependant variable. Only covariates found to be significant at the .05 level were included in the final models. Quadratic terms were only considered as covariates when the corresponding linear term(s) had been accepted for a model. When a higher order term was accepted, the linear term was still included in the model regardless of its significance level. The values used for the explained variance were the adjusted values (R^2adj), in order to take into account the chance contributions of each variable in the model being tested. Some of the models of the elderly subjects ($n_1 = 67$) were compared with those of the young adult subjects ($n_2 = 48$: 8 subjects at 6 walking speeds) using an adaptation of Student's *t*-test, following formula (1) (Dagnelie, 1986). The significance level used was $<.05$.

$$t_v = \frac{B_1 - B_2}{\sqrt{S^2_{Y.X}\left\{\frac{1}{S^2_{X_1}(n_1 - 1)} + \frac{1}{S^2_{X_2}(n_2 - 1)}\right\}}} \quad (1)$$

where $v = n_1 + n_2 - 2k$, k is the number of coefficients in the equation, B_1 and B_2 are the coefficients to be compared, S^2x_1 and S^2x_2 are the variances of the x_1 and x_2 distributions, and $S^2{}_{Y.X}$ is a composite residual variance calculated as follows:

$$S^2_{Y.X} = \frac{S^2_{y1}(1 - r^2_{y1.x1})(n_1 - k) + S^2_{y2}(1 - r^2_{y2.x2})(n_2 - k)}{n_1 + n_2 - 2k} \quad (1a)$$

Results and discussion

Effect of age, sex, and height on kinematic parameters of gait

Stride length and cycle duration can be considered as the fundamental parameters of locomotion because they are a synthesis of its spatio-temporal characteristics (Inman et al., 1981). Velocity is calculated by dividing stride length by cycle duration. Double support duration can be considered as an index of stability and control of balance (Gabell & Nayak, 1984). Swing phase is the complement of the double support phase in stride duration.

We first checked for a possible relationship between age and height, which gave no significant result, R^2adj = 21%, $F(1, 65) = 3.11$. The

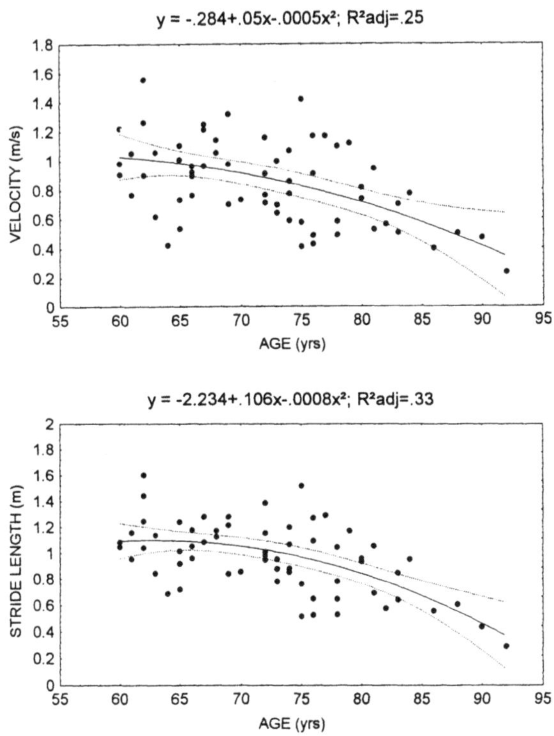

FIGURE 1. Relationship between age and two locomotor parameters (velocity and stride length) in elderly subjects. The solid lines represent the best fit of the second order polynomial equations, and the dotted lines represent the limits of the 95% confidence interval of the model.

models were tested on velocity, stride length, swing duration, and double support duration with age, sex, and height using forward stepwise multiple regressions.

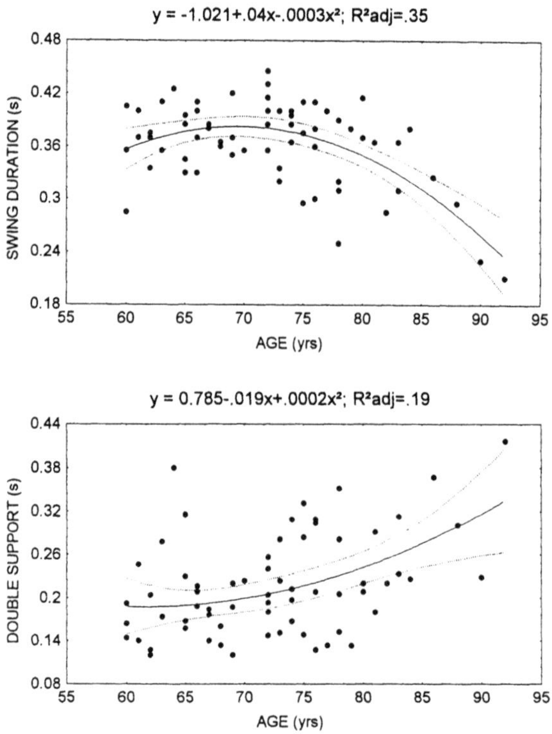

FIGURE 2. Relationship between age and two locomotor parameters (swing duration and double support duration) in elderly subjects. The solid lines represent the best fit of the second order polynomial equations, and the dotted lines represent the limits of the 95% confidence interval of the model.

For all of the dependent variables considered, height was not accepted in the model; only age (in its quadratic form) and sex were accepted in the models. The variances explained by the final models were 37% for velocity, 50% for stride length, 25% for double support duration, and 46% for swing duration. The variance explained by age alone is given in Figure 1 and Figure 2 for each locomotor parameter considered. Figures 1 and 2 show the best fit for a polynomial equation (second order) for age and a given locomotor parameter.

Effects of velocity on stride length, swing duration, and double support duration in elderly gait

Velocity was added as an independent variable to the previously tested models (see previous section). For stride length, velocity (in linear and quadratic form) and sex were accepted in the final model, which explained 92% of the variance. For double support duration, velocity (linear and quadratic) and age (quadratic form) were accepted, and the model explained 90%. For swing duration, velocity (linear and quadratic) and age (quadratic) were accepted and the model explained 30%.

The models with velocity alone (linear and quadratic) were tested on stride length, double support duration, and swing duration. The variance explained by velocity alone is given in Figure 3 for each locomotor parameter considered (see filled-in dots; elderly population).

These results demonstrate that velocity plays a major role in elderly gait, because it determines several locomotor parameters, such as double support duration. This is important to note, since many studies have emphasized the increase in double support duration with advancing age (see for example Murray, Kory, & Clarkson, 1969). The fear of falling is often thought to explain increased double support duration (Murray et al., 1969). The results presented above show that, although age can be considered responsible for a long double support duration, slowing explains most of the increase in double support duration.

However, the fact that a large part of the variance was explained by velocity in these results has to be taken with caution, since velocity is functionally linked to all other kinematic parameters. Obviously, variations in velocity are explained in part by variations in the distance or duration parameters.

Effects of velocity on other parameters in elderly and young adults

In order to deepen our understanding of the effects of age on the relationships between velocity and other parameters, the relationship between velocity and a given parameter (stride length, double support duration, and swing duration) was also tested for a population of young adults walking at a wide range of speeds, including those spontaneously adopted by the elderly subjects. These data are presented in Figure 3 (see empty dots; young adults). For the young adult group, each dot represents one of the 8 subjects in one of the 6 walking speed conditions.

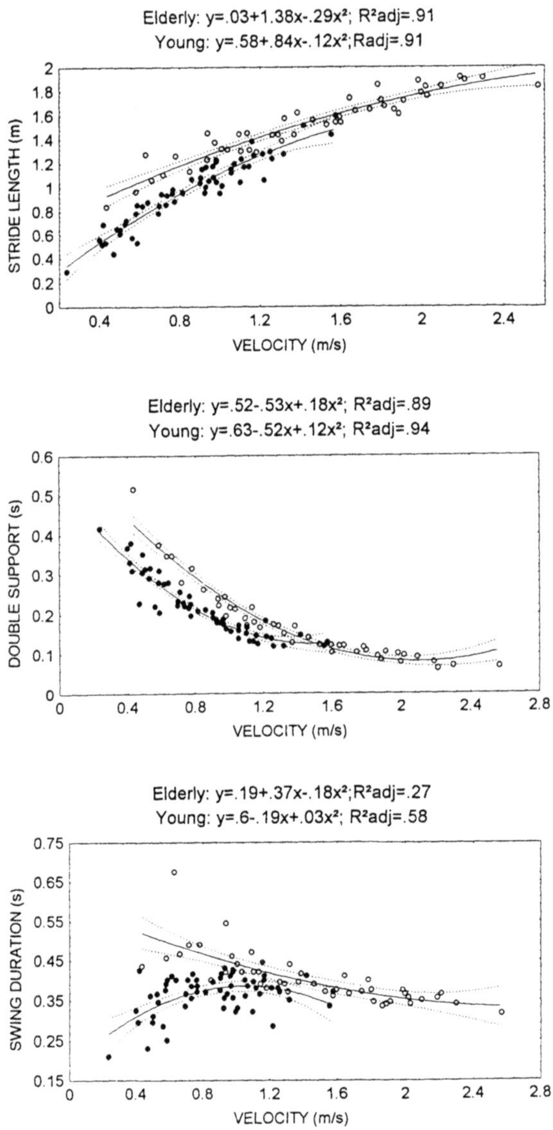

FIGURE 3. Relationship between velocity and three locomotor parameters in elderly subjects (filled-in dots) and young adult subjects (empty dots). The solid lines represent the best fit of the second order polynomial equations, and the dotted lines represent the limits of the 95% confidence interval of the model.

For each of the parameters considered, what is striking is the fact that the two models do not overlap. The differences between the young- and old-adult models were tested using the adaptation of Student's *t*-test (see formula 1), and were found to be significant for stride length, $t(110) = 8.45$, double support duration, $t(110) = -9.78$, and swing duration, $t(110) = 22.34$. The difference between the models is particularly surprising for swing duration, which is a negative function of velocity in the elderly subjects, and a positive one in the young adult subjects. However this finding has to be considered with caution, since double support duration was a greater determinant of velocity than was swing duration. The complete model with double support and swing durations explained 90% of the velocity variance in the elderly, and 77% in the young adults. The model with double support duration alone explained 87% (elderly) and 75% (young adults) of the variance, although the model with swing duration alone explained 14% (elderly) and 26% (young adults).

These results are complementary to previously reported data (Ferrandez, Pailhous, & Durup, 1990), but contradict the previous finding that the relative double support duration (ratio of double support duration to total cycle duration) is a complex function of velocity (ax^b) which looks very similar in the young and elderly populations. Actually, although walking speed largely determines double support duration and stride length, slowing alone is insufficient for explaining the long double support phase, the short steps, and swing phase duration in the elderly gait.

We then investigated how young adult subjects and elderly subjects modulate their walking speed in an attempt to determine whether or not the ways of modulating kinematic parameters between free walking and fast walking differ in the two populations.

EXPERIMENT 2: INTENTIONAL MODULATIONS OF WALKING SPEED

Methods

Subjects. The subjects were the same elderly persons as in Experiment 1. A young adult group composed of nine male university students served as controls. They were 20 to 28 years old (mean: 25; median: 25; SD: 2.3). For this group, three sessions were held one week apart. Each subject signed an informed consent form, in compliance with university rules.

Instructions. The subjects were told they would be members of a control group in an experiment on pathological locomotion. Each subject was led into the experiment room (8 meters long and 6 meters wide), and after being fitted with the apparatus un Experiment 1, was asked to walk to the assistant standing at the other end of the room. The task was performed under two conditions, one with the instructions "Go over to that person", and one with the instructions "Go over to that person as fast as possible". Thus, all subjects first walked with a free gait and then with a fast gait. Subjects walked barefoot.

Data analysis. Effects of condition (free walking vs. fast walking) and group (young vs. elderly) were tested on velocity, stride length, stride duration, double support duration, and swing duration using a MANOVA (with condition as a repeated measure). Five separate two-way ANOVAs with one repeated measure were conducted for analyzing the effects of condition and group, for each locomotor parameter. The modulation mechanisms were analyzed by testing the regression of the stride length ratio (stride length in fast walking/stride length in free walking) on the duration ratio (stride duration in fast walking/stride duration in free walking). The regressions obtained for the elderly population ($n_1 = 67$) and the young adult population ($n_2 = 27$: 9 subjects in 3 sessions) were compared using formula (1). The significance level was $<.05$.

Results and discussion

The MANOVA on velocity, stride length, stride duration, double support duration, and swing duration yielded significant effects of group and walking condition, and an interaction between the two (Wilk's Lambda (5,88) was .14 for group, .19 for walking condition, and .46 for interaction). Separate ANOVAs for each of the 5 locomotor parameters considered are given in Table 1. Figure 4 presents the data for each locomotor parameter, with the fast walking data plotted against the free walking data.

For each parameter, the diagonal (from 0 to the maximum) in Figure 4 represents no difference between the free walking condition and the fast walking condition. The condition effects yielded by the ANOVAs are indicated by the position of the data above or below the diagonal. In the fast walking condition, the subjects increased their velocity and their stride length, and decreased their stride duration, double support duration, and swing duration.

TABLE 1.
Results of five separate ANOVAs (velocity, stride length, stride duration, double support duration, swing duration), with repeated measures for age group and walking condition (free gait vs. fast gait).

	df effect	MS effect	df error	MS error	*F*	*p*-level
Velocity						
Age group (AG)	1	29.465	92	.131	224.693	.000*
Walking condition (W)	1	5.497	92	.014	378.506	.000*
AG × W	1	.406	92	.014	28.017	.000*
Stride length						
Age group	1	19.584	92	.100	195.379	.000*
Walking condition	1	1.490	92	.006	223.749	.000*
AG × W	1	.061	92	.006	9.162	.003*
Stride duration						
Age group	1	.686	92	.024	27.790	.000*
Walking condition	1	.719	92	.003	195.525	.000*
AG × W	1	.012	92	.003	3.466	.065
Double support duration						
Age group	1	.188	92	.004	43.885	.000*
Walking condition	1	.088	92	.000	118.634	.000*
AG × W	1	.003	92	.000	4.570	.035*
Swing duration						
Age group	1	.001	92	.003	.606	.438
Walking condition	1	.013	92	.000	60.211	.000*
AG × W	1	.000	92	.000	.150	.699

* $p < .05$.

The separate distributions of the two populations indicate the differences between the groups: the elderly differed from the young adults on every parameter considered, except swing duration, where the young adult distribution is included in the elderly's. For velocity and stride length, the difference between the two populations is striking: the two distributions do not overlap at a confidence interval of 95%.

The interaction is expressed by the difference in distance between the two populations along the x-axis and the y-axis. Elderly subjects increased their velocity and stride length less than did young adult subjects, and decreased their double support duration more than did young adult subjects.

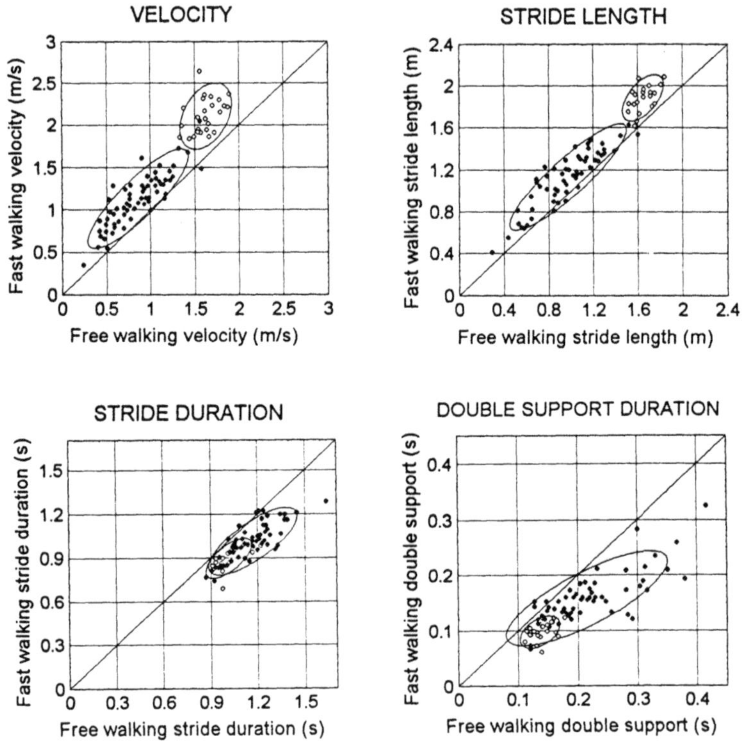

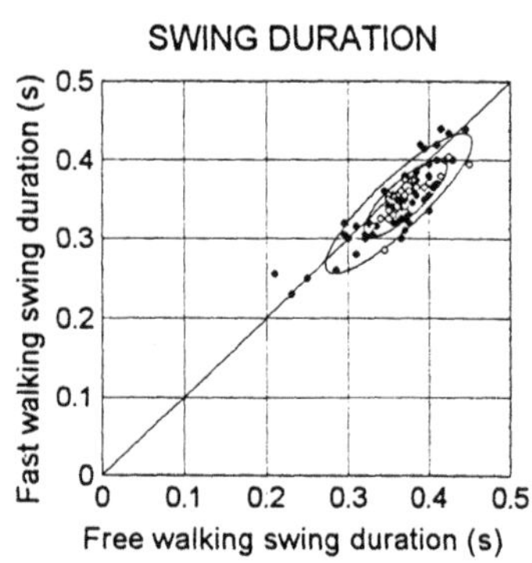

FIGURE 4. Free-walking values plotted against fast-walking values of five locomotor parameters, for elderly subjects (filled-in dots) and young adult subjects (empty dots). The ellipses delimit the 95% confidence intervals for each group.

Thus, elderly subjects like young adult subjects modulated their velocity by increasing stride length and decreasing stride duration. For a better understanding of these modulation mechanisms, we attempted to determine how modifications on stride length are related to modifications on stride duration. After having calculated the ratio of fast walking to free walking for each of the two parameters, the stride length ratio was plotted against the stride duration ratio. Regressions were calculated for each population, and the equations were compared using formula (1). These data are presented in Figure 5.

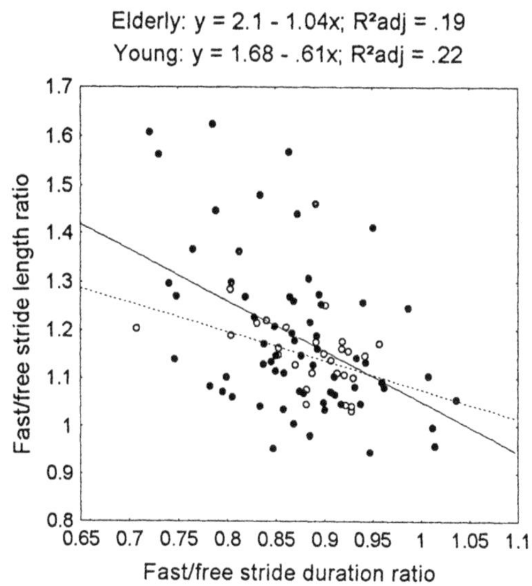

FIGURE 5. Linear regressions of fast/free stride length ratio on fast/free stride duration ratio, for elderly subjects (filled-in dots and solid line) and young adult subjects (empty dots and dotted line).

The regressions were significant for both populations: $F(1, 65) = 16.54$, for the elderly group, and $F(1, 25) = 8.5$, for the young adult group. There was no significant difference between these two regressions, $t(90) = .45$.

This xy distribution can be represented in three dimensions, representing the impact of stride-length and stride-duration modulations (ratio between free walking and fast walking) on the increase in velocity. The xy distribution is thus a plane with no dispersion under or above it. The orientation of this plane for each group is shown in Figure 6. The lack of a significant difference between the two regressions tested above appears here in the surprising proximity of the two planes, showing how the velocity modulation mechanisms of young adult subjects and elderly subjects were identical.

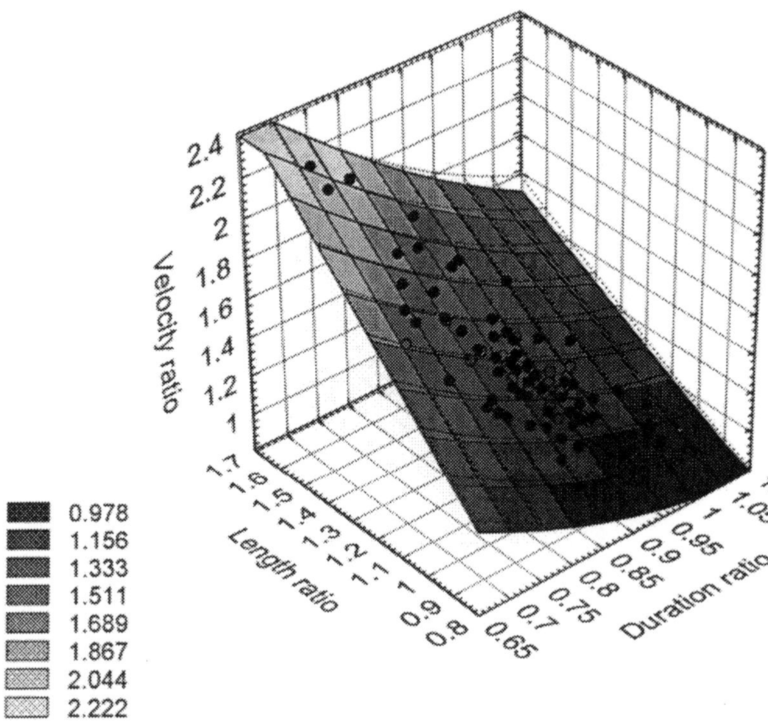

FIGURE 6. Impact of stride duration ratio and stride length ratio on velocity ratio, for elderly subjects (filled-in dots and solid lines) and young adult subjects (empty dots and dotted lines).

Stride duration is the sum of swing duration and double support duration. To test for a possible difference between the two populations in the way stride duration was decreased between free walking and fast walking, the same ratio as above was calculated, i.e., the fast walking/free walking ratio for swing duration and double support duration. The regression of the swing duration ratio on the double support duration ratio for each population was then calculated. For the young adults, the regression was significant, $F(1, 25) = 14.41$, and the equation was $y = .77+.25x$. For the elderly, the regression was not significant, $F(1, 65) = .63$, so the way the subjects in the two groups modulated duration could not be compared.

This study thus showed that intentional modulations of walking speed do not alter with age. Some questions remain however: Is the spontaneous low velocity of elderly people also compatible with more sophisticated modulations? Is the gait of older adults more variable (intra-individual variability) than that of younger adults? From a functional point of view, stability of gait is of great importance because it is the starting point for further adaptation of motor output to meet environmental constraints and comply with the subject's intentions (Pailhous & Bonnard, 1992). Is the elderly gait more variable from one stride to the next, and are elderly people as efficient as younger individuals at walking over an irregular terrain? The next section was designed to answer these questions.

EXPERIMENTS 3 AND 4: VARIABILITY AND SLOWNESS IN MODULATIONS OF STRIDE LENGTH

Subjects

The experiments were conducted in the laboratory. Two groups of subjects took part: one group of 14 elderly subjects and one group of 8 young adult subjects.

All of the elderly subjects could perform daily life activities normally. Most were retired teachers aged 61 to 80 (mean: 70). They were healthy and did not suffer from disorders causing pain in walking (bone or joint disorders in the lower limbs or spinal cord, or vascular or neuromuscular disorders in the lower limbs); deficits of motor, sensory, cerebellar or vestibular origin; major visual deficits; severe heart or breathing malfunctions; or asthenia or depressive tendencies. None were taking sedative medication. For the young adult control group, the subjects were students or members of the laboratory staff. The age range was 25 to 48 (mean: 35). The experiment lasted nearly half an hour. Each subject signed an informed consent form, in compliance with university rules.

Materials and procedure

Constraint experiment. Subjects walked with their usual comfortable shoes on a motor-driven treadmill (Gymroll BRL 1800; 0-10 km/h; walking surface: .60 m wide and 1.80 m long with lateral protective bars). In order to become familiar with walking on the treadmill before the experiment, the elderly subjects walked naturally for at least 30 seconds at 2.5 km/h, without any precise instructions. The practice walking time could be extended if the subject wished. All of the young subjects had already had experience walking on the treadmill.

The imposed (required) walking speeds were 2, 2.5, and 3 km/h. The order of the required walking speeds was counterbalanced. One minute of walking at each speed was recorded. A two-minute break was allowed between each session. For the elderly group, the duration of the break was increased if the subject desired. Subjects were requested to put their feet on the marks as they walked. The marks formed a pattern composed of 9 strides (alternating left and right strides), shown in Figure 7.

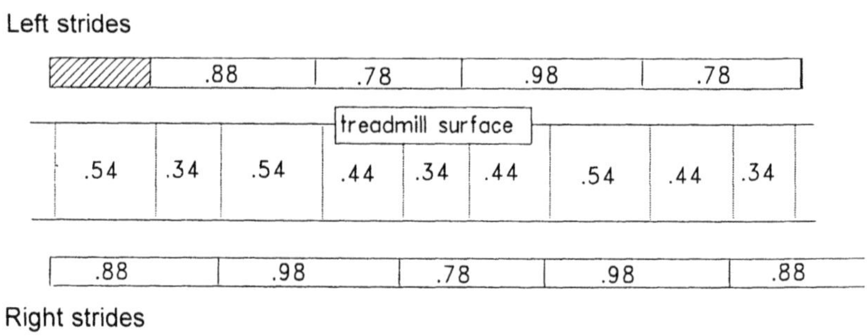

FIGURE 7. Pattern of required stride lengths.

Regular experiment. Thirteen young adult subjects aged 21 to 48 (mean: 31), including the 8 subjects of the constraint experiment, participated in this experiment, where they had to walk on the treadmill (no marks were drawn on the surface) at 6 required velocities (ranging from 2 to 4.5 km/h .5 km/h steps) without any instructions other than to walk as naturally as possible. One minute of walking at each speed was recorded. For the elderly subjects, data for only one speed (2.5 km/h) was available from the familiarization trial (see constraint experiment).

Data collection. The longitudinal foot displacement was recorded by connecting each foot via a thread to a precision potentiometer (10 Kohms, 10 revolutions) which transduced a displacement in voltage, moving clockwise when the foot moved forward and counterclockwise when the foot moved backward. The potentiometer signals were digitized at a sampling rate of 100 Hz. Before being analyzed they were filtered using a Finite Impulsive Response filter with a 33-point window and a 10-Hz cutoff (-3 dB).

Data analysis

Comparison of intra-individual variability. For analyzing intra-individual variability in the regular experiment, the instantaneous fluctuations of a subject were calculated for each trial, i.e., the difference in percentage between two consecutive strides, using formula (2):

$$fr_j = ((n_i - n_j)/n_i) \times 100 \tag{2}$$

where n_i is the length of stride i, and n_j is the length of stride $i - 1$. This formula gives the relative values. The average of these values on one trial was always close to zero, and was never greater than 1 or less than -1. This means that the total increase and the total decrease in a given trial compensated for each other. These data were then transformed into absolute values using the following formula:

$$fa_j = |((n_i - n_j)/n_i) \times 100| \tag{3}$$

For each trial, at least 20 consecutive strides were available. Each elderly subject performed one trial (at 2.5 km/h), and each young adult subject performed 6 trials, one per speed (2, 2.5, 3, 3.5, 4, & 4.5 km/h).

In the constraint experiment, the pattern of marks where subjects had to put their feet was composed of 9 strides (alternating left and right). The fluctuations were calculated using formulas (2) and (3), where n_i is the length of stride i, and n_j is the length of stride $i - 9$. The average of these values on one trial was always close to zero, and was never greater than 1 or less than -1. For each trial, 63 consecutive strides were available. Each subject performed 3 trials (at 2, 2.5, and 3 km/h).

ANOVAs were used to analyze the effects of group or walking speed (as a repeated measure) on intra-individual variability. The effects of walking condition were also tested by comparing data in the regular experiment with data for the same subjects in the constraint experiment, using an ANOVA with repeated measures. The significance level used was $<.05$.

Comparisons of accuracy. An error index was calculated to assess stride length regulation accuracy using formula (4):

$$e = (n_i - A)/n_i \times 100 \qquad (4)$$

where e is the error index, n_i is the length of stride *i*, and A, the required length for that stride (.78, .88, or .98). This index was treated as a relative value, and then transformed into its absolute value when necessary. ANOVAs with repeated measures served to analyze the effects of group, required walking speed, and required stride length on accuracy. They also were used to compare accuracy and intra-individual variability. The significance level was $< .05$.

Results and discussion

Intra-individual variability. Standard deviation is generally used to express the dispersion of a distribution. However it is not a relevant index for assessing fluctuations over time. In locomotion, which is a repetitive movement, the instantaneous fluctuations of a given parameter (for example, stride length), i.e., the difference in percentage between two consecutive strides, is a useful measure.

The means and standard deviations are given in Figure 8 for the young adults at 6 walking speeds in the regular experiment.

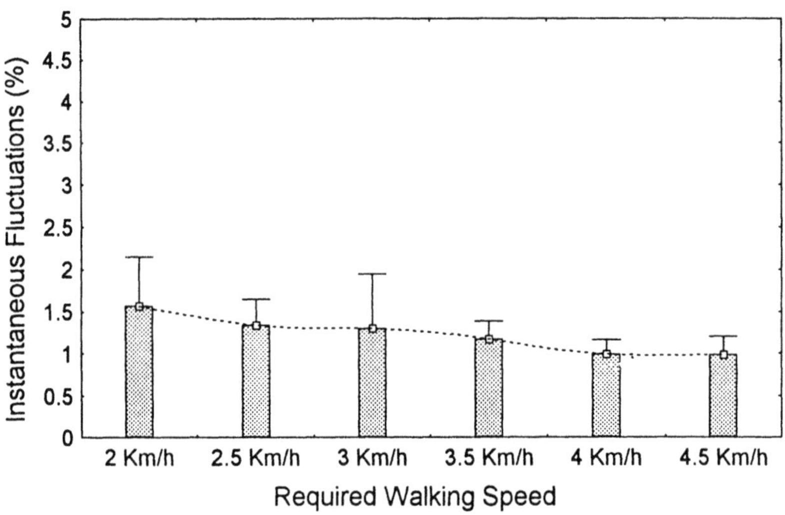

FIGURE 8. Means (bars) and standard deviations (whiskers) of instantaneous fluctuations (in %) for young adults at each required walking speed.

Intra-individual variability in young adults was greater at high walking speeds, as assessed by a two-way ANOVA with one repeated measure, $F(5, 60) = 6.75$.

The instantaneous fluctuations of elderly subjects in the regular experiment were compared to those of the young adult subjects at the slowest walking speed (and thus the one with the greatest instantaneous fluctuations). This comparison revealed greater intra-individual variability in elderly subjects than in young adults, $F(1, 25) = 20.65$. The means and standard deviations are given in Figure 9 (see light gray boxes).

In the constraint experiment, intra-individual variability was assessed by the difference in percentage between stride *i* and stride *i* - 9. A two-way ANOVA with one repeated measure on the two groups (elderly and young) and the three walking speeds (2, 2.5, and 3 km/h) revealed no significant effects, $F(1, 20) = .58$ for group, $F(2, 40) = .04$ for walking speed, and $F(2, 40) = 2.41$ for the interaction. The mean, standard error, and standard deviation of each group for all three walking speeds pooled are given in Figure 9 (see dark gray boxes).

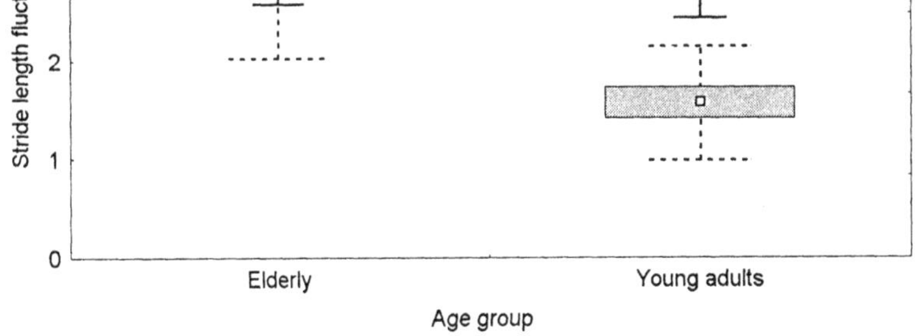

FIGURE 9. Means, standard errors of means (boxes), and standard deviations (whiskers) for stride length fluctuations (in %) in regular walking (light-gray boxes) and in constraint walking (dark-gray boxes) by elderly and young adults.

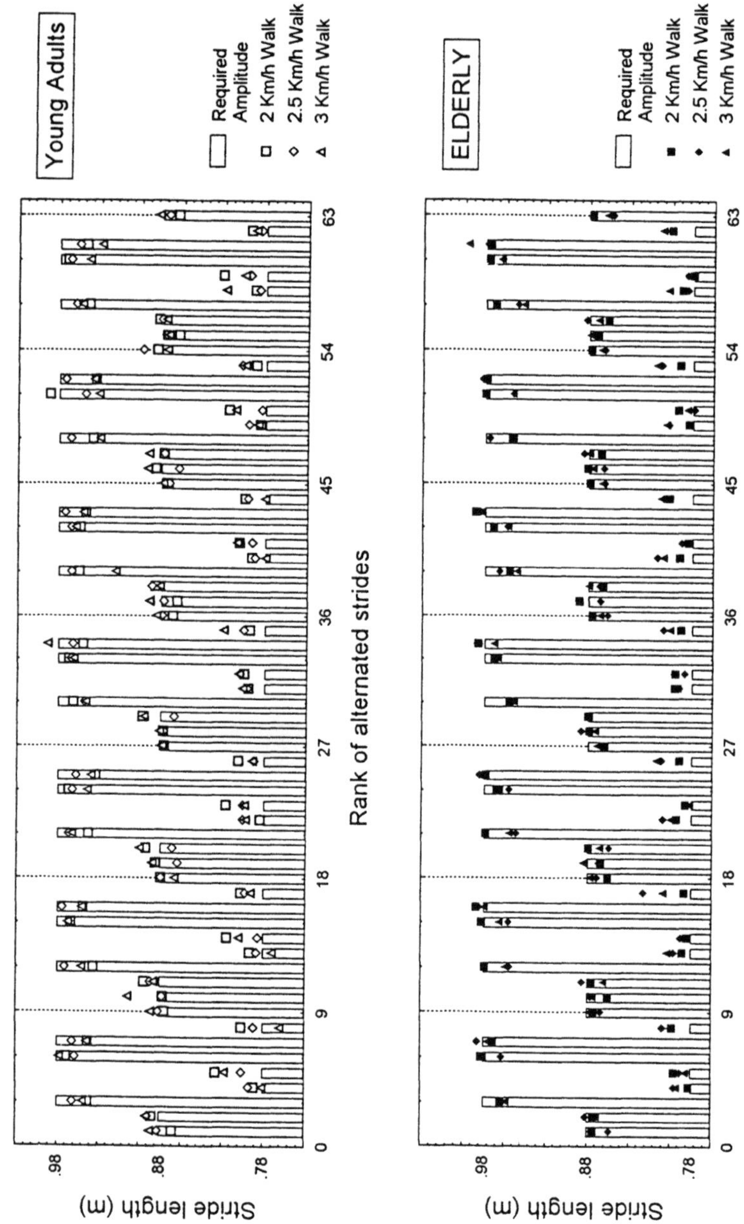

FIGURE 10. Mean actual stride lengths and required stride lengths for the 9 alternated strides on 7 consecutive patterns at each required speed, for young adults (upper) and elderly (lower).

The intra-individual variability values in the regular experiment and the constraint experiment were compared for the subjects who participated in both experiments (14 elderly subjects and 8 young adult subjects) using a two-way ANOVA with one repeated measure. This analysis yielded a significant effect of group, $F(1, 20) = 8.58$, and condition, $F(1, 20) = 8.94$, and a group-by-condition interaction, $F(1, 20) = 12.19$. Separate analyses for each group yielded a significant effect of condition for the young adults only, $F(1, 7) = 58.49$. These analyses confirmed what can be seen in Figure 9, i.e., intra-individual variability was greater in the elderly than in the young adults, and in the constraint experiment than in the regular experiment. However, only the young adult subjects exhibited greater intra-individual variability in the constraint experiment than in the regular experiment, which reached a level similar to that of the elderly subjects.

Stride length regulation accuracy. The subjects intentionally varied their stride length in an attempt to accurately place their feet on the marks drawn on the treadmill surface, as requested by the experimenter. The required stride lengths were .78, .88, and .98. The means of the 63 consecutive, alternated strides (7 patterns of 9 strides) is shown in Figure 10.

Upon first glance, subjects seem to have a tendency to lengthen the short strides and shorten the long strides. Strides 3, 6, and 7 were often underestimated, and strides 4, 5, and 8, overestimated. To check this point, an error index was calculated for each stride. This index was the difference between the actual stride length and the required stride length (formula (4) described in the method section). Figure 11 shows the mean error index for each group of subjects at the 3 required walking speeds and the 3 required stride lengths.

The first striking result is the accurate performance in both groups. The largest error, only 3%, was obtained by the young adult group at the shortest length (.78) and the fastest walking speed (3 km/h). This is no more than the intra-individual variability measured in the same experiment (see dark gray box in Figure 9). Two-way ANOVAs with one repeated measure were computed to compare the error values at the shortest length (.78) (all walking speeds pooled) with intra-individual variability obtained in the same experiment (constraint experiment) and in the regular experiment. The analysis comparing error with intra-individual variability in the same experiment demonstrated a significantly lower error value than intra-individual variability value, $F(1, 20) = 23.57$, and no effect of group. When comparing error with intra-individual variability in the regular experiment, the analysis demonstrated only an effect of group, $F(1, 20) = 7.12$, and an interaction, $F(1, 20) = 6.8$. This interaction means that the intra-individual variability was significantly higher than the error value for the elderly group, $F(1, 13) = 5.04$, whereas these two values cannot be considered as different for the young adults, $F(1, 7) = 3.66$.

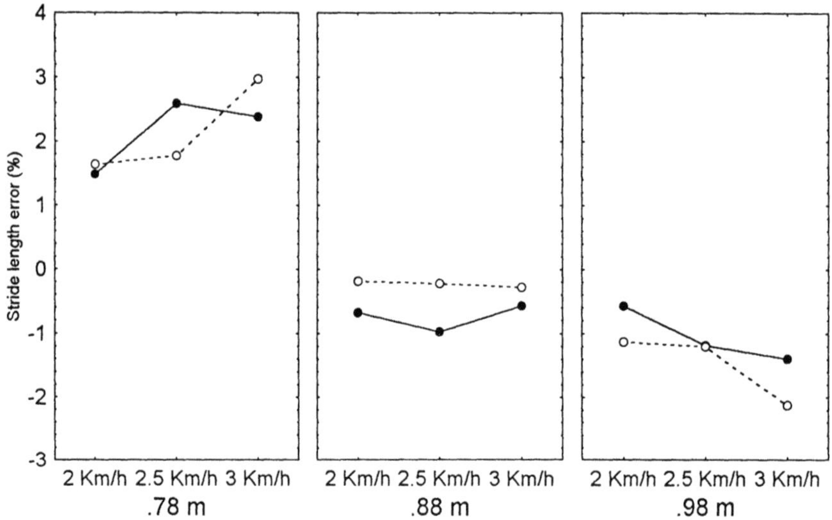

FIGURE 11. Mean error values at each required walking speed and required stride length, for elderly subjects (filled-in dots and solid line) and young adult subjects (empty dots and dotted line).

Thus, the young adults as well as the elderly adults can be considered to have performed this task accurately, since the error indices were lower than the intra-individual variability observed in the same experiment. Moreover, the elderly adults, who performed as accurately in this task as the young adults, had an error index which was lower than their regular intra-individual variability.

To more deeply analyze the factors influencing accuracy, the effects of group, required walking speed, and required stride length were tested using a three-way ANOVA with two repeated measures. The results are given in Table 2.

A striking result of this analysis is the fact that a large part of the variance was explained by the required stride length factor. This factor, when considered alone, accounted for 95% of the total variance explained by the experiment, and 99.9% when associated with all interactions in which it was involved. The same analysis was conducted on the absolute values of this index, and once again, required stride length accounted for a large part of the variance (78% when associated with interactions). The main points demonstrated by this analysis are thus a tendency to overestimate the shorter length (required at .78) by about 2%, and underestimate the longer length (required at .98) by about 1%. This tendency towards stride length uniformity was more pronounced at higher walking speeds. However the interaction was less pronounced in elderly subjects.

TABLE 2.
Results of an ANOVA with repeated measures, to test the effects of group (AG), required walking speed (RWS), and required stride length (RSL), on accuracy.

	df effect	MS effect	df error	MS error	*F*	*p*-level
AG	1	.011	20	.041	.268	.609
RWS	2	.090	40	.006	13.575	.000*
RSL	2	194.578	40	2.911	66.838	.000*
AG × RWS	2	.026	40	.006	3.888	.028*
AG × RSL	2	3.440	40	2.911	1.181	.317*
RWS × RSL	4	5.437	80	.631	8.609	.000*
AG × RWS × RSL	4	1.803	80	.631	2.854	.028*

* $p < .05$.

Swing phase modulations in regulating stride length. Subjects had to regulate their stride length by walking with short, medium, and long strides. Succeeding at such regulation (which the previous section showed to be highly accurate) requires modulating duration on the swing phase. Swing duration can be split into two phases: the acceleration phase before the peak velocity, and the deceleration phase after the peak velocity.

Figure 12 shows the acceleration and deceleration durations at the three required walking speeds and stride lengths for the two groups of subjects.

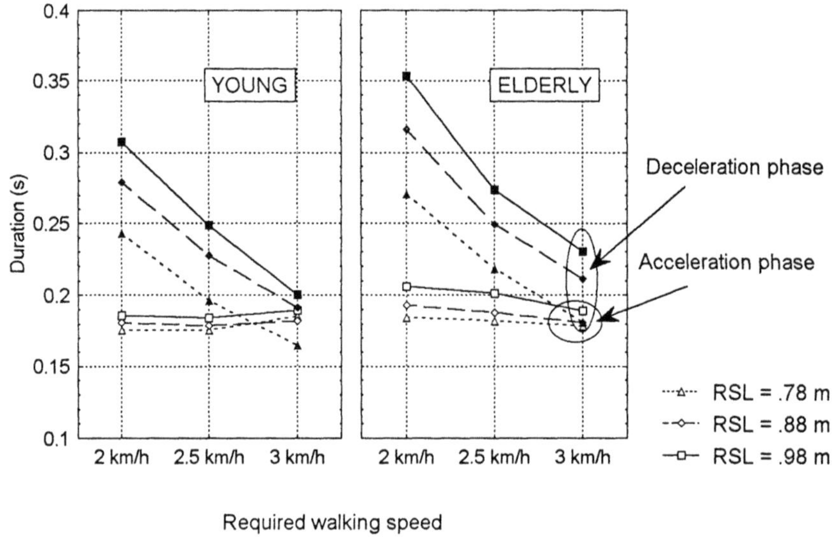

FIGURE 12. Duration of deceleration phase (filled-in symbols) and acceleration phase (empty symbols) for the three required walking speeds, the three required stride lengths (RSL) and the two groups.

To test for equal or different acceleration and deceleration phases, a 4-way ANOVA with 3 repeated measures was conducted (phase [acceleration and deceleration] with 2 categories, walking speed with 3 categories, and required stride length with 3 categories) on the two groups of subjects (elderly and young adults). The results of this ANOVA are given in Table 3.

The most notable result observed here was the relative constancy of the acceleration phase duration compared to the variation in deceleration phase duration with both walking speed and required stride length: the slower the walking speed and the greater the required stride length, the greater the difference between the duration of the acceleration and deceleration phases.

Another noticeable effect was revealed by the interaction between group and walking speed: duration decreased as walking speed increased. This effect occurred for both groups, but was more pronounced for the elderly.

Thus, stride length was modulated in this experiment mainly during the deceleration phase of the swing. What is surprising is that elderly subjects behaved the same way as young adult subjects. Is this consistent with what happens in regular walking? To answer this question, the durations of the acceleration and deceleration phases were measured in the regular experiment at a walking speed adapted to the subject (2.5 km/h for the elderly group, and 4 km/h for the young adult group). The age effects on these durations were assessed using a 2-way ANOVA with one repeated measure (phase). This analysis revealed a significant age-group effect, $F(1, 25) = 8.4$, and a significant phase effect, $F(1, 25) = 24.38$, but no interaction.

TABLE 3.
Results of an ANOVA with repeated measures, to test the effects of group (AG), phase (acceleration and deceleration, P), required walking speed (RWS), and required stride length (RSL), on duration.

	df effect	MS effect	df error	MS error	*F*	*p*-level
AG	1	.027	20	.003	7.620	.012*
P	1	.295	20	.004	66.714	.000*
RWS	2	.080	40	.000	255.549	.000*
RSL	2	.036	40	.000	225.335	.000*
AG × P	1	.009	20	.004	2.137	.159
AG × RWS	2	.001	40	.000	6.025	.005*
P × RWS	2	.069	40	.000	143.816	.000*
AG × RSL	2	.000	40	.000	5.290	.009*
P × RSL	2	.015	40	.000	54.106	.000*
RWS × RSL	4	.001	80	.000	26.374	.000*
AG × P × RWS	2	.000	40	.000	.747	.480
AG × P × RSL	2	.000	40	.000	.031	.969
AG × RWS × RSL	4	.000	80	.000	1.106	.359
P × RWS × RSL	4	.000	80	.000	3.934	.005*
AG × P × RWS × RSL	4	.000	80	.000	.588	.671

* $p<.05$.

Thus, in regular walking as well as in walking at required stride lengths, the elderly subjects differed from the young adult subjects in their swing duration (exhibiting a shorter swing duration at their preferred walking speed and a longer swing duration at matched walking speeds), but did not differ in their acceleration-deceleration ratio.

GENERAL DISCUSSION

This series of experiments sheds some light on several aspects of age-related changes in gait modulation.

In particular, the effects of slowness were investigated. In several cases, a young adult group matched on walking speed (the same as elderly people) was used to determine whether the observed effects of aging can be attributed – at least in part – to slowness.

The effects of age and slowness on kinematic gait patterns and modulations were studied by looking at intentional speed modulations (free walking vs. fast walking) and spatial modulations (modulating one's stride length). The question of intra-individual variability of stride length and its possible increase with age was addressed more specifically by studying instantaneous fluctuations (variation from one stride to the next) in regular walking, and fluctuations from one stride to the same stride of the following pattern in constraint walking, where subjects were requested to accurately modulate their stride length. The relationships between intra-individual variability and accuracy were investigated in the two age groups, as well as potential age-related changes in modulation capabilities.

Kinematic gait patterns in healthy elderly: effects of slowing with age

Walking becomes slower with age; age-related decreases in stride length and increases in double support duration were found, but these modifications could be largely explained by velocity in both the elderly and the young adults. However, the links between velocity and other kinematic parameters (stride length, double support duration, or swing duration) were found to be different in elderly and young adults.

The long duration of the double support phase encountered in elderly people can be largely understood as a biomechanical consequence of low velocity. Within the cycle the double support phase has the highest velocity at the subject's center of gravity (Inman et al., 1981). This is in line with Tuller et al.'s (Tuller, Turvey, & Fitch, 1982) interpretation, which suggests that subjects control their velocity mainly during the double support phase (by acting upon propulsive forces). The results obtained here also show that the swing phase is less related to velocity than the double support phase. This finding is consistent with Goslow, Reinking, and Stuart's data (1973) which showed that in quadrupeds, the support phase decreases markedly as the velocity increases, while the swing phase remains essentially unchanged.

It is well known that one of the main characteristics of the elderly is the general slowing down of cognitive and sensorimotor activity (see, for example, Salthouse, 1985; and Welford, 1988). We can see from the present results on the kinematic patterns of gait in healthy elderly that slowing can explain a large part of their gait characteristics. This finding suggests that future studies on elderly gait should take the velocity of elderly people into account, for example, by matching walking speeds of young adults with those of elderly adults.

Modulations of walking speed

Intentional modulations of walking speed, as defined in the present experiment, did not alter with age. Elderly subjects brought about increases in speed with very similar participations of stride length and duration as those achieved by young adults. Thus, the ability to intentionally increase walking speed is still intact in elderly subjects.

The walking speed spontaneously adopted by the elderly might be called a "cruising speed" or also an "economic speed", since these subjects are able to increase it in fair proportions, with the same modulation capabilities as young adults. Although slow, this cruising speed provides elderly subjects with several alternatives for coping with progressive internal changes (lower sensory acuity in several modalities, disorders in control of balance, low overall motivation) and gradual or sudden modifications in their environment (variations in light or slope, obstacles). The walking speed used by elderly adults may provide the greatest number of solutions for adapting at several levels (e.g., biomechanical, forces, or control of dynamic equilibrium), and thus ensures their safety to the greatest possible extent.

Intra-individual variability of gait: effects of aging and slowing

In walking on a treadmill without specific instructions, intra-individual variability in stride length was low in young adult subjects (less than 2%), as previously demonstrated (Pailhous & Bonnard, 1992). When walking regularly at their preferred speed of about 2.5 km/h, elderly subjects' stride length was more variable than that of young adults. To our knowledge, no previous studies on locomotion have shown any sign of increasing variability with age in healthy people. The results obtained here are consistent however with those obtained by Cooke et al. (1989) on upper limb movements. Worringham (1991) reported that young adults show more variable upper limb movements at high velocities than at low velocities. The opposite result was obtained here, i.e., increasing variability was observed with decreasing velocity in young adult locomotion.

Effects of aging, slowing, and intra-individual variability on spatial modulations

When instructed to regulate their stride length, elderly subjects exhibited as accurate spatial modulations as young adult subjects. The same tendency towards uniformity of locomotion (i.e., underestimating long strides and overestimating short strides) was observed in both groups of subjects, and this tendency was more pronounced at higher velocities.

To perform spatial modulations successfully, one has to modulate velocity and duration, especially during the swing phase. The swing duration can be split in two phases: the acceleration phase before the peak velocity, and the deceleration phase after the peak velocity. At matched velocities and stride lengths, no age-related effects were observed in the spatial modulations: the subjects mainly achieved these modulations by varying the duration of the deceleration phase, while keeping the acceleration phase approximately constant regardless of the required stride length or walking speed. Age did not have any effect on this strategy.

Conclusions

Slowness is one of the main characteristics of elderly people's sensorimotor behavior. Despite their slowness (or thanks to it?) and the variability of their movement parameters, elderly people are still able to use efficient strategies to cope with sudden changes in the environment, and are capable of adapting well to their own decline. The issue of age-related variability has not been extensively investigated, and the question as to its origin has not yet been solved. In particular, should increasing variability in the elderly be viewed simply as increasing "noise" in the system, or as an emergent property of the decline process affecting the various subsystems in a non-linear way?

To be able to gain a better understanding of these basic questions about the aging process (especially as regards the heterogeneity and non-linearity involved), researchers certainly need new theoretical and paradigmatic approaches to investigate a wide variety of subsystems in a large range of age groups. Life-span studies and longitudinal studies are particularly suitable for this purpose.

ACKNOWLEDGMENTS

This work was supported by the Centre National de la Recherche Scientifique, and by a grant from the "PIR-Villes research program" (n° 94N84/0052). The authors thank the two anonymous reviewers for their helpful and constructive critiques on a first draft of this chapter. They are grateful to the "Mutuelle Générale de l'Education Nationale des Bouches-du-Rhône" for having provided help in recruitment of elderly subjects for one of the experiments. They also thank Vivian E. Waltz for her assistance in the English version of the chapter. Part of the results reported here were published previously (Ferrandez, Pailhous, & Durup, 1990, and Ferrandez, 1993).

REFERENCES

Bernstein, N. (1967). *The co-ordination and regulation of movements.* Oxford: Pergamon.

Bessou, P., Dupui, P., Montoya, R., & Pages, B. (1989). Simultaneous recording of longitudinal displacements of both feet during human walking. *Journal of Physiology, 83,* 102-110.

Birren, J. E., Woods, A. M., & Williams, M. V. (1980). Behavioral slowing with age: causes, organization and consequences. In L.W. Poon (Ed.), *Aging in the 1980's* (pp. 293-308). Washington, DC: American Psychological Association.

Blin, O., Ferrandez, A.-M., & Serratrice, G. (1990). Quantitative analysis of gait in Parkinson patients: increased variability of stride length. *Journal of the Neurological Sciences, 98,* 91-97.

Clarkson, P. M. (1978). The relationships of age and level of physical activity with the fractionated components of patellar reflex time. *Journal of Gerontology, 33,* 650-656.

Cooke, J. D., Brown, S. H., & Cunningham, D. A. (1989). Kinematics of arm movements in elderly humans. *Neurobiology of Aging, 10,* 159- 165.

Curcio, C. A., Buell, S. J., & Coleman, P. D. (1982). Morphology of the aging central nervous system: not all downhill. In F. G. Pirozzolo & G. J. Maletta (Eds.), *The aging motor system* (pp. 7-35). New York: Praeger.

Dagnelie, P. (1986). *Théorie et méthodes statistiques* (2nd edition). Gembloux: Presses agronomiques de Gembloux.

Diggles-Buckles, V. (1993). Age-related slowing. In G. E. Stelmach & V. Hömberg (Eds.), *Sensorimotor impairment in the elderly* (pp. 73-87). Dordrecht: Kluwer Academic Publishers.

Dobbs, R. J., Charlett, S. G., Bowes, C. J. A., Weller, C., Hughes, J., & Dobbs, S. M. (1993). Is this walk normal? *Age and Ageing, 22,* 27-30.

Ferrandez, A.-M. (1993). Modulations of gait in normal aging an in Parkinson's disease. In G. E. Stelmach & V. Hömberg (Eds.), *Sensorimotor impairment in the elderly* (pp. 209-230). Dordrecht: Kluwer Academic Publishers.

Ferrandez, A.-M., Pailhous, J., & Durup, M. (1990). Slowness in elderly gait. *Experimental Aging Research, 16* (2), 79-89.

Gabell, A., & Nayak, U. S. L. (1984). The effects of age on variability in gait. *Journal of Gerontology, 39,* 662-666.

Goslow, G. E., Reinking, R. M., & Stuart, D. G. (1973). The cat step cycle: hind limb joing angles and muscle lengths during unrestrained locomotion. *Journal of Morphology, 141,* 1-42.

Hatze, H. (1986). Motion variability - its definition, quantification, and origin. *Journal of Motor Behavior, 18* (1), 5-16.

Imms, F. J., & Edhlom, O. G. (1981). Studies of gait and mobility in the elderly. *Age and Ageing, 10,* 147-156.

Inman, V. T., Ralston, H. J., & Todd, F. (1981). *Human walking.* Baltimore, MD: Williams and Wilkins.

Jänicke, B., Coper, H., & Schulze, G. (1988). Adaptivity as a paradigm for age-dependent changes exemplified by motor behavior. In J. J. Joseph (Ed.), *Central determinants of age-related declines in motor function* (pp. 18-32). New York: The New York Academy of Sciences.

Kay, B. A., Saltzman, E. L., & Kelso, J. A. S. (1991). Steady-state and perturbed rhythmical movements: a dynamical analysis. *Journal of Experimental Psychology: Human Perception and Performance, 17,* 183-198.

Kenshalo, D. (1977). Age changes in touch, vibration, temperature, kinesthesis and pain sensitivity. In J. E. Birren & K. W. Schaie (Eds.), *Handbook of the psychology of aging* (2nd Edition, pp. 562-579). New York: Van Nostrand Reinhold.

Kenshalo, D. (1979). Changes in the vestibular and somesthetic systems as a function of age. In J. Ordy & K. Brizzee (Eds.), *Sensory systems and communications in the elderly*. New York: Raven Press.

Kline, D., Schieber, F., & Coyne, A. (1983). Age, the eye and the visual channels: contrast sensitivity and response speed. *Journal of Gerontology, 38,* 211-216.

Kokmen, E., Bossemeyer, R., & Williams, W. (1978). Quantitative evaluation of joint motion sensation in a aging population. *Journal of Gerontology, 33,* 62-67.

Larsson, L. (1983). Histochemical characteristics of human skeletal muscle during aging. *Acta Physiologica Scandinavica, 117,* 469-471.

Lexell, J. (1993). What is the cause of the ageing atrophy? Assessment of the fiber type composition in whole human muscles. In G. E. Stelmach & V. Hömberg (Eds.), *Sensorimotor impairment in the elderly* (pp. 143-153). Dordrecht: Kluwer Academic Publishers.

Lupien, S., & Lecours, A. R. (1993). Toutes choses n'étant pas égales par ailleurs : réflexion sur l'accroissement des différences interindividuelles avec l'âge. *Revue de Neuropsychologie, 3* (1), 3-35.

McClellan, J. H., & Parks, T. W. (1973). A unified approach to the design of optimum FIR linear phase digital filters. *IEEE Transactions on Circuit Theory, CT-20,* 697-701.

Morse, C. K. (1993). Does variability increase with age? An archival study of cognitive measures. *Psychology and Aging, 8* (2), 156-164.

Murray, M. P., Kory, R. C., & Clarkson, B. H. (1969). Walking patterns in healthy old men. *Journal of Gerontology, 24,* 169-178.

Nelson, E. A., & Dannefer, D. (1992). Aged heterogeneity: fact or fiction? The fate of diversity in gerontological research. *The Gerontologist, 326,* 17-23.

Pailhous, J., & Bonnard, M. (1992). Steady-state fluctuations of human walking. *Behavioural Brain Research, 47,* 181-190.

Patla, A. E. (1985). Some characteristics of EMG patterns during locomotion: implications for locomotor control processes. *Journal of Motor Behavior, 17,* 443-461.

Phillips, J. G., Stelmach, G. E., & Teasdale, N. (1991). What can indices of handwriting quality tell us about Parkinsonian handwriting? *Human Movement Science, 10,* 301-314.

Salthouse, T. A. (1985). *A theory of cognitive aging.* Amsterdam: North-Holland.

Scholtz, J. P., & Kelso, J. A. S. (1990). Intentional switching between patterns of bimanual coordination depends on the intrinsic dynamics of the patterns. *Journal of Motor Behavior, 22,* 98-124.

Scholtz, J. P., & Kelso, J. A. S. (1989). A quantitative approach to understanding the formation and change of coordinated movement pattern. *Journal of Motor Behavior, 21,* 122-144.

Schöner, G., & Kelso, J. A. S. (1988). A dynamic pattern theory of behavioral change. *Journal of Theorical Biology, 135,* 501-524.

Sekuler, R., & Hutman, C. (1980). Human aging and spatial vision. *Science, 209,* 1255-1256.

Stelmach, G. E., & Sirica, A. (1986). Aging and proprioception. *Age, 9,* 99-103.

Teasdale, N., Stelmach, G. E., & Breuning, A. (1991). Postural sway characteristics of the elderly under normal and altered visual and support surface conditions. *Journal of Gerontology: Biological Sciences, 46,* B238-244.

Teulings, H. L., & Stelmach, G. E. (1993). Signal-to-noise ratio of handwriting size, force, and time: cues to early markers of Parkinson's disease? in G. E. Stelmach & V. Hömberg (Eds.), *Sensorimotor impairment in the elderly* (pp. 311-327). Dordrecht: Kluwer Academic Publishers.

Tuller, B., Turvey, M. T., & Fitch, H. L. (1982). The Bernstein perspective: II. The concept of muscle linkage or coordinative structure. In J. A. S. Kelso (Ed.), *Human motor behavior* (pp. 253-270). Hillsdale, NJ: Lawrence Erlbaum Associates.

Visser, H. (1983). Gait and balance in senile dementia of Alzheimer's type. *Age and Ageing, 12,* 296-301.

Welford, A. T. (1988). Reaction time, speed of performance, and age. In J. J. Joseph (Ed.), *Central determinants of age-related declines in motor function* (pp. 1-17). New York: The New York Academy of Sciences.

Williams, H. G. (1990). Aging and eye-hand coordination. In C. Bard, M. Fleury, & L. Hay (Eds.), *Development of eye-hand coordination across the life span* (pp. 327-357). Columbia, SC: University of South Carolina Press.

Winter, D. A. (1987). *The biomechanics and motor control of human gait: normal, elderly and pathological.* Waterloo: University of Waterloo Press.

Winter, D. A. (1984). Kinematic and kinetic patterns in human gait: variability and compensating effects. *Human Movement Science, 3,* 51-76.

Winter, D. A., Patla, A. E., Frank, J. S., & Walt, S. E. (1990). Biomechanical walking pattern changes in the fit and healthy elderly. *Physical Therapy, 70,* 340-347.

Worringham, C. J. (1991). Variability effects on the internal structure of rapid aiming movements. *Journal of Motor Behavior, 23* (1), 75-85.

Changes in sensory motor behavior in aging
A.-M. Ferrandez and N. Teasdale (Editors)

AGING AND COORDINATION FROM THE DYNAMIC PATTERN PERSPECTIVE

Laurence S. GREENE[1] and Harriet G. WILLIAMS[2]

University of Colorado and University of South Carolina

Abstract

Movement scientists have extensively described aging effects on speeded unilateral and uniarticular movements; in contrast, little is known about how aging affects the coordination of multi-degree of freedom actions. In this chapter we review studies on motor coordination (defined in terms of the spatio-temporal patterning among multiple elements of motor systems) in older adults. In addition, we present our application of the principles and research strategy of 'dynamic pattern theory' (Kelso & Schöner, 1988) to the study of aging and coordination. Findings on the older mover's ability to coordinate intra- and inter-limb movements are equivocal. Older subjects exhibit (a) loose spatio-temporal coupling of muscle synergies during postural responses and (b) temporal asynchrony in discrete bimanual reaching. However, locomotor coordination (based on temporal relative phasing between the legs) appears to be well-maintained with age. We observed similar bimanual phasing relationships in 23-78 year olds during cyclical in-phase (IP) and anti-phase (AP) movements. In a phase shift experiment, where movement frequency was systematically increased, older subjects exhibited abrupt transitions from AP to IP cycling at significantly lower

1. University of Colorado at Boulder, Department of Kinesiology, Boulder, CO 80309-0354, U.S.A. (e-mail: greenel@horton.Colorado.edu).
2. University of South Carolina, Department of Exercise Science, Columbia, SC 29208, U.S.A. (e-mail: n520020@univscvms).

frequencies than young counterparts. We present findings which suggest that older subjects experience critical fluctuation and phase shifts at lower frequencies than young subjects due to fundamental deficiencies in coordination. We also present evidence for a deficiency in the speed at which older subjects execute voluntary changes from IP to AP movements. An explanation for age-related incoordination may be the relative strengthening of intrinsic dynamics. We discuss our findings in relation to conceptualizations of aging as a thermodynamic process whereby constraints are altered in ways that reduce behavioral stability and the ability to adapt to environmental challenges (Yates, 1988).

Key words: Aging, coordination, bimanual coordination, dynamic systems, synergetics.

INTRODUCTION

Many older persons judge the quality of life by their ability to carry out everyday tasks, social activities, and recreational pursuits with independence, effectiveness, and vigor. Age-related declines in physical functioning diminish such abilities and attributes, and often lead to negative outcomes such as dependence on others, social isolation, hypokinetic disease, depression, and injuries (and related sequelae) suffered in accidents like falling (Cummings & Nevitt, 1989; Katz, 1983; Rowe & Kahn, 1987). In contrast, the maintenance of physical, and specifically *movement-related*, capacities with age is associated with independence in daily living, an active social life, physical fitness, and other markers of functional health (Jette, Branch, & Berlin, 1990; Ostwald, Snowdon, Rysavy, Keenan, & Kane, 1989).

The association between the quality of life and movement efficacy puts a premium on new research and expansion of the knowledge base in gerontological motor behavior. Presently, the literature is replete with descriptions of age-related psychomotor slowing in simple movement tasks (see reviews by Welford, 1984, and Williams, 1990). These descriptions have led many gerontologists to agree that declines of reactive (i.e., information processing) and movement speed are among the most ubiquitous markers of the aging process (Birren, Woods, &

Williams, 1980). The implications of behavioral slowing with age are indeed significant. For example, slow performance of manual tasks is a better predictor of the loss of functional independence in daily living skills than abnormalities on medical examinations, number of chronic diseases, and social network strength (Williams, Gaylord, & McGahie, 1990).

In contrast to our well-documented understanding of the age-related decline in psychomotor speed and its functional implications, the degree to which problems of *motor coordination* affect the elderly is largely unknown. The two primary purposes for this chapter are to (a) review findings from studies which have examined aging effects on coordination and (b) present findings from our own research in which we have applied the principles and operational strategy of 'dynamic pattern theory' (Jeka & Kelso, 1989; Kelso & Schöner, 1988) to studying bimanual coordination in older persons. We define coordination in terms of *the spatio-temporal relationships among multiple (i.e., at least two) elements of motor systems* (see Turvey, 1990 for a more comprehensive definition); thus, we restrict our review of studies to those (few) which examined spatial and temporal patterns of muscle and limb activity during multi-degree of freedom actions. Our review of literature will specifically address the unresolved question of whether the ability to coordinate intra- and inter-limb actions declines with age; moreover, we will speculate on underlying neural mechanisms and non-neural control parameters which may explain contradictory evidence for both reduced and maintained coordination across the adult life span.

Our research on coordination dynamics in older persons follows the precedent of movement scientists who have applied dynamic systems theory to studying the development of coordinated behavior in infants and children (e.g., Roberton & Halverson, 1985; Thelen & Ulrich, 1991; Whitall & Clark, 1994). Using the principles and tools of non-linear dynamics and synergetic physics, these researchers have successfully characterized emergent phenomena in movement behavior and identified some of the parameters which engender behavioral changes in developing organisms. We feel strongly that dynamic systems approaches will afford significant new insights into aging and movement behavior. Our sentiment has been inspired by Yates' (1988) argument for a *physical and dynamical* rather than *information processing* view of the aging process and its behavioral consequences. Accordingly, aging may be viewed as a non-linear, thermodynamic process whereby constraints are altered in ways that affect behavioral stability and the ability to adapt to environmental challenges. The background and justification for the dynamical perspective are presented in the following section.

THEORETICAL ORIENTATIONS TO AGING AND MOTOR BEHAVIOR

Our contemporary understanding of the aging mover is inextricably tied to the theoretical perspectives and research paradigms that have traditionally characterized motor behavior research. Specifically, movement scientists have relied heavily on information processing and motor programming theories to describe movement-related capacities in older persons (e.g., Spirduso, 1980; Welford, 1988). Within these frameworks, researchers have generally used experiments involving simple uniarticular movements where the observables of interest are maximum speed and accuracy. The outcomes of these research approaches have clearly influenced our impressions of the 'movement problems' faced by the elderly and even the mechanisms underlying declines in sensorimotor functioning. For example, we commonly view the age-related slowing of reaction time as a key determinant of deficits in common movement skills such as driving a car or restoring upright posture following an unexpected perturbation of balance (Stelmach & Nahom, 1992; Stelmach & Worringham, 1985). Moreover, we may attribute senescent declines in motor skill to impairments in specific stages of information processing (e.g., Larish & Stelmach, 1982; Salthouse & Somberg, 1982; Simon & Pouraghbagher, 1978). Reaction time paradigms, which have been used to isolate processing stages leading to the formation of motor programs, are commonly viewed as "behavioral windows" into CNS functioning in the elderly.

The value and productivity of traditional approaches to studying aging and motor behavior are not debatable; however, their limitations should be considered. In the general context of movement science, these limitations have been raised in discussions which contrast *information processing* and *dynamic systems* as appropriate models for theory building (see Kelso & Tuller, 1984; Kelso, Holt, Rubin, & Kugler, 1981; Kugler, Kelso, & Turvey, 1980). However, such discussions generally have not considered special populations such as older adults. By definition, information processing models have generally restricted our description of aging effects to central processes leading up to, but not including, movement execution. In many cases, these models do not account for the peripheral consequences and sensory regulation of "programmed" movements. As argued by Bernstein (1967) in his influential treatise, movement outcomes cannot be solely determined by centrally-executed commands because they are shaped by the mutual and dynamic influences of neural and non-neural factors (e.g., anatomical and biomechanical constraints, information flow fields, environmental obstacles, gravity and inertial forces). Given the peripherally-based structural and functional changes (e.g., loss of strength, flexibility, and somatosensation) which occur during the aging process, a centralist approach inordinately narrows the scope of inquiry. Indeed, these structural and functional changes may act as constraints which influence the older person's ability to perform movements directly and by altering perception of environmental affordances (Konczak, Meeuwsen, & Cress, 1992). Thus, we would argue that a complete description of the aging mover particularly requires a model that accounts for the dynamic cooperation between central and peripheral processes, and the coupling of perception and action (e.g., Reed, 1982; Turvey, Shaw, & Mace, 1978).

As a consequence of using reaction time paradigms to isolate specific central processing operations, researchers may inappropriately characterize the aging mover as a "passive processor" of information. To make generalizations concerning movement behavior based on reaction time experiments requires assumptions that motor output depends on sensory input and that organisms are inactive until responses are triggered by external stimulation. These outdated assumptions clearly have not been substantiated; following Gibson (1979) and Bernstein (1967), we have come to understand that effective goal-directed movement depends on active exploration of environmental affordances and preparatory postural adjustments which provide stability for voluntary movements. We would argue that older persons are *particularly* active (and even efficacious) in exploring environmental affordances and in achieving preparatory postural stability. For example, the 'speed/accuracy trade-off' phenomenon, which partly explains behavioral slowing with age (Rabbitt, 1979), may be interpreted to suggest that older persons act with particular awareness of environmental information (i.e., accuracy demands of a given task). Moreover, as demonstrated by Konczak et al. (1992), older persons are considerably *more accurate* than young counterparts in matching their perception of environmental affordances with their actual capacity to execute goal-directed movements effectively. With regard to preparatory postural adjustments, the interval between postural and (upper extremity) voluntary muscle activity is disproportionately longer in older versus young subjects (Inglin & Woollacott, 1988); thus, older persons appear to be particularly sensitive to establishing stability in preparation for focal movements.

By adopting the focus of most branches of the natural sciences, researchers who study the aging mover have generally exploited a 'reductionist' style of inquiry; as previously addressed, most studies have examined elderly subjects' performance on unilateral and often single-joint movements. A consequence of reductionism is the inability to describe coordination, or cooperative phenomenon among elements of perception-action systems. Moreover, by fractionating a system the researcher may overlook the dynamical properties which define that system (see Haken, 1985). In movement science, this point was illustrated in a classical experiment by Kelso, Southard, and Goodman (1979). They demonstrated that laws explaining the control of a single limb do not hold when individuals perform bilateral movements. Thus, coordinated motor behavior cannot be understood by "re-assembling" information gained from studies on single-degree of freedom movements. Movement abilities which young adults may take for granted are essential to the quality of life of older adults. Thus, the importance of studying coordination in the elderly is manifested in daily tasks and recreational activities such as walking, using the hands and fingers to manipulate objects, playing a musical instrument, exercising, and so forth. A richer interpretation of declines in movement-related capacities with age may be derived from studies on coordination. For example, slowing in manual dexterity tasks may be due to deficiency in coupling or decoupling the hands and fingers. Poor balance in older persons may be attributable to dysfunction in organizing and executing synergistic muscle responses, rather than (or in addition to) declines in the speed of muscle activation. Our understanding of factors which contribute to age-related declines in performing such activities is essential to helping older persons maintain functional independence and quality of life. Accordingly, researchers are challenged to (a) describe the degree to which multi-degree of freedom movements are coordinated in older adults, (b) identify constraints and control parameters which engender changes in coordinated behavior across the adult life span, and (c) devise strategies and interventions for retarding age-related declines in complex movements.

AGING EFFECTS ON INTRA- AND INTER-LIMB COORDINATION

Nearly all published studies on coordinated behavior in the elderly have been empirical in nature, describing performance differences between young and older subjects on postural, locomotor, and bimanual actions. Few studies have *directly* investigated aging effects on coordination as we have defined it (i.e., in terms of the spatio-temporal relationships between multiple elements of motor systems); moreover, research findings on the degree to which coordination declines with age have been equivocal. In this section, we review studies in which researchers, regardless of their intent to study coordination *per se*, measured spatial and/or temporal responses in more than one muscle and/or effector system in older subjects who performed multi-degree of freedom tasks. Specifically, we synthesize separate lines of evidence which either support or refute the argument that coordination declines with age.

Evidence for declines in coordination with age

Some studies suggest that coordination, like other dimensions of motor behavior (e.g., reactive/movement speed, strength), becomes deficient with age. For example, Woollacott, Shumway-Cook, and Nashner (1986) observed abnormal muscle activation patterns in the intra-limb postural responses of 61-78 year olds. Perturbations to a support platform were delivered to subjects during static stance, and EMG activity was recorded in the gastrocnemius (G), tibialis anterior (TA), hamstrings (H), and quadriceps (Q) muscles of the left leg. In response to forward translation of the support platform (causing backward sway), all young subjects (19-38 years) exhibited consistent temporal sequencing of synergistic muscles on the anterior side of the body (cf. Nashner, 1977); specifically, distal musculature (TA) was always activated prior to proximal musculature (Q). Onset latencies were tightly coupled in young subjects (TA was activated at around 100 ms and was followed by Q activity 10-20 ms later). In contrast, five of 12 older subjects exhibited intermittent reversals of the classical disto-proximal postural synergy. As illustrated in Figure 1, in this abnormal pattern Q responded prior to TA. Some of the older subjects were also characterized by looser kinetic coupling of the postural muscle synergy. They had relatively low correlations between distal and proximal EMG amplitude across trials (r = .12 to .86); in contrast, young adults had relatively high correlations (r = 82 to 1.00) for this measure of kinetic coupling. Whereas older subjects had abnormal EMG patterns, they did not lose their balance during perturbations under normal visual conditions.

Stelmach, Phillips, DiFabio, and Teasdale (1989) presented evidence for age-related deficiency in organizing intra- and inter-limb postural muscle activity. These researchers delivered platform perturbations to young (19-33 years) and older (64-76 years) adults during static stance and voluntary sway. During voluntary forward sway when the support platform was also moved forward, the appropriate stabilizing response activated muscle groups on the posterior part of the legs (G and H). However, forward movement of the platform initially caused backward sway and inappropriately activated TA (via reflex pathways) in both young and older subjects. Young subjects effectively compensated for the perturbation by inhibiting TA and activating G significantly faster than older subjects (94 ms vs 154 ms). As observed by Woollacott et al. (1986) and Stelmach et al. (1989), the abnormal sequencing and looser temporal coupling of intra-limb postural muscles in older persons may contribute to their postural instability and relatively high risk of falling.

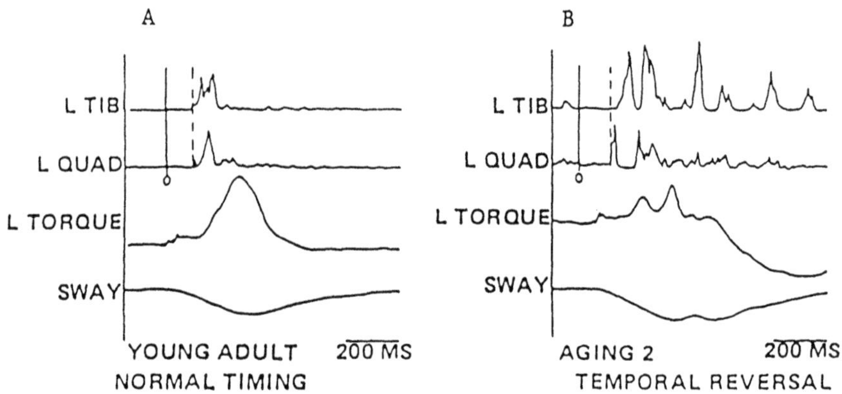

FIGURE 1. Characteristic muscle activation patterns of young and older subjects in response to platform perturbations which caused posterior sway. Graph A shows distal-to-proximal sequencing in a young subject. The left tibialis anterior (L TIB) is activated before the left quadriceps (L QUAD). Graph B shows a temporal reversal in an older subject. (Adapted from Woollacott et al., 1986, with permission from Baywood Publishing Company.)

Stelmach et al. (1989) also reported marked asynchrony of inter-limb postural muscle activation in older subjects. For example, during backward sway, the mean difference in onset latencies of the right and left TA for older subjects (M = 25 ms) was over two times greater than for young subjects (M = 10 ms). A similar pattern of inter-limb temporal decoupling in older adults was revealed by Stelmach, Amrhein, and Goggin (1988) in a study on bimanual coordination. These researchers adopted the bimanual movement paradigm introduced by Kelso et al. (1979); in this paradigm, subjects perform discrete two-handed movements to targets of similar (symmetric condition) and disparate (asymmetric condition) amplitudes and widths. (Kelso et al. observed that in young adults there was a strong coupling of the two limbs such that they initiated and terminated movement nearly simultaneously.) As expected, older subjects (67-75 years) were significantly slower than young subjects (21-25 years) in initiating and executing bimanual movements. With respect to inter-limb coordination, RT differences between the hands were significantly greater in elderly (M = 30 ms) than young (M = 16 ms) subjects. Moreover, response time (i.e., RT + MT) differences between the hands were significantly greater for older (M = 44 ms) than young subjects (M = 21 ms). Older subjects were less able to compensate for temporal asynchrony in movement initiation. This finding was supported by lower negative correlations for RT differences between hands and MT differences between hands in older subjects. In contrast, young subjects had high negative correlations; that is, they compensated for a slow RT of one hand by moving that hand faster to the target, thereby reducing asynchrony in movement termination.

The apparent age-related incoordination of intra- and inter-limb muscle activity and movement has important functional and mechanistic implications. Many common motor skills require a high degree of spatial symmetry and temporal synchrony in controlling opposite limbs. In some tasks, such as lifting a heavy object, the lower and upper limbs must produce actions which are coupled, or closely linked in space and time. In posture control, synergistic activation of intra- and inter-limb muscle groups ensures stabilization of joints and reduces body sway. Synergistic coupling may reflect optimization of movement organization (Kelso et al., 1979); that is, the unitary action of muscle collectives and limbs reflects a solution to the essential problem of coordination by compressing degrees of freedom. Thus, a breakdown of coupling in tasks that exploit tight linkages suggests deficiencies in movement organizational processes and perhaps deterioration of the neuromuscular mechanisms that underlie coordinative structures.

Whereas some tasks are best performed with a high degree of spatio-temporal coupling, others require that effectors function independently of one another (e.g., tying shoe laces, typing on computer keyboard, and driving a car). In such tasks, the intrinsic tendency to link limb movements in space and time must be inhibited. Spirduso and Choi (1992) reported that older subjects had particular difficulty in *decoupling* coordinated patterns and force production. Subjects used the index finger(s) and/or thumb(s) to apply force to spring levers; there were two levers for each hand. Pressure on one lever moved a computer screen cursor in the horizontal direction; pressure on the other lever moved the cursor vertically. The task involved tracing the sides of a triangle template on the computer screen using different digit combinations, one of which was a pinching action with the right index finger and thumb.

In the pinch condition, the first side of the triangle could be traced by producing flexion of the index finger and thumb; in this movement, which replicates the natural pinching synergy, homologous muscle groups (i.e., flexors) are active simultaneously. To trace the other two sides required the subject to decouple the flexion synergy. For example, to trace side 2 the subject gradually released the levers by extending both the finger and thumb. Tracing side 3 required an asymmetric pattern using non-homologous muscles: the index finger and thumb performed flexion and extension movements, respectively. Older subjects (61-81 years) were significantly slower and less accurate than young subjects (18-23 years) in all conditions. Interestingly, in most conditions older subjects were only slightly less accurate while tracing side 1 of the triangle. However, older subjects had marked difficulty decoupling the flexion synergy. For example, in the pinching task, they demonstrated 100 and 50% more error than young counterparts in tracing sides 2 and 3 of the triangle, respectively. These findings suggest that both coupling *and* decoupling capacities decline with age; however, older persons are particularly less able to decouple functional synergies.

The evidence presented above for age-related incoordination comes from studies which examined intra- and inter-limb movements restricted to the upper or lower extremities. Inglin and Woollacott (1988) observed deficient coordination of remote muscle groups in older persons during whole-body actions. These researchers examined activation patterns of postural and upper extremity muscles in young (19-33 years) and older (64-76 years) subjects who made rapid pushing or pulling movements on a handle while they stood on a stable platform. In such a task, anticipatory postural adjustments serve to compensate for destabilizing forces which occur during voluntary movements of the upper extremities (Cordo & Nashner, 1982). For example, when the arm is extended to push an object using the triceps (TRI), the body will move backwards; to compensate for this destabilizing force requires anticipatory forward displacement of the center of gravity. This action is accomplished by contraction of postural muscles on the anterior side of the body (e.g., TA, Q).

Inglin and Woollacott (1988) observed longer coupling latencies between postural and voluntary muscles in older than young subjects (Figure 2). This response in older subjects was attributed to an inordinate delay in activating voluntary muscles. This delay is probably not simply caused by the slowing of the voluntary muscle system; older persons may exhibit longer anticipatory adjustments as result of the slowing of postural reflexes and because they are particularly concerned with establishing stability prior to potentially destabilizing voluntary movements. Some older subjects exhibited abnormal postural and voluntary muscle activation sequences. In pushing movements, young subjects typically showed a distal-proximal pattern on the same side of the body (e.g., TA, Q, and TRI were activated sequentially). In contrast, older subjects had a high incidence of activating muscles on opposite sides of the legs; for example, 40% of the older subjects activated TA and H during postural adjustments for pushing movements. These findings are indicative of an age-related breakdown in the ability to coordinate responses of different muscle systems which are involved in common goal-directed tasks.

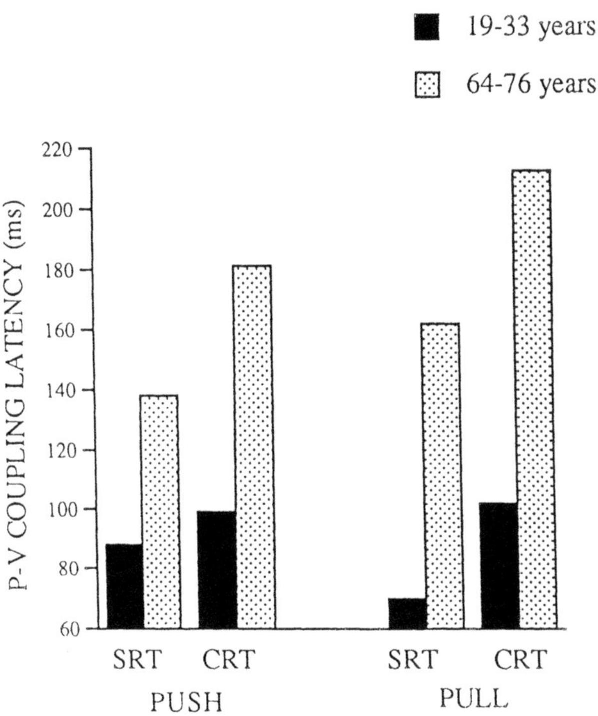

FIGURE 2. Differences between onset latencies for distal postural muscles and voluntary muscles during pushing movements (causing backward sway) and pulling movements (causing forward sway) in simple and choice RT conditions. For pushing movements, anticipatory postural adjustments were in tibialis anterior and quadriceps; voluntary movements were in triceps. For pulling movements, anticipatory postural adjustments were in gastrocnemius and hamstrings; voluntary movements were in biceps. Note that coupling latencies are particularly longer for older subjects during the CRT task. (Figure produced using data from Inglin & Woollacott, 1988.)

Evidence for the maintenance of coordination with age

In contrast to the studies reviewed above, several studies have demonstrated little difference in coordinated movement patterns between young and older adults. For example, Rothstein, Larish, Pretruzzello, Crews, and Nahom (1989) found that older subjects demonstrated tight interlimb coupling during rapid bimanual movements to targets placed at equal and unequal distances. Specifically, in performing a task based on the Kelso et al. (1979) paradigm, older subjects (M = 78.1 years) did not differ from young subjects (M = 19.7 years) with respect to latency differences between the hands in initiating and terminating bimanual movements. Warabi, Noda, and Kato (1986) observed tight temporal coupling of eye and hand movements in older subjects during a unimanual pointing task; whereas ocular and hand EMG responses were significantly slower in 60-69 than 20-29 year olds, differences in onset latencies between muscles were similar across age groups. It is noteworthy that the neural pathways innervating these two muscle systems are not linked anatomically. Moreover, the regulating mechanisms for effective eye and hand movement are markedly different. Thus, the high degree of eye-hand coordination in the older subjects suggests that the coupling mechanism for these two systems is well-maintained with age.

Williams and Bird (1992) presented evidence for the maintenance of locomotor coordination with age. The authors, working within the dynamic systems framework, asked whether manipulations of 'constraints on action' would differentially affect coordinative patterns in young (20-29 years) versus healthy and active older (50-59, 60-69, 70-80 years) women. Two constraints, movement pace (preferred versus fast) and terrain (level ground versus stairs), were used in the study. Coordination was assessed by the temporal phasing relationship between the limbs. This measure was calculated as the relative point in time at which one foot contacted the ground within the other foot's cycle (beginning with ground contact). For mean relative phase, there were no main effects of age, terrain, or pace; values were approximately 50% across age groups and conditions. It is of interest, however, that the variability (i.e., standard deviation) of relative phase did distinguish between age groups when subjects walked at fast speeds. The youngest women were significantly less variable than all other groups in this condition; no statistical differences in phasing variability were detected among the three oldest groups. Given the lack of an age effect for the mean temporal phasing of the limbs, Williams and Bird (1992) concluded that coordinative patterns in locomotor tasks of varying difficulty are not changed by aging in healthy persons up to 80 years.

What factors explain incoordination in older adults?

There has been little systematic investigation of the mechanisms and control parameters which might explain the age-related differences in motor coordination reviewed above. Thus, scientists can only speculate on reasons for why coordination in some movements becomes deficient with age. Such speculation is limited by the complexity of factors involved in the execution and regulation of coordinated behavior (e.g., Swinnen, Massion, & Heuer, 1994). Clearly, reductionist approaches which attempt to isolate factors underlying incoordination in the elderly are limited because senescent changes occur in many different systems (e.g., neural, skeletal-muscular, cognitive) and at different hierarchical levels of given systems; the *dynamic interaction* of factors within and across systems likely explains senescent changes in sensori-motor function. Compounding the problem of explaining age-related incoordination is evidence for the highly coordinated performance of some tasks by subjects in their 70s and 80s (e.g., Warabi et al., 1986; Williams & Bird, 1992). Beginning with the essential problem of coordination – *that of compressing degrees of freedom to achieve functional outcomes* – we address the following question in this section: Why do older adults exhibit abnormalities (compared to young counterparts) in executing synergistic muscle and limb activity? In keeping with the principles of dynamic systems theory, we do not seek to identify a single "mechanism;" rather, we offer a number of possible contributing factors to senescent changes in coordinated behavior.

Senescent changes in the CNS likely contribute to deficiencies in coordination. However, given the distributed nature of the CNS structures which control specific movements, it is difficult to link neural alterations in a single area of the brain to age-related breakdowns in synergistic behavior. As determined by studies examining multi-limb performance in patients with brain lesions, areas and structures which contribute to cooperative bilateral actions include the following: premotor, supplementary motor, parietal association, corpus callosum and anterior commisures, cerebellum, and basal ganglia (Wiesendanger, Wicki, & Rouiller, 1994). In the spinal cord neural networks control rhythmical limb movements. Post-mortem studies indicate that aging results in significant decreases in neuronal density, dendritic density, and synaptic connections in many of these areas (Berlin & Wallace, 1975; Coleman & Flood, 1987). In addition, old brains are characterized by accumulation of lipofuscin and neurofibrillary tangles (Dayans, 1970; Timiras & Vernadakis, 1972). Whereas changes in given CNS structures have been hypothesized to be mechanisms underlying behavioral slowing and variable movement timing with age (e.g., Ivry & Keele, 1989; Spirduso, 1982), they have not been directly linked to incoordination in older persons.

Movement studies on patients with brain lesions provide some insight into how senescent neural deterioration might affect coordination. Wiesendanger et al. (1994) observed several characteristics of inter-limb movements which are typical of patients with brain lesions in the premotor cortex, supplementary motor area, parietal association cortex, and the commisures. These patients have particular difficulty performing asymmetric bilateral movements which involve non-homologous muscle groups; in contrast, they are able to produce symmetrical patterns involving homologous muscles with a higher degree of coordination. In addition, patients with brain lesions maintain the capacity to perform well-practiced, everyday movements, but have difficulty performing new "abstract" bilateral tasks. These two characteristics describe the performance of older persons during inter-limb movements. First, as observed by Spirduso and Choi (1992), older subjects had inordinate difficulty decoupling homologous muscle activity in the natural pinching synergy. (In the following section we present evidence which shows that older persons are deficient in producing bimanual out-of-phase patterns using non-homologous muscles.) Second, the high degree of locomotor and eye-hand coordination which has been observed in older adults suggests that movements which have been practiced across the life span are more resistant to aging effects. Thus, we would argue in a general manner that neural deterioration is partly responsible for age-related changes in coordinated behavior.

Older persons may have difficulty decoupling symmetric upper limb movements (involving homologous muscles) due to an inability to differentiate neural output to the limbs/effectors. Neuroanatomical and behavioral evidence suggests that the difficulty of performing asymmetric patterns may be partly attributed to 'interference' effects at higher CNS levels (e.g., callosum) and across cortico-spinal pathways which originate at cortical and brain stem levels (Brinkman & Kuypers, 1972, 1973; Marteniuk, MacKenzie, & Baba, 1984; Swinnen, Young, Walter, & Serrien, 1991). These pathways project contralaterally *and* ipsilaterally to control upper extremity muscles. Successful performance of asymmetric bimanual tasks might thus require inhibition of conflicting neural commands at common spinal neurons (Swinnen et al., 1991). With age the capacity to inhibit patterns of interference may become deficient (cf. Spirduso & Choi, 1992). However, there is no direct evidence to support such speculation.

Evidence supports a considerable contribution of sensory processes to the control of coordinated movements. For example, several researchers have argued that multi-limb actions can be conceived as multiple tasks which require adequate attention and regulation based on sensory input (Baldissera, Cavallari, & Tesio, 1994; Cohen, 1970; Peters, 1994); incoordination in such actions may thus be due to deficient sensory processes. Given the well-documented senescent deterioration of sensory structures and function (e.g., Horak, Mirka, & Shupert, 1989; Skinner, Barrack, & Cook, 1984; Stelmach & Sirica, 1986), one might argue that the age-related deficiency in coordination is partly due to sensory declines. For example, incoordination of postural muscles during perturbation of balance may be related to the inability to organize or "perceptually re-weight" sensory input from visual, proprioceptive, and vestibular sources (Woollacott, 1989). Moreover, deficient somatosensation may be partly responsible for the older individual's difficulty in fine control of the pinching synergy. Cole (1991) reported that grasp force while pinching an object was 143% greater in older than young subjects; he argued that the excessive force in the former group was due to their attempt to enhance reduced tactile signals.

A final hypothesized mechanism which might account for age-related declines in motor coordination is enhanced lateralization effects. As argued by Peters (1994), handedness plays a role in the performance of everyday bimanual skills (e.g., peeling an apple) and laboratory tasks (e.g., polyrhythmic finger tapping). It is possible that aging may have differential effects on the cerebral hemispheres and/or peripheral response mechanisms; thus, performance decrements on inter-limb tasks may be due to lateral asymmetry. Weller and Latimer-Sayer (1985) presented evidence for increasing right-hand dominance (in right-handed subjects) with age on manual dexterity tasks. These authors compared right- and left-hand skill in 16-87 year olds on a pegboard task. Age was significantly correlated with the time taken to complete the task in both unimanual and bimanual conditions. The difference between movement times for the right and left hands increased significantly as a function of age, reflecting progressive declines in left hand skill among older subjects. Ferron (1992, cited by Peters, 1994) found that older subjects had marked difficulty producing a 2:1 rhythm in finger tapping; specifically, they were unable to lift the left index finger when the right hand performed the faster rhythm.

Equivocal findings of the degree to which coordination declines with age warrants consideration of reasons for the *maintenance* of spatio-temporal patterns in some multi-limb actions such as walking. Citing research on the development of locomotion in infants, Williams and

Bird (1992) concluded that "the coordinative structure governing locomotion is present even in very young infants and does not change throughout the life span in relatively healthy adults" (p. 252). This suggests that the neural substrates for locomotion, such as central pattern generators in the spinal cord, are relatively unaffected by the aging process. Later in this chapter, we present evidence for the maintenance of "intrinsic dynamics" or congenital coordination tendencies with age.

DYNAMIC PATTERN APPROACH TO AGING AND MOTOR COORDINATION

As reviewed above, studies on intra- and inter-limb coordination have enhanced our understanding of how older persons organize and execute movements; findings from these studies nicely complement insights gained from experiments on speeded single-degree of freedom movements. However, our description of coordinated motor behavior in the elderly is still incomplete and somewhat limited by previous conceptual and experimental approaches. Consider, for example, those studies in which the temporal synchrony among muscle responses and limb movements was measured during discrete reaching or postural stabilization; whereas the difference in onset or response latencies between two muscles or limbs reflects their spatio-temporal relationships, the measure captures coordination at only a brief instant in time. Differences in onset or movement latencies do not characterize coordination in its truest sense – as an evolving process of pattern formation where change is a function of the system's current state (i.e., its stability) and coordination tendencies (Kelso, 1994). In the language of dynamic systems theory, previous studies have reduced the *state space* through which the older mover's coordination dynamics can be observed. In addition, variables that might alter coordinated patterns (e.g., the speed of translating a support platform during postural perturbations) have not been systematically explored in older persons.

In this section we present a brief overview of the research strategy of dynamic pattern theory (Jeka & Kelso 1989; Kelso & Schöner, 1988) and a summary of experiments in which we applied the strategy to study coordinated behavior and movement pattern formation in the elderly. Essentially, dynamic pattern theory applies synergetic physics (Haken, 1977) to movement science. Synergetics exploits the propensity for self-organization in physical and biological systems wherein the patterns that emerge cannot be predicted by the movement of individual elements of the system. In the dynamic pattern or synergetic strategy, the scientist seeks to characterize existing and emerging coordinated patterns in systems composed of numerous elements. As summarized below the experimental approach allows the scientist to *quantify*, using the tools of non-linear dynamics, aspects of stability and fluctuation in complex systems as they respond to a wide range of external perturbations.[1]

Dynamic pattern research strategy

A preliminary step in the dynamic pattern research strategy is to identify macroscopic measures which characterize patterns of interacting elements in the system; these macroscopic or collective variables, also termed *order parameters*, provide low-dimensional descriptions of high-dimensional systems. Order parameters reduce surface complexity by compressing degrees of freedom; they also reflect changes in pattern configuration when systems lose and gain stability. A common example of an order parameter in limb movement is the temporal phase relation between the legs in human locomotion: In walking, the two legs are approximately one-half cycle (or 180 degrees) out of phase. Relative phase defines the relationship and, specifically, the coordination between the two legs; moreover, if the walking pattern is perturbed (e.g., if the individual steps in a pothole), the value for relative phase would reflect the observed change in pattern.

Having identified the order parameter(s) in the system of interest, the researcher observes how it changes over time. In observing and mapping order parameter dynamics, researchers are particularly interested in aspects of *stability* and *fluctuation*. Stable systems are characterized by their propensity to maintain preferred patterns or *attractor states*, particularly in response to external perturbations; an attractor is a region in the system's state space to which independent trajectories naturally converge over time (Jeka & Kelso, 1989). Fluctuation refers to variability in a coordinated pattern over time (e.g., from cycle to cycle in repetitive movements). To quantify the stability/fluctuation of a system the researcher may calculate the variability of the order parameter on a

1. An essential goal of the dynamic pattern approach has been to characterize the "landscape" of coordinated movement patterns by developing equations of motion (e.g., Haken et al., 1985). In our application of the principles and research strategy of dynamic pattern theory we assume that the same mathematical models which have been developed for characterizing synergetic phenomena in younger adults are appropriate for experiments on older adults.

cycle-to-cycle basis. For example, in human adult locomotion, 180 degree phasing is a highly stable attractor because the cycle-to-cycle standard deviation of this value is relatively low (Whitall & Clark, 1994). As we discuss below, some dynamic systems are characterized by stability at more than one attractor, a condition called *multi-stability*.

Perhaps the most distinguishing feature of synergetic systems is that they *self-organize*; that is, the patterns formed by the system change without specific instructions or code from a central organizing mechanism (e.g., a motor program). Self-organization is emergent as a product of dynamic interactions among system components, energy flux through the system, and physical constraints on the system. To observe self-organized pattern formation in the synergetic strategy, the researcher scales non-specific variables, called *control parameters*, which vary energy flux through the system. Heat is an example of a control parameter which engenders pattern formation in liquids; for example, when water at room temperature is heated to a critical temperature the physical process of convection causes the pattern formed by water molecules to change from one characterized by a smooth surface to one of hexagonal rolls as the water starts to boil. In the language of synergetics, the sudden (non-linear) change in system configuration is called a *phase shift* (see Haken, 1983 for examples of non-linear phase transitions in physical, biological, and social systems).

Coordinated phenomena in dynamic systems are predictable under various conditions of perturbation. For example, as a control parameter is scaled to critical values, variability of the order parameter will increase systematically. This phenomenon, called *critical fluctuation*, underscores the synergetic concept that order (i.e., stability) emerges out of conditions of disorder (i.e., fluctuation or instability) in open, complex systems. Stability, or reduced variance, will emerge again once a new stable mode is established. Another prediction of synergetics is linked to the concept of *critical slowing*. This concept predicts that external perturbations will have a weak effect on the order parameter when the system is in a highly stable state; however, as a critical transition point is approached, the system is more likely to undergo change when externally perturbed. A measure of critical slowing is *relaxation time*, or the duration from the onset of a perturbation (which causes the system to deviate from its attractor state) to the point in time when the order parameter returns to its pre-perturbation value. Synergetic theory predicts that relaxation time will increase as the system loses stability.

Evidence for synergetic phenomena in human movement

Kelso and co-workers have revealed that synergetic phenomena underlie coordinated motor behavior in humans (e.g., Haken, Kelso, and Bunz, 1985; Kay, Kelso, Saltzman, & Schöner, 1987; Kelso, 1984; Kelso & Scholz, 1985; Scholz & Kelso, 1990; Scholz, Kelso, & Schöner, 1987). These researchers used a common methodology in their experiments. Subjects performed cyclical wrist or finger movements with the hands moving in one of two coordinative modes: in-phase (IP) and anti-phase (AP). For wrist movements in the IP mode, homologous muscles are active simultaneously to bring the hands toward (flexion) and away from (extension) the body's midline together. In the AP mode, non-homologous muscles are active simultaneously in an alternating pattern to produce flexion in one wrist and extension in the other.

In this paradigm, the temporal relative phase between the two hands is an appropriate order parameter because it (a) adequately "captures" all coordinative states, including attractor states, and (b) reflects changes in coordinative pattern with a high degree of resolution (Jeka & Kelso, 1989). A point estimate of relative phase may be calculated as the relative point in time at which left (right) peak flexion occurs within a cycle of right (left) hand movement. In AP movement, relative phase is approximately .5 (or 180 degrees when expressed on a unit circle) because one limb reaches a landmark position (e.g., peak flexion) half-way through the other limb's cycle (starting and ending with peak flexion). In the IP mode, relative phase is 0 degrees because the two limbs share the same spatio-temporal pattern and reach the landmark position nearly simultaneously. As explained below, a *continuous* estimate of relative phase can be calculated with a phase plane technique by taking the difference in phase angles between the right and left hands for each data sample. Yaminishi, Kawato, and Suzuki (1980) and Haken et al. (1985) have shown that 0 and 180 degree phasing may be conceived as attractor states for the two hands in cyclical movements: While other coordinative modes are possible, they cannot be produced with a high degree of stability over cycles, as evidenced by high phase variability (but see Zanone & Kelso, 1992 for remarkable learning effects which enable subjects to produce patterns other than 0 and 180 degrees).

Kelso (1984) observed phase shifts in bimanual coordination during cyclical wrist movements when a control parameter, movement frequency, was experimentally scaled. Subjects' hands were initially prepared in the AP mode. Over the course of a trial, frequency was increased from 1 to 5 Hz by systematic decreases in the inter-pulse interval of a metronome to which movements were synchronized. In unresisted (i.e., low torque) and resisted (i.e., high torque) conditions, all subjects exhibited a spontaneous phase shift to the IP mode at a critical frequency; relative phase changed abruptly from approximately 180 to 0 degrees. A comparison of the last five AP cycles before the phase shift to the first five cycles in the IP mode revealed an increase in movement frequency and a decrease in amplitude. Mechanical energy expenditure and cycle-to-cycle variability (i.e., standard deviation of relative phase) decreased across the phase shift, indicating greater stability in the IP mode. The frequency at which a phase shift occurred varied among subjects from approximately 1.75 to 3.25 Hz for unresisted movements and 1.3 to 2.4 Hz for resisted movements. However, when each subject's switching frequency was expressed relative to his natural or "preferred" frequency, there was little variation.

Kelso and Scholz (1985) presented evidence to support predictions of critical fluctuation and critical slowing in inter-limb coordination. Subjects performed rhythmical index finger movements in the horizontal plane. Metronome frequency was increased from 1.25 to 3.5 Hz in .25 Hz steps every 4 seconds. As illustrated in Figure 3, when the fingers were initially prepared in the IP mode, no changes were observed in mean relative phase or its variability with increased movement frequency. However, when the fingers were prepared in the AP mode, a phase shift to IP movement occurred at a critical frequency. Variability of relative phase increased as movement frequency approached the transition value, confirming the synergetic prediction of critical fluctuation. Following the phase shift to the IP pattern, variability decreased markedly. The synergetic prediction of critical slowing down was tested using a measure of relaxation time; this measure was calculated as the time required for relative phase to stabilize following perturbations caused by increases in metronome frequency. It was observed that relaxation time increased as frequency approached critical transition values and stabilized thereafter.

Stable patterns in synergetic systems reveal their *intrinsic dynamics*, or the attractor states to which they naturally gravitate under given conditions. In the experiments reviewed above, two stable coordinative modes (AP and IP) reflect the intrinsic dynamics of the two hands when cycling below a critical frequency; above the individual's switching frequency, only IP movements are stable. Scholz and Kelso (1990) illustrated that, paradoxically, intrinsic dynamics can be suppressed by *intention* and, *at the same time*, can constrain intentional changes in coordinated behavior. In this experiment, subjects intentionally switched (finger movements) from the IP to AP mode, or vice versa. The researchers hypothesized that the ease with which subjects switched patterns would depend on the stability of the initial pattern. For example, switching from the IP to AP mode would be more difficult (and would take longer) than switching in the opposite direction because the IP mode is more stable (i.e., it can be easily maintained in the face of perturbations such as increased movement frequency). The relative instability of the AP attractor, particularly at frequencies approaching critical transition values, would predict faster intentional switching from the AP to IP mode. As predicted, across frequency levels, intentional switching time was significantly greater when subjects switched from IP to AP (M = 583 ms, SD = 153 ms) than AP to IP (M = 337 ms, SD = 33 ms) movement.

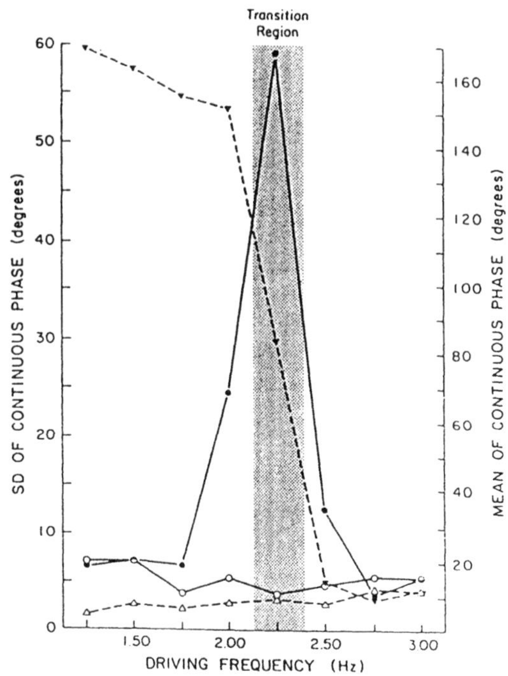

FIGURE 3. Mean relative phase and its standard deviation for the IP and AP modes across increased pacing frequencies. When the hands are prepared in the AP mode (closed triangles), mean relative phase changes abruptly from around 180 to 0 degrees at approximately 2 Hz. Note the increase in the standard deviation of relative phase (closed circles) prior to the phase shift; also, note the reduced standard deviation after the shift to the IP mode is completed. When the hands are prepared in the IP mode (open triangles), there is no change in mean relative phase and its standard deviation (open circles). (From Kelso & Scholz, 1985 with permission from Springer-Verlag Publishers.)

Application of the dynamic pattern research strategy to studying coordination in plder adults

In a preliminary effort to apply the dynamic pattern research strategy (Greene, Williams, Burke, & Lopez, 1993), we examined order parameter dynamics during cyclical bimanual movements (wrist flexion-extension) performed by young (M = 25.9 years; n = 6) and older adults (M = 66.9 years; n = 6). Relative phase values in the AP and IP modes were not different between the two groups when moving at a preferred frequency. However, with systematic increases in movement frequency older subjects exhibited spontaneous phase shifts (from AP to IP coupling) at significantly lower rates than young subjects (phase shift frequencies were 1.51 and 2.16 Hz for older and young subjects, respectively). This evidence for critical fluctuation and phase shifts at lower frequencies in older subjects motivated a series of four more comprehensive experiments on cyclical bimanual coordination in the elderly (Greene, 1994); below, we summarize the purposes, methodology, and findings of these studies.

Experiment 1: Kinematics and order parameter dynamics

In our first study we examined (a) kinematic characteristics and (b) order parameter dynamics of bimanual coordination in right-handed 23-32 (n = 10), 60-68 (n = 10), and 70-78 (n = 10) year olds (Greene, Williams, & Lopez, submitted). Using handles situated in the vertical plane, subjects performed cyclical, wrist flexion-extension movements. Unimanual and bimanual movements were performed at preferred and maximum frequencies; movement amplitude was freely-chosen by the subject. Bimanual movements were executed in the IP and AP coordinative modes. A potentiometer was secured in the joint axis of each handle; the potentiometer spindle rotated with hand movement, allowing kinematic measures to be derived through analog-to-digital conversion at a sampling rate of 250 Hz. To study movement kinematics, we measured the frequency and amplitude of the two hands during unimanual and bimanual movements. Given the conflicting evidence for aging effects on inter-limb coupling during discrete movements in previous studies, we were particularly interested in the degree to which cyclical bimanual movements were spatio-temporally coupled across the three age groups.

Our investigation of order parameter dynamics was motivated by questions of whether aging might alter coordination tendencies for upper extremity inter-limb movements; specifically, we were interested to learn whether, like young adults, older adults would demonstrate relative phase values of approximately 0 (IP) and 180 (AP) degrees. To answer these questions, we obtained point *and* continuous estimates of relative phase.[2] The point estimate was defined in terms of the latency to left hand peak flexion within a cycle of right hand movement (cf. Kay et al., 1987); this ratio was then expressed relative to the unit circle. The continuous estimate was derived using a phase plane technique (cf. Scholz & Kelso, 1989). Individual phase angles (Φ) for the right and left hand were calculated using the following equation:

$$\Phi = \mathrm{ARCTAN}\frac{\dot{x}}{x}$$

where $\dot{x}$ is normalized instantaneous velocity and x is position rescaled to the interval (-1,1). For a given cycle, continuous relative phase was the mean of absolute differences in phase angles for the two hands over each data point (250 Hz). For each of 10 trials, point and continuous estimates of mean relative phase were determined by averaging values for eight movement cycles; phase variability was calculated as the standard deviation of absolute differences in phase angles across those eight cycles.

The analysis of movement frequency revealed three noteworthy results: (a) 60-68 and 70-78 year olds had significantly lower frequencies than 23-32 year olds (Table 1) ; (b) there was no main effect or interaction for hand; and (c) there was a significant Age × Pace interaction which indicated that 23-32 year-olds had proportionally greater differences in frequency between maximum and preferred paced movement compared to 60-68 and 70-78 year-olds (Figure 4). The age-related reduction in movement frequency was predictable based on previous studies and common observation; senescent slowing is particularly evident in repetitive tasks which require movement reversals, or changes in limb direction (Williams, 1990). That there was no main effect or interaction for hand indicates a high degree of frequency coupling of the limbs regardless of age. Finally, the significant Age × Pace interaction indicated that older subjects increased frequency *proportionally less* than young subjects in maximum- versus preferred-paced conditions.

---

2. In a study on locomotor patterns in older adults, Williams and Bird (1992) used a *point* estimate of relative phase. That is, coordination was assessed at only one point in each locomotor cycle. It is conceivable that a more resolute measure of relative phase, one that captured the coordinative pattern throughout the locomotor cycle, would have revealed significant age differences. Thus, we chose to use the continuous estimate in addition to the point estimate to determine whether the former would produce different results.

Williams and Bird (1992) also found that older subjects had a smaller range of speeds in preferred– and fast-paced walking and stair-climbing. It is conceivable that older subjects do not achieve their true maximum speed in laboratory testing due to experiential factors. Specifically, in daily activities older persons generally are not required, nor do they frequently choose, to perform movements at maximum rates under test conditions.

TABLE 1.
Frequency (Hz) means and standard deviations.[a]

Coordinative mode	Age (years) 23-32		60-68		70-78	
	Left	Right	Left	Right	Left	Right
Preferred pace						
Unimanual	1.46	1.46	.96	.97	.92	.92
	(.45)	(.41)	(.26)	(.29)	(.35)	(.36)
Bimanual-IP	1.52	1.52	.98	.98	.88	.89
	(.43)	(.43)	(.21)	(.21)	(.32)	(.32)
Unimanual-AP	1.32	1.32	.84	.84	.80	.79
	(.35)	(.43)	(.17)	(.17)	(.27)	(.26)
Maximum pace						
Unimanual	3.57	3.74	2.29	2.49	2.08	2.12
	(1.02)	(1.18)	(.68)	(.91)	(.54)	(.62)
Bimanual-IP	3.76	3.76	2.28	2.28	2.07	2.07
	(1.02)	(1.01)	(.90)	(.91)	(.55)	(.55)
Bimanual-AP	2.44	2.46	1.63	1.63	1.59	1.59
	(.72)	(.73)	(.32)	(.31)	(.31)	(.31)

[a] Standard deviations are given in parentheses.

For movement amplitude, a non-significant trend revealed higher values in 70-78 year olds than 60-68 and 23-32 year olds (Table 2). Of particular interest was a significant interaction for Age × Pace × Hand. As illustrated in Figure 5, the interaction was due to two factors characterizing the performance of 70-78 year-olds during maximum-paced movement: (a) large increases in amplitude compared to the other two groups, and (b) marked differences in amplitude between the two hands. The former result may be attributable to an age-related deficiency in the ability to reverse movement direction rapidly (i.e., from flexion to extension, and vice versa). In repetitive tasks, limb acceleration in one direction must be overcome by the precisely-timed activation of antagonist muscles during deceleration at movement end-points. With age, there is an increase in the duration of the deceleration phase of rapid movements (Darling, Cooke, & Brown, 1989). Thus, older persons may take longer to change movement directions, and may thereby move through excessive amplitudes. The ability to finely grade muscle force in agonist and antagonist muscles, as is required in movement reversals, may be diminished in older adults due to the loss of motor units (see Kamen & DeLuca, 1989).

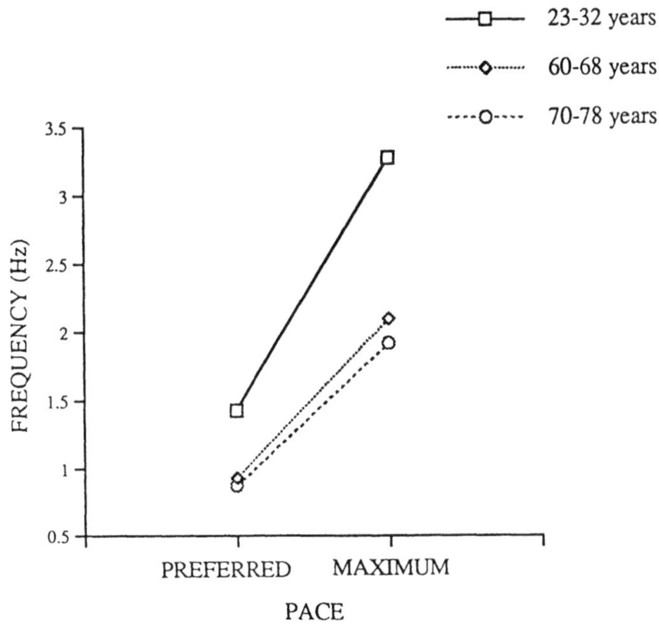

FIGURE 4. Age × Pace interaction for movement frequency of unimanual and bimanual movements; for bimanual movements frequencies are averaged across hands (see text for discussion).

For the 70-78 year olds, amplitude differences *between* the hands indicate a breakdown in spatial coupling: the right hand moved farther and faster (based on measures of average velocity) than the left hand during maximum-paced movement. Reasons for the hand asymmetry are unclear; however, as we previously reviewed, there is evidence for age-related declines in left-hand skill in right-handed individuals (e.g., Weller & Latimer-Sayer, 1985).

TABLE 2.
Amplitude (degrees) means and standard deviations.[a]

Coordinative mode	Age (years)					
	23-32		60-68		70-78	
	Left	Right	Left	Right	Left	Right
Preferred pace						
Unimanual	177.52	179.99	184.15	180.53	168.78	170.08
	(34.03)	(37.91)	(55.18)	(63.96)	(35.50)	(33.64)
Bimanual-IP	160.93	168.56	174.09	183.30	162.79	180.63
	(26.98)	(28.49)	(50.03)	(56.66)	(26.06)	(23.81)
Unimanual-AP	176.50	180.20	174.09	175.63	183.64	191.33
	(31.83)	(34.03)	(57.76)	(73.90)	(37.28)	(40.60)
Maximum pace						
Unimanual	161.49	158.79	165.51	147.64	174.93	191.25
	(38.77)	(40.71)	(63.59)	(80.13)	(34.92)	(41.99)
Bimanual-IP	149.91	151.97	162.75	165.94	168.94	193.76
	(35.05)	(38.91)	(51.41)	(57.95)	(34.83)	(32.94)
Bimanual-AP	181.93	178.45	166.47	170.44	197.73	219.42
	(37.84)	(32.57)	(50.24)	(64.09)	(34.48)	(45.19)

[a] Standard deviations are given in parentheses.

The analysis of point and continuous estimates of relative phase and phase variability revealed no main effects or interactions involving age. Across age groups mean phasing patterns at preferred and maximum speeds approximated attractor state values for IP and AP movements (Table 3). Moreover, the analysis of phase variability indicates that regardless of age subjects produced equally stable phasing patterns over cycles (Table 4). These results suggest that the mechanisms which link cyclical interlimb actions are relatively unaffected by age (cf. Williams & Bird, 1992). However, a speed-accuracy trade-off for coordination may have contributed to the stability of relative phase in 60-68 and 70-78 year-olds during maximum paced movement. That is, under instructions to move at maximum rates, older subjects may have compromised speed in order to maintain the coordinative pattern. As discussed above, the Age × Pace interaction for frequency supports this conclusion: older subjects did not increase frequency as much as young subjects across pace conditions. Although speed-accuracy trade-offs in the elderly have been demonstrated in numerous studies on discrete unilateral movements

(Welford, 1984), the phenomenon has not been noted in multi-limb coordination tasks.

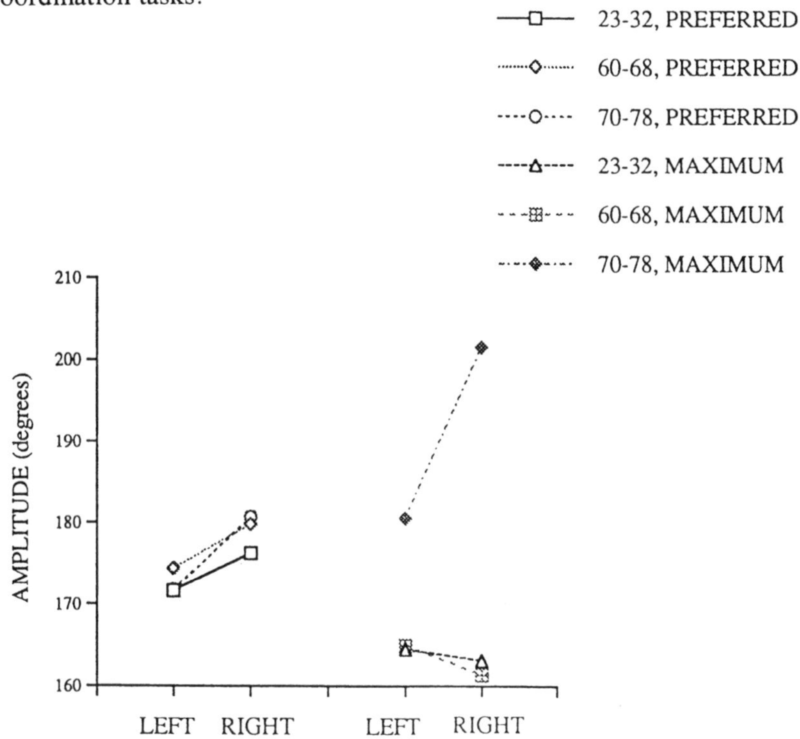

FIGURE 5. Age × Pace × Hand interaction for movement amplitude of unimanual and bimanual movements (see text for discussion).

Experiments 2 and 3: Phase shifts and phase maintenance/stability

To follow up the descriptive study on kinematics and order parameter dynamics, we conducted experiments on *phase shifts* (experiment 2) and *phase maintenance/stability* (experiment 3) using the same subjects who participated in experiment 1. Experiments 2 and 3 were motivated by our pilot work which revealed that older subjects exhibited phase transitions from the AP to IP mode at lower frequencies than young counterparts. One can argue that this result is a marker of deficient coordination in older subjects because they are, relatively speaking, unsuccessful at maintaining the intended AP mode. Thus, we sought to answer the question of whether aging diminishes coordination *per se*, or if senescent incoordination is attributable to some other factor(s), such as

TABLE 3.
Means and standard deviations for point and continuous estimates of relative phase (degrees).

	Age (years) 23-32		60-68		70-78	
	M	SD	M	SD	M	SD
	Point relative phase					
Preferred pace						
In-phase	9.39	2.94	9.33	3.88	10.44	3.00
Anti-phase	179.54	9.01	188.78	11.19	181.58	11.09
Maximum pace						
In-phase	14.37	3.56	14.23	4.58	15.31	170.08
Anti-phase	189.36	11.24	180.98	16.52	187.64	180.63
	Continuous relative phase					
Preferred pace						
In-phase	12.79	2.35	15.39	3.53	16.19	170.08
Anti-phase	179.16	4.39	178.15	5.62	176.16	180.63
Maximum pace						
In-phase	159.90	5.10	16.30	3.74	16.30	16.30
Anti-phase	175.21	5.98	179.09	5.93	175.90	175.90

TABLE 4.
Means and standard deviations for point and continuous estimates of phase variability (degrees).

	Age (years) 23-32		60-68		70-78	
	M	SD	M	SD	M	SD
	Point relative phase variability					
Preferred pace						
In-phase	6.10	1.72	6.14	2.49	6.97	1.56
Anti-phase	9.57	2.83	10.44	3.28	12.18	3.90
Maximum pace						
In-phase	9.19	2.55	7.76	2.63	7.97	2.54
Anti-phase	14.98	3.59	13.50	6.35	14.71	3.82
	Continuous relative phase variability					
Preferred pace						
In-phase	3.50	.75	3.09	.75	4.13	1.61
Anti-phase	5.12	1.35	4.93	1.25	6.55	2.85
Maximum pace						
In-phase	6.23	2.63	5.42	3.49	5.27	2.47
Anti-phase	6.86	2.28	6.39	.69	6.74	1.26

behavioral slowing: That is, do older persons exhibit critical fluctuation and phase shifts at lower frequencies (than young persons) because they are "less coordinated," or simply because they are slower? Because older persons have slower *maximum* movement frequencies, it is not appropriate to compare their absolute values of phase shift frequency to those of young persons; it is possible that regardless of age, individuals exhibit spontaneous phase shifts at some constant frequency relative to maximum. A key principle of the dynamic systems approach to movement science is that control parameters such as frequency and force are distinct from coordinative structures or the formation of muscle synergies (Kelso & Tuller, 1984). Thus, it is possible that in previous "perturbation" studies older persons exhibited abnormal spatio-temporal patterns of coordination because control parameters (e.g, the speed of platform translation in posture control studies) exceeded critical values.

To answer to whether older persons experience deficits in coordination *per se*, we first conducted an experiment to replicate our previous findings on age-related differences in the frequency at which phase shifts occur. We asked subjects to synchronize movements in the AP mode to a metronome set at 80% of their maximum AP rate (as determined in experiment 1); subjects performed eight trials. In each trial, after a synchronization period which consisted of 10-15 cycles at the base rate, metronome frequency increased by 20% of the base rate after every five tones until reaching 120% of the individual's maximum rate.[3] Subjects were instructed to allow changes in coordinative pattern to occur without resistance if they could no longer maintain the AP pattern (cf. Kelso, 1984). The algorithm used to determine movement frequency at the phase shift identified the point when relative phase first deviated from 180 degrees by more than 40 degrees. The dependent measure, *phase shift frequency*, was obtained by averaging frequencies of the right and left hands over the five movement cycles leading up to the phase shift, which we designated as the point when relative phase was less than 140 degrees or greater than 220 degrees.

As illustrated in Figure 6, phase shift frequency was significantly lower in 60-68 and 70-78 year olds than in 23-32 year olds. To offer

---

3. Pilot testing indicated that with practice all subjects could move at frequencies considerably higher than their maximum frequency in preliminary testing. Thus, in experiment 2 we were confident that subjects could move at 120% of the maximum frequency they performed in experiment 1. Pilot data also indicated that 5 cycles at each frequency plateau were sufficient for stabilizing the movement pattern before the frequency was increased.

insight into whether this result could be explained by age-related declines in coordination, we conducted an experiment to examine subjects' ability to *maintain* AP and IP movement patterns with increasing *normalized* frequencies. Frequencies were normalized to 80, 100, and 120% of maximum movement rates (as determined in experiment 1) in order to eliminate the bias of maximum speed.[4] To determine phase maintenance over the increasing normalized frequencies, we calculated the absolute deviation of observed point relative phase values ($RP_{observed}$) from attractor state ($RP_{attractor}$) values. The attractor states for IP and AP movements were set at 0 and 180 degrees, respectively. Within a trial the dependent measure, *phase deviation*, was calculated by taking the mean absolute difference between $RP_{observed}$ and $RP_{attractor}$ across 10 cycles of movement. Subjects performed eight trials in each normalized frequency/coordination mode condition. To assess *phase stability*, we measured the standard deviation of relative phase values across the 10 cycles for each trial.

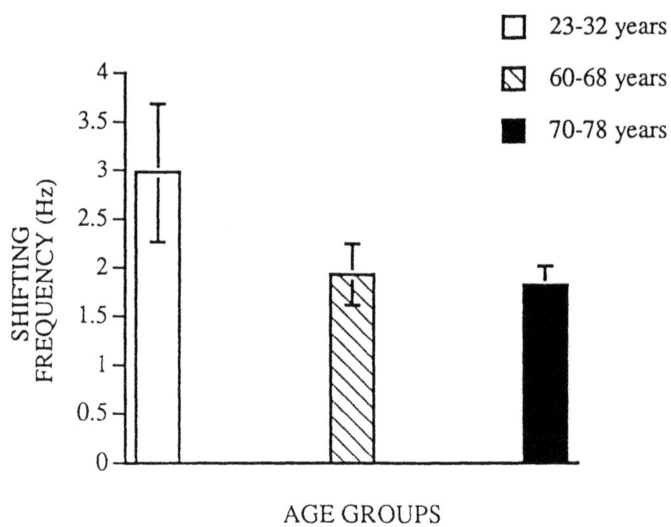

FIGURE 6. Age differences in phase shift frequency. Error bars show standard deviations.

---

4. To ensure that subjects moved at a rate close to their prescribed normalized frequency, the duration of each movement cycle was subtracted from the duration of the metronome cycle. The absolute difference between these values was averaged for the 10 cycles in each trial. If the average temporal deviation exceeded 50 ms, the trial was re-done.

We reasoned that if older subjects had poorer coordination than young counterparts, they would have greater difficulty maintaining the coordinated pattern and would demonstrate a proportionally greater rate of deviation from attractor state values (specifically in the AP mode) as normalized pacing frequency increased; in contrast, if older subjects were not deficient in coordination, the effect of increased normalized frequency on relative phase deviation would be similar to that observed for young subjects. Thus, a significant Age × Frequency Level interaction would provide support for an age-related decline in coordination. Figure 7 illustrates a trend toward this interaction, although it was not statistically significant (p = .3367). Regardless of age, the effect of increased frequency on phase deviation was slight in the IP mode; however, both older groups deviated considerably from 180 degrees in the AP mode. Collapsed across coordinative modes, means for phase deviation increased by 12.08 and 9.41 degrees for 60-68 and 70-78 year-olds, respectively, across frequency levels (80-120%). For young

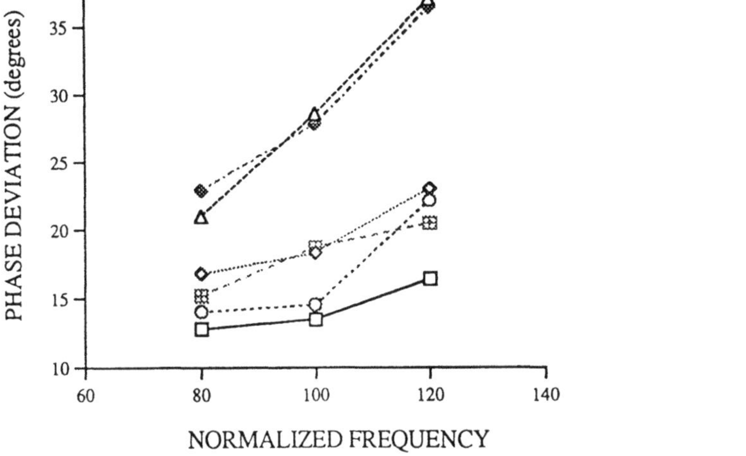

FIGURE 7. Age differences in relative phase deviation across normalized metronome frequency levels in in-phase (IP) and anti-phase (AP) modes. Phase deviation was calculated as the absolute deviation of relative phase from 'attractor state values' (0 and 180 degrees for IP and AP modes, respectively.)

subjects the increase was only 4.92 degrees. High inter-individual variability among older subjects may have accounted for the lack of a significant interaction. Collapsed across coordinative modes and frequency levels, the standard deviations of phase deviation values were 7.23, 17.26, and 13.50 degrees for the 23-32, 60-68, and 70-78 year-olds, respectively.

Our analysis of phase stability (i.e., the standard deviation of relative phase values) indicated that older subjects had considerable difficulty in producing consistent AP movements. There was a significant Age X Coordinative Mode interaction for relative phase variability collapsed across frequency levels. As illustrated in Figure 8, the two oldest age groups had similar values to young subjects in the IP mode. However, older subjects exhibited proportionally larger increases in phase variability when performing in the AP mode. These results indicate that for upper extremity movements the coordination of asymmetric patterns involving non-homologous muscles is impaired with age; in contrast, symmetric patterns involving homologous muscles are not affected by aging. As discussed below, these differential aging effects on IP and AP movement may influence the efficacy with which older persons can intentionally change coordinated patterns.

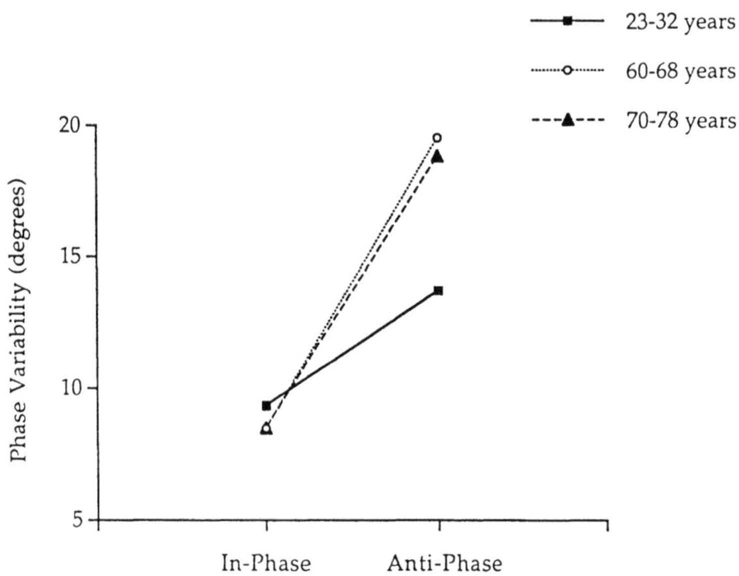

FIGURE 8. Age × Coordinative Mode interaction for relative phase variability (see text for discussion).

Experiment 4: Intentional and intrinsic dynamics

To complete our survey of coordination dynamics in older adults, we conducted an experiment on intentional changes in coordinated pattern. This experiment was intended to determine whether aging affects the degree to which intrinsic dynamics, or preferred behavioral modes, influence or constrain intentional movement processes. Adapting the paradigm of Scholz and Kelso (1990), we compared young and older subjects on the time taken to execute voluntary phase shifts from the IP to AP mode, and vice versa. Subjects performed 10 trials in each of six switching direction (AP to IP, IP to AP)/frequency level (80, 100, 120%) conditions, where frequency was normalized to phase shift frequency from experiment 2.

At the start of a trial subjects synchronized movements in a given coordinative mode to a computer-generated metronome. After 15 tones, the metronome ceased and the subject continued moving, attempting to match the same rate. A stimulus LED, the cue to switch from one mode to the other, was illuminated at random intervals of 500, 1000, or 1500 ms after the metronome was turned off. Upon the appearance of the light, subjects changed patterns from the IP to AP mode, or vice versa. Subjects were instructed to switch as rapidly as possible and, once having switched, to re-establish synchrony with the metronome in the new pattern. The trial ended after 10 post-switch cycles had been completed. The dependent measure, *switching response time*, was defined as the latency (in ms) between presentation of the light stimulus to the point when continuous relative phase stabilized in the intended coordinative mode (Figure 9). The criterion for stabilization was that post-stimulus relative phase values had to be maintained within 40 degrees of attractor state values for at least 1 s. For example, for changing from the IP to AP mode, the switch would be completed when relative phase values exceeded 140 degrees for at least 1 s.

As reviewed above, Scholz and Kelso (1990) demonstrated that intrinsic dynamics, which reflect coordination tendencies, act to constrain intentional changes in coordinated patterns. These authors found that the duration of intentional switching from the IP to AP mode was significantly longer than switching in the opposite direction. Because the IP mode is more stable, it is more difficult to decouple when switching to the new mode.

We hypothesized that intrinsic dynamics might have a *greater* constraining influence on older versus young subjects; if so, one would expect that switching from the IP to AP mode would be slowed proportionally more for older than young subjects; accordingly, we hypothesized that there would be a statistical interaction between age and switching direction. As illustrated in Figure 10, this interaction was significant. Compared to the 23-32 year-olds, both older groups demonstrated greater proportional increases in switching time in the IP to AP mode. However, our conclusion that intrinsic dynamics have a greater constraining influence on older adults is limited by our failure to replicate the result of Scholz and Kelso (1990), who found that, in young adults, switching time in the IP to AP mode was nearly two times slower than in the opposite direction.[5]

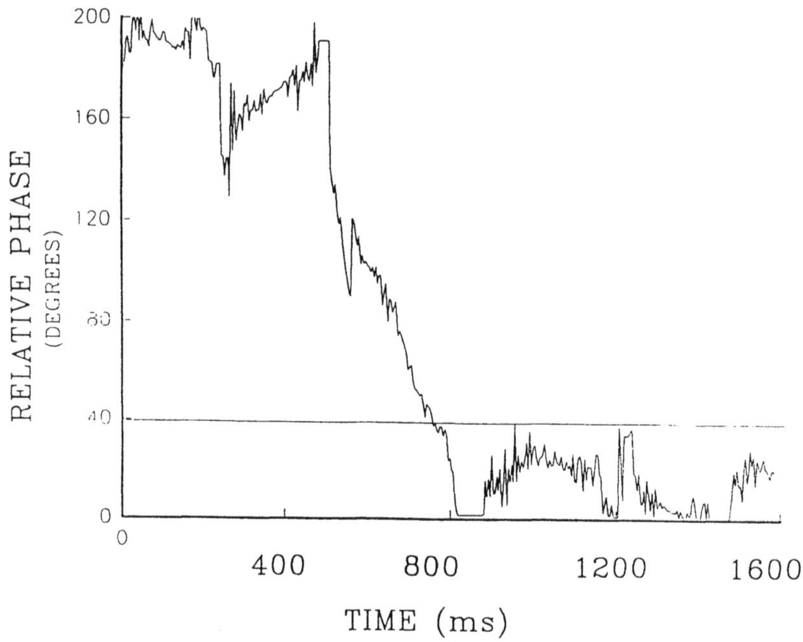

FIGURE 9. Continuous relative phase values during the intentional switching experiment. In this example, the subject is cycling in the AP mode upon presentation of the response stimulus (at time zero). Note that relative phase values approximate 180 degrees until the switch to the IP mode is initiated (at approximately 500 ms). The switch was designated to be completed when relative phase values were within 40 degrees of the intended mode (in this case 0 degrees) for at least 1 s.

5. Thanks to an anonymous reviewer for bringing this point to our attention.

Nevertheless, in the context of theoretical models of bimanual pattern dynamics (Haken et al., 1985, Kelso & Schöner, 1988), the aging process may be said to strengthen the attractor state (and "deepen" the attractor basin) for symmetrical upper extremity movement. Thus, breaking the symmetry during switching from the IP to AP mode would require greater energy and thereby a longer latency response for older individuals. Switching from the AP to IP mode would be easier and faster for older persons, in a relative sense, because the former pattern is more unstable and the latter pattern is a strong attractor. Further research might test the notion that increasing bias toward symmetrical movements has a constraining influence on the elderly individual's capacity to perform and learn more complex motor skills than simple cyclical bimanual movements.

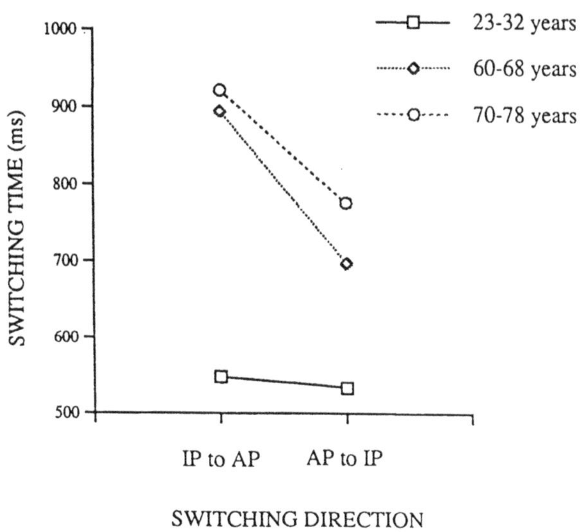

FIGURE 10. Age × Switching Direction interaction for intentional switching response time (see text for discussion).

CONCLUSION

In this chapter, we have reviewed literature on coordinated behavior in older adults and presented our application of the principles and research strategy of dynamic pattern theory to studying aging and coordination. We have stressed the importance of understanding how aging affects the ability to organize and execute complex movements as functional units. For effective outcomes, most daily tasks require highly ordered interactions among muscle groups and effectors which span multiple joints; effectiveness in daily living skills largely determines the quality of life for older adults. Although few studies have been conducted on aging and coordination, we can offer several preliminary conclusions which might be further investigated in the future. First, in laboratory testing involving bilateral upper extremity movements and responses to postural perturbations, many (but not all) older subjects exhibit temporal asynchrony of limb movements and abnormal muscle activation patterns. Second, common movements such as walking and pointing to targets in space (i.e., eye-hand coordination) are slowed with age, but their spatio-temporal patterns of muscle activation and effector movement are not degraded. Third, older adults have marked difficulty decoupling movement synergies, especially synergies involving homologous muscle groups in the upper extremities. Asymmetric patterns, as in cyclical anti-phase movements, are produced with greater variability (than in young adults) and tendencies toward critical fluctuation and shifts to symmetric patterns at relatively low frequencies. Fourth, intrinsic dynamics appear to have a greater constraining influence on intentional behavior in older versus young adults. That is, strong coordination tendencies such as in-phase coupling during cyclical bimanual movements reduce the efficacy with which older adults can form new and more difficult movement patterns.

The application of dynamic systems approaches to the study of aging and coordination is somewhat unique; however, there are strong experimental and theoretical precedents for such application. The research strategy discussed herein provides quantitative measures of coordination which allow scientists to identify control parameters that influence behavioral change over diverse time scales. In fact, the most promising application of the dynamic pattern approach may prove to be in longitudinal studies in which researchers identify and map the trajectories of candidate control parameters which may underlie senescent declines in movement control (see Thelen & Ulrich, 1991 for discussion of such an approach in studying motor development in infants). Dynamic systems theory also provides a conceptual backdrop for developing a unifying theory of aging. As argued by Yates (1988), aging can be thought of as an entropic process whereby instability emerges and constraints are formed "with a loss of dynamical range and richness" (p. 104). The aging organism illustrates the great paradox of many physical and natural systems which are characterized by flux between order and disorder. For example, whereas older persons demonstrate marked variability in many behaviors, they are also highly stable (and "set in their ways") in others. Dynamic issues such as stability, fluctuation, and change may thus be central to advancing our understanding of the aging mover.

REFERENCES

Baldissera, F., Cavallari, P., & Tesio, L. (1994). Coordination of cyclic coupled movement of hand and foot in normal subjects and on the healthy side of hemiplegic patients. In S. Swinnen, H. Heuer, J. Massion, & P. Casaer (Eds.), *Interlimb coordination: Neural, dynamical, and cognitive constraints* (pp. 230-242). San Diego: Academic Press.

Berlin, M., & Wallace, R. B. (1975). Aging and the central nervous system. *Experimental Aging Research, 2,* 125-164.

Bernstein, N. A. (1967). *The coordination and regulation of movements.* Oxford, England: Pergamon Press.

Birren, J. E., Woods, A. M., & Williams, M. V. (1980). Behavioral slowing with age: Causes, organization and consequences. In L.W. Poon (Ed.), *Aging in the 1980's* (pp. 293-308). Washington, D.C.: American Psychological Association.

Brinkman, J., & Kuypers, H. G. J. M. (1972). Splitbrain monkeys: Cerebral control of ipsilateral and contralateral arm, hand, and finger movements. *Science, 176,* 536-538.

Brinkman, J., & Kuypers, H. G. J. M. (1973). Cerebral control of contralateral and ipsilateral arm, hand, and finger movement in the split-brain rhesus monkey. *Brain, 96,* 653-674.

Cohen, L. (1970). Interaction between limbs during bimanual voluntary activity. *Brain, 93,* 259-272.

Cole, K. J. (1991). Grasp force control in older adults. *Journal of Motor Behavior, 23,* 251-258.

Coleman, P. D., & Flood, D. G. (1987). Neuron number and dendritic extent in normal aging and Alzheimer's disease. *Neurobiology of Aging, 8,* 521-545.

Cordo, P. J., & Nashner, L. M. (1982). Properties of postural adjustents associated with rapid arm movements. *Journal of Neurophysiology, 47,* 287-302.

Cummings, S. R., & Nevitt, M. C. (1989). A hypothesis: the causes of hip fractures. *Journal of Gerontology, 44,* M107-111.

Darling, W. G., Cooke, J. D., & Brown, S. H. (1989). Control of simple arm movements in elderly humans. *Neurobiology of Aging, 10,* 149-157.

Dayan, A. D. (1970). Quantitative histological studies on the aged human brain. II. Senile plaques and neurofibrillary tangles in senile dementia. *Acta Neuropathologica, 16,* 85-94.

Ferron, D. (1992). Changes with aging in right hemisphere activation as reflected in bimanual and dihaptic task performance. Unpublished doctoral dissertation, Carelton University, Ottawa, Canada.

Gibson, J. J. (1979). *The ecological approach to visual perception.* Boston: Houghton-Mifflin.

Greene, L. S. (1994). *Interlimb coordination in older adults: Application of dynamic pattern theory.* Doctoral dissertation, University of South Carolina. University Microfilms No. 94-10005.

Greene, L. S., Williams, H. G., & Lopez, O. A. (submitted). Kinematics and relative phase patterns of bimanual coordination in older adults.

Greene, L. S., Williams, H. G., Lopez, O. A., & Burke, J. R. (1993). Interlimb coordination in older adults: Application of dynamic pattern theory. *Journal of Sport and Exercise Psychology, 15* (Suppl.), S36.

Haken, H. (1977). *Synergetics – an introduction.* Heidelberg: Springer.

Haken, H. (1983). *Synergetics, an introduction: Non-equilibrium phase transitions and self-organization in physics, chemistry, and biology.* Heidelberg: Springer-Verlag.

Haken, H. (1985). Operational approaches to complex systems: An introduction. In H. Haken (Ed.), *Complex systems - operational approaches in neurobiology, physics, and computers* (pp. 1-13). Berlin: Springer-Verlag.

Haken, H., Kelso, J. A. S., & Bunz, H. (1985). A theoretical model of phase transitions in human hand movements. *Biological Cybernetics, 51,* 347-356.

Horak, F. B., Mirka, A., & Shupert, C. L. (1989). The role of peripheral vestibular disorders in postural dyscontrol in the elderly. In M. H. Woollacott & A. Shumway-Cook (Eds.), *Development of posture and gait across the life span* (pp. 253-279). Columbia, SC: University of South Carolina Press.

Inglin, B., & Woollacott, M. (1988). Age-related changes in anticipatory postural adjustments. *Journal of Gerontology, 43,* M105-113.

Ivry, R. B., & Keele, S. W. (1989). Timing functions of the cerebellum. *Journal of Cognitive Neuroscience, 1,* 136-152.

Jeka, J. J., & Kelso, J. A. S. (1989). The dynamic pattern approach to coordinated behavior: A tutorial review. In S. A. Wallace (Ed.), *Perspectives on the coordination of movement*. Amsterdam: North-Holland/Elsevier.

Jette, A. M., Branch, L. G., & Berlin, J. (1990). Musculoskeletal impairments and physical disablement among the aged. *Journal of Gerontology, 45,* M203-208.

Kamen, G., & DeLuca, C. J. (1989). Unusual motor unit firing behavior in older adults. *Brain Research, 482,* 136-140.

Katz, S. (1983). Assessing self-maintenance: Activities of daily living, mobility, and instrumental activities of daily living. *Journal of the American Geriatrics Society, 31,* 721-726.

Kay, B. A., Kelso, J. A. S., Saltzman, E. L., & Schöner, G. (1987). Space-time behavior of single and bimanual rhythmic movements: Data and limit cycle model. *Journal of Experimental Psychology: Human Perception and Performance, 13,* 178-192.

Kelso, J. A. S. (1984). Phase transitions and critical behavior in human bimanual coordination. *American Journal of Physiology, 240,* R1000-1004.

Kelso, J. A. S. (1994). Elementary coordination dynamics. In S. Swinnen, H. Heuer, J. Massion, & P. Casaer (Eds.), *Interlimb coordination: Neural, dynamical, and cognitive constraints* (pp. 301-317). San Diego: Academic Press.

Kelso, J. A. S., Holt, K. G., Kugler, P. N., & Turvey, M. T. (1980). On the concept of coordinative structures as dissipative structures: II. Empirical lines of convergence. In G. E. Stelmach & J. Requin (Eds.), *Tutorials in motor behavior* (pp. 3-48). Amsterdam: North-Holland.

Kelso, J. A. S., & Scholz, J. P. (1985). Cooperative phenomena in biological motion. In H. Haken (Ed.), *Complex systems: Operational approaches in neurobiology, physics and computers* (pp.124-149). Heidelberg: Springer-Verlag.

Kelso, J. A. S., & Schöner, G. (1988). Self-organization of coordinative movement patterns. *Human Movement Science, 7,* 27-46.

Kelso, J. A. S., Southard, D. L., & Goodman, D. (1979). On the coordination of two-handed movements. *Journal of Experimental Psychology: Human Perception and Performance, 5,* 229-238.

Kelso, J. A. S., & Tuller, B. H. (1984). A dynamical basis for action systems. In M. S. Gazzaniga (Ed.), *Handbook of cognitive neuroscience* (pp.321-356). New York: Plenum Press.

Konczak, J., Meeuwsen, H. J., & Cress, M. E. (1992). Changing affordances in stair climbability in young and older adults. *Journal of*

Experimental Psychology: Human Perception and Performance, 18, 691-697.

Kugler, P. N., Kelso, J. A. S., & Turvey, M. T. (1980). On the concept of coordinative structures as dissipative structures: I. Theoretical lines of convergence. In G. E. Stelmach & J. Requin (Eds.), *Tutorials in motor behavior* (pp. 49-70). Amsterdam: North-Holland.

Larish, D. D., & Stelmach, G. E. (1982). Preprogramming, programming and reprogramming of aimed hand movements as a function of age. *Journal of Motor Behavior, 14,* 322-340.

Marteniuk, R. G., MacKenzie, C. L., & Baba, D. M. (1984). Bimanual movement control: Information processing and interaction effects. *The Quarterly Journal of Experimental Psychology, 36A,* 335-365.

Nashner, L. M. (1977). Fixed patterns of rapid postural responses among leg muscles during stance. *Experimental Brain Research, 30,* 13-24.

Ostwald, S. K., Snowdon, D. A., Rysavy, S. D. M., Keenan, N. L., & Kane, R. L. (1989). Manual dexterity as a correlate of dependency in the elderly. *Journal of the American Geriatrics Society, 37,* 963-969.

Peters, M. (1994). Does handedness play a role in the coordination of bimanual movement? In S. Swinnen, H. Heuer, J. Massion, & P. Casaer (Eds.), *Interlimb coordination: Neural, dynamical, and cognitive constraints* (pp. 595-615). San Diego: Academic Press.

Rabbitt, P. (1979). How old and young subjects monitor and control responses for accuracy and speed. *British Journal of Psychology, 70,* 305-311.

Reed, E. S. (1982). An outline of a theory of action systems. *Journal of Motor Behavior, 14,* 98-134.

Roberton, M. A., & Halverson, L. E. (1988). The development of locomotor coordination: Longitudinal change and invariance. *Journal of Motor Behavior, 20,* 197-241.

Rothstein, D., Larish, D., Petruzzello, S., Crews, D., & Nahom, A. (1989). Bimanual coordination in the healthy old. *The Gerontologist, 29,* 258A-259A.

Rowe, J. W., & Kahn, R. L. (1987). Human aging: Usual and successful. *Science, 237,* 143-149.

Salthouse, T. A., & Somberg, B. L. (1982). Isolating the age deficit in speeded performance. *Journal of Gerontology, 37,* 59-63.

Scholz, J. P., & Kelso, J. A. S. (1989). A quantitative approach to understanding the formation and change of coordinated movement patterns. *Journal of Motor Behavior, 21,* 122-144.

990). Intentional switching between
n depends on the intrinsic dynamics
Behavior, 22, 98-124.
Schöner, G. S. (1987). Non-equili-
dinated biological motion: Critical
. *Physics Letters A, 123,* 390-394.
& Cook, S. (1984). Age-related
inical Orthapaedics and Related

ι. R. (1978). The effect of aging on
ιoice reaction time task. *Journal of*

ıl fitness, aging and psychomotor
5, 850-865.
se and the aging brain. *Research*
, 54, 208-218
2). Age and practice effects on force
fingers in precision pinching and
Stelmach & V. Homberg (Eds.),
he elderly (pp. 393-412). Boston:

ξ Goggin, N. L. (1988). Age differ-
ournal of Gerontology, 43, P18-23.
bio, R. P., & Teasdale, N. (1989).
s, and voluntary sway. *Journal of*

92). Cognitive-motor abilities of the
4, 53-65.
6). Aging and proprioception. *Age,*

C. J. (1985). Sensorimotor deficits
plications for falls in the elderly.
679-694.
ıer, H. (1994). Topics on interlimb
. Heuer, J. Massion, & P. Casaer
Neural, dynamical, and cognitive
: Academic Press.
ter, C. B., & Serrien, D. J. (1991).
ıl movements. *Experimental Brain*

Thelen, E., & Ulrich, B. D. (1991). Hidden skills: A dynamic systems analysis of treadmill stepping during the first year. *Monographs of the society for research in child development, 56,* 1-104.

Timiras, P. A., & Vernadakis, A. (1972). Structural, biochemical, and functional aging of the nervous system. In P. S. Timiras (Ed.), *Developmental physiology and aging.* New York: Macmillan.

Turvey, M. T. (1990). Coordination. *American Psychologist, 45,* 938-953.

Turvey, M. T., Shaw, R. E., & Mace, W. (1978). Issues in the theory of action. In J.Requin (Ed.), *Attention and performance VII* Hillsdale, NJ: Erlbaum.

Warabi, T., Noda, H., & Kato, T. (1986). Effects of aging on sensorimotor functions of eye and hand movements. *Experimental Neurology, 92,* 686-697.

Welford, A. T. (1984). Psychomotor performance. In C. Eisdorfer (Ed.), *Annual review of gerontology and geriatrics* (pp. 237-273). New York: Springer.

Welford, A. T. (1988). Reaction time, speed of performance and age. In J. A. Joseph (Ed.), *Central determinants of age-related declines in motor function* (pp. 1-16). New York: New York Academy of Sciences, Vol. 515.

Weller, M. P. I., & Latimer-Sayre, D. T. (1985). Increasing right hand dominance with age on a motor skill task. *Psychological Medicine, 15,* 867-872.

Whitall, J., & Clark, J. E. (1994). The development of bipedal interlimb coordination. In S. Swinnen, H. Heuer, J. Massion, & P. Casaer (Eds.), *Interlimb coordination: Neural, dynamical, and cognitive constraints* (pp. 395-412). San Diego: Academic Press.

Wiesendanger, M., Wicki, U., & Rouiller, E. (1994). Are there unifying structures in the brain responsible for interlimb coordination? In S. Swinnen, H. Heuer, J. Massion, & P. Casaer (Eds.), *Interlimb coordination: Neural, dynamical, and cognitive constraints* (pp. 180-207). San Diego: Academic Press.

Williams, H. G. (1990). Aging and eye-hand coordination. In C. Bard, M. Fleury, & L. Hay (Eds.), *Development of eye-hand coordination across the life span* (pp. 327-357). Columbia, SC: University of South Carolina Press.

Williams, K., & Bird, M. (1992). The aging mover: A preliminary report on constraints to action. *International Journal of Aging and Human Development, 34,* 241-255.

Williams, M. E., Gaylord, S. A., & McGahie, W. C. (1990). Timed manual performance in a community elderly population. *Journal of the American Geriatrics Society, 38,* 1120-1126.

Woollacott, M. H. (1989). Aging, posture control, and movement preparation. In M. H. Woollacott, & A. Shumway-Cook (Eds.), *Development of posture and gait across the lifespan* (pp. 155-175. Columbia, SC: University of South Carolina Press.

Woollacott, M. H., Shumway-Cook, A., & Nashner, L. M. (1986). Aging and posture control: Changes in sensory organization and muscular coordination. *International Journal of Aging and Human Development, 23,* 97-114.

Yaminishi, J., Kawato, M., & Suzuki, R. (1980). Two coupled oscillators as a model for the coordinated finger tapping by both hands. *Biological Cybernetics, 37,* 219-225.

Yates, F. E. (1988). The dynamics of aging and time: How physical action implies social action. In J. E. Birren & V. L. Bengtson (Eds.), *Emergent theories of aging* (pp. 90-117). New York: Springer.

Zanone, P. G., & Kelso, J. A. S. (1992). Evolution of behavioral attractors with learning: non-equilibrium phase transitions. *Journal of Experimental Psychology: Human Perception and Performance, 18,* 403-421.

Changes in sensory motor behavior in aging
A.-M. Ferrandez and N. Teasdale (Editors)

POSTURE CONTROL AND MUSCLE PROPRIOCEPTION IN THE ELDERLY

Laurette HAY

Université de Provence, Marseille

Abstract

This chapter first reviews some studies on the decrease in the static and dynamic control of posture in the elderly. The various components (spinal reflexes, postural responses, sensory integration mechanisms) of postural control are examined, together with their respective contributions to the ageing process. The notion of a hierarchical impairment of this multilevel system has been documented in a series of studies in the literature, and in a study on proprioceptive function in the elderly, using the tendon vibration technique. Three levels at which the propriomuscular afferents generated by this type of vibratory stimulation are processed were studied in the elderly by analyzing the latency and the intensity of the reflex, postural, and perceptual responses. The results showed that the greatest age-related impairment occurred at the highest perceptual level, since both the latency and the intensity of the vibration-induced postural illusions were affected. The postural responses (Vibration-Induced Falling) decreased in intensity with age, whereas the latency remained unchanged. No significant age-related changes were detected at the reflex level (Tonic Vibration Reflex). These data are in agreement with hierarchical models of postural dyscontrol in the elderly.

Key words: Posture, balance, proprioception, ageing.

Correspondence should be sent to Laurette Hay, Laboratoire de Neurobiologie Humaine, URA CNRS 372, Université de Provence, Centre Saint Jérôme, Avenue Escadrille Normandie-Niemen, 13397 Marseille Cedex 20, France (email: lnh@phocea.univ-mrs.fr).

STATIC AND DYNAMIC POSTURE CONTROL IN THE ELDERLY

One of the main problems besetting ageing adults is that they have an increasing tendency to fall, which is a common cause of injury and even mortality, and generally leads to reduced mobility. This is of course strongly related to the ability to control balance, i.e. to maintain the center of gravity of the body over its supporting base. This relationship can be disrupted unexpectedly by environmental events or continuously by locomotor activity involving changes in the supporting base and therefore introducing periodic destabilization of postural balance. Since their bipedal stance and locomotion give humans both a simpler locomotor cycle and a more complex system of static and dynamic equilibrium control than those of other species (Pailhous & Clarac, 1984), it is not surprising that keeping one's balance becomes an important and constant problem in the elderly. The need to focus on balance seems to greatly alter the ageing locomotor pattern. This has led some authors to hypothesize that deterioration of the balance control system may be one of the main causes of the changes in locomotion that occur in the elderly.

Even while standing quietly, many individuals in the elderly population encounter various balance problems. Such problems have generally been studied by examining the amplitude or velocity of the postural sway under static conditions. Our knowledge in this area is based on comparisons of populations of different ages, and not on longitudinal studies, so that talking about ontogenetic changes in balance control during ageing requires some circumspection. One of the first authors to investigate the control of quiet stance across the life-span found that postural stability was directly proportional to age until adolescence; then postural sway appeared to be very similar across age groups in adults up to the fifth decade, at which point postural stability became inversely proportional to age (Sheldon, 1963). These age-related differences were observed whatever the visual conditions, but were more pronounced when visual feedback was not available. The results of another investigation of a similar kind on a wide range of age groups (Pyykkö, Aalto, Hytönen, Starck, Jäntti, & Ramsay, 1988) confirmed Sheldon's data (Figure 1), and showed that postural stability when the eyes are open, as indicated by sway velocity, was approximately the same in children under ten and in adults over sixty. But with the eyes closed, the older people showed a higher sway velocity than the children.

Moreover, a number of studies have pointed out differences between the sexes in ageing balance control. Overstall, Exton-Smith, Imms, and Johnson (1977) showed that the age-related increase in the postural sway

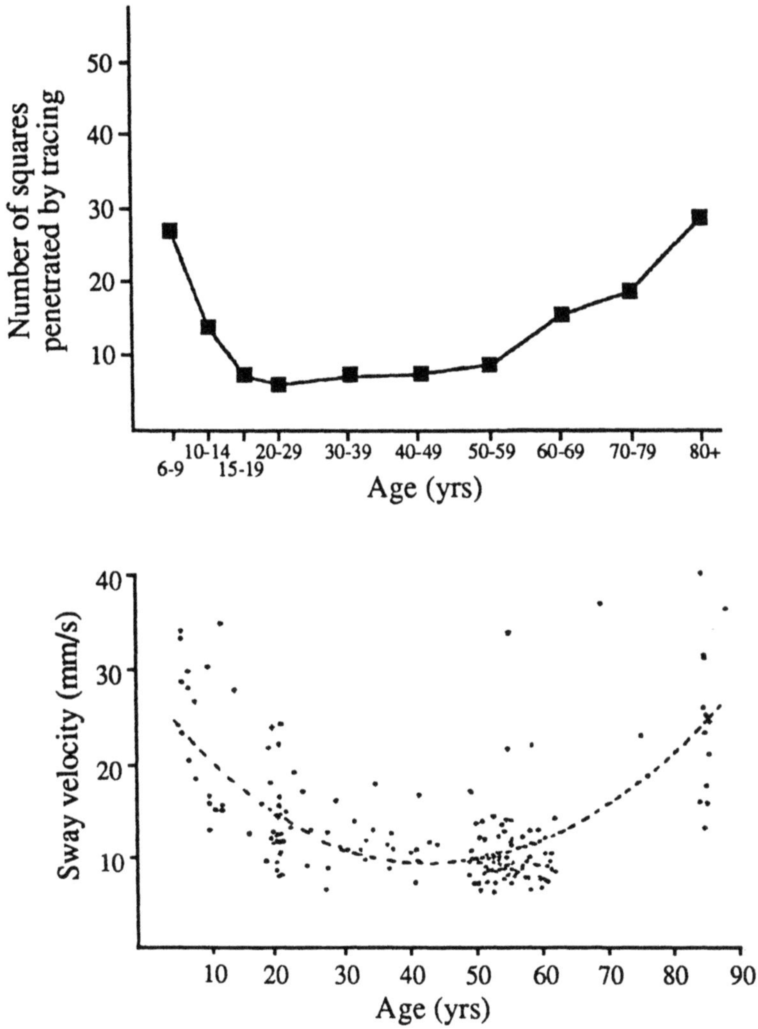

FIGURE 1. Postural sway measured during quiet stance, with the eyes open. Adapted from Sheldon (1963) (upper part) and Pyykkö et al. (1988) (lower part).

was identical in men and women, but that the women in all age groups studied exhibited lower equilibrium performance. In general, prior studies on the effects of gender have revealed that performance on balance tests is lower in women than in men (Hinchcliffe, 1983; Ochs, Newberry, Lenhardt, & Harkins, 1985; Overstall, 1983), or that sex-related differences cannot be observed (Brocklehurst, Robertson, & James-Groom, 1982), whereas more recent data indicate higher performance in women (Ekdahl, Jarnlo, & Andersson, 1989; Juntunen, Matikainen, Ylikoski, Ylikoski, Ojala, & Vaheri, 1987; Pyykkö, Jäntti, & Aalto, 1990). These contradictory findings make it difficult to conclude from the literature that men and women behave differently when facing balance problems. Other authors (Hasselkus & Shambes, 1975; Era & Heikkinen, 1985) confirmed the effects of age on the stability of stance, and observed, moreover, that practice affects postural control. Female subjects placed in a forward leaning position had more difficulty controlling their postural sway than when placed in an upright position. However, they were able to increase their stability with repeated trials in this more stress-inducing position (Hasselkus & Shambes, 1975).

Changes in balance control in older people are not always the consequence of ageing itself, which leads to the gradual and widespread degeneration of the neural and musculoskeletal systems and may vary according to the subjects' activity history and genetic factors. These changes also often result from specific diseases frequently encountered among the elderly, and sometimes from side effects of the drugs associated with those diseases. This pathological feature of the age-related decrease in postural balance was even used by Horak, Shupert, and Mirka (1989) as the basis for explaining most of the ageing process. These authors proposed that the effect of age per se on postural control deficits is of minor importance, and that the main factor is the fact that the development of specific pathologies becomes increasingly probable with age, which explains the increased variability of postural control in the elderly population. In a study on balance control in a group of older adults without any history of neurological dysfunction (Manchester, Woollacott, Zederbauer-Hylton, & Marin, 1989), a neurological examination revealed that half of the subjects had a borderline case of a disease of the peripheral or central nervous system which correlated with loss of balance.

The dynamic balance problems encountered during locomotion or during any other balance disturbance in the elderly have generally been studied under restricted, standardized conditions where a moveable platform is shifted to induce postural responses simulating those which might occur during perturbed walking. The patterns of muscular activity involved in the postural responses to unexpected platform perturbations have been analyzed extensively in elderly and young adults by Woollacott and her colleagues (Woollacott, Shumway-Cook, & Nashner, 1982,

1986; Woollacott, 1988), who described the changes in the dynamic postural response patterns reflecting the age-related impairment of the postural control system. In particular, these changes affected the temporal sequencing and relative consistency of distal and proximal synergist activation (Woollacott et al., 1982). However, 50% of the subjects were able to adapt their postural responses when repeating the trials (Woollacott et al., 1986). The absolute latency of these responses does not seem to be strongly affected by age, since only slight increases were reported (Woollacott, 1986; 1988), and other authors have even reported that the latencies remained unchanged (Manchester et al., 1989; Panzer, Kaye, Edner, & Holme, 1992).

These deficits in the corrective responses that serve to restore balance are not the only ones responsible for the equilibrium problems encountered by the elderly during locomotor activity, since the so-called long latency automatic postural responses are part of a posture control hierarchy, which also comprises a lower reflex level and higher integrative mechanisms. In this chapter, the involvement of these various levels of posture control in the ageing process will be examined in the light of some studies in the literature. This question will be addressed with particular reference to a study on proprioceptive function, known to be strongly involved in postural control. This study aimed to analyze how propriomuscular input generated by vibratory muscle tendon stimulation is processed by the elderly nervous system at various levels. For this purpose, reflex, postural, and perceptual responses to vibration-induced propriomuscular messages will be described. The notion of a multilevel control of posture, which diverges strongly from the purely reflexological approach (Sherrington, 1906, 1910) and originates from theories on the hierarchical organization of motor control (Paillard, 1960; Bernstein, 1967), has greatly influenced the thinking and hypotheses of research on the elderly. The various features of impaired postural control have been described in terms of multilevel or multicomponent models in some excellent reviews in the literature (see for example Woollacott et al., 1982, 1986; Stelmach & Worringham, 1985; Horak et al., 1989).

SPINAL REFLEXES

Stretch reflex

In addition to higher level mechanisms which integrate convergent visual, vestibular, and somatosensory input in order to control posture and locomotion, the lower level sensorimotor loops (monosynaptic stretch reflex and long latency automatic postural responses) involved in the regulation of both of these activities are proprioceptive in origin. The literature on senescence is quite prolific as regards the monosynaptic stretch reflexes involved in the lowest level of control, but it includes some rather contradictory data. Milne and Williamson (1972) and McLennan, Timothy, and Hall (1980), for example, described an age-related decrease in the ankle jerk reflex, whereas other authors (Appenzeller, Imarisio, & Gilbert, 1966; Clarkson, 1978; Carel, Korczyn, & Hochberg, 1979; DeVries, Wiswell, Romero, Heckathorne, 1985) did not observe any significant changes in the Achilles tendon reflex of elderly subjects. Nor did comparisons of the Hoffman and T reflexes reveal the existence of any changes in the specific involvement of the fusimotor system (DeVries et al., 1985). After having carefully examined the results of several studies on tendon reflexes and compared them with their own data, Bryndum and Marquardsen (1964) concluded that when the absence of ankle jerk was relatively frequent, it was mainly due to factors that were not linked to ageing, such as peripheral neuropathy.

On the basis of contradictory data on age-related effects on the Achilles tendon reflex, DeVries et al. (1985) suggested that the subjects' activity history may be a powerful factor, and that there may have been a link in their study between the fact that their elderly population consisted of vigorous, active individuals, and the observed lack of any age-related changes in the T reflex. The importance of this factor was also exemplified in a recent study using a young adult group and an elderly population (60 to 86 years) consisting of 28 healthy, independent and active individuals subjected to a preliminary clinical examination of the Achilles and knee reflexes. No major changes were observed across age groups; only one 80-year-old subject did not show any reflex at all, and the Achilles reflex was not present in one young and another 80-year old subject (Quoniam, Hay, Roll, & Harlay, 1995).

Tonic vibration reflex

In the above study by Quoniam et al. (1995), another method using tendon vibration was applied to test for the effects of proprioceptive input at the reflex level. It has now been clearly established that a vibratory stimulation applied to the tendon of a muscle mainly activates the muscle spindle primary endings (Roll, Vedel, & Ribot, 1989a) and gives rise to an afferent pattern similar to that which would have resulted from an actual movement stretching the muscle being vibrated.

A motor response was induced, the Tonic vibration reflex or TVR (Eklund & Hagbarth, 1966; Hagbarth, 1973), which is an involuntary contraction of the vibrated muscle. This response mainly corresponds to the high frequency activation of the myotatic loop via the same pathways as those involved in the stretch reflex, and is also influenced by various afferents, via polysynaptic pathways. This tonic muscular response to vibration can be observed under isometric conditions when muscle shortening is prevented by joint fixation, and is facilitated when the stimulated limb is kept relaxed and within sight. The influence of visual information about the fixed state of the limb suggests the possible functional reorganization at the spinal level of the excitatory effect of the muscle spindle afferents on either the agonist or the antagonist motoneuron pool (Roll, Gilhodes, & Tardy-Gervet, 1980). Since the occurrence of a vibration-induced muscular response shows that at least some proportion of the Ia-afferent fibers and the alpha motoneurons of the stimulated muscle are functioning normally, the vibration technique can be applied to ageing people to determine whether the reflex pathways are intact. Guieu and his colleagues (Guieu, Ribot-Ciscar, Tardy-Gervet, Dano, & Roll, 1990) established that this method is more sensitive than the clinical method of examination, where a reflex-hammer is used to quantify reflexes. These authors succeeded in activating the myotatic loop by applying vibrations, and in inducing TVR responses in patients who suffered from polyradiculoneuropathy and showed no deep reflex at all. They suggested that the continuous propriomuscular message induced by the vibration may have led to a temporal summation which enabled the alpha motoneurons to reach their discharge threshold, contrary to what occurs with a single tendon tap, which generates only a short lasting spindle message. Moreover, vibratory stimulation induces an afferent pattern which is similar to the sustained inflow arising from the stretch receptors during continuous muscle stretching (Hagbarth, 1973), so this type of activation may more faithfully mimic that elicited by balance disturbances in the sensorimotor loops than the massive synchronous afferent volley evoked by a single tendon tap.

In the present study, 28 healthy male and female subjects ranging in age from 60 to 86 participated in the experiments. They were autonomous, active individuals selected among volunteers without neurological disorders. They were divided into three age groups corresponding to the sixth, seventh, and eighth decades (10, 10, and 8 subjects, respectively). A control group of nine young adults (mean age: 29) was also tested. In the first two experiments, two subjects had to be dropped for technical reasons, one in the seventh-decade group and one in the eighth-decade group.

The TVR was tested here in the upper limb, since the conditions required to obtain this response (the stimulated limb must be relaxed and within sight) are more easily achieved when testing the upper than the lower limbs. Although some differences in somatic sensitivity exist between the upper and lower limbs, both show a similar age-related pattern of decline (Rosenberg, 1958; Kokmen, Bossemeyer, & Williams, 1977; McLennan et al., 1980; Brocklehurst et al., 1982). The subjects, seated on a chair with their forearms laying comfortably on horizontal supports, were asked to keep their arm relaxed and look at their hand, while a vibrator was applied for eight seconds to the distal tendon of either the biceps or the triceps brachii muscle. In all age groups tested, the vibration induced an increase in the electromyographic activity of the vibrated muscle. Neither the latency nor the amplitude of this activity varied significantly across the age groups (Quoniam et al., 1995). This absence of age-related changes in the TVR responses of the biceps and triceps brachii muscles is in line with clinical data on the tendon reflexes induced by a reflex-hammer in the lower limbs, which were present in all of the age groups in this population.

Another question that arises concerning the lower level sensorimotor loops is their functional contribution to corrective reactions to postural perturbations, and the extent to which impairment of reflex activity may account for dynamic imbalance problems in the elderly. Is it possible, for example, that fast, transient changes in the muscle stretching induced at the ankle joint by a postural perturbation might be compensated for by spinal reflex activity if the myotatic loop is still efficient? In fact, the stretch reflex does not seem to be very involved in balance control in young adults (Gurfinkel, Lipshits, Mori, & Popov, 1976). This reflex may be of secondary importance in correcting unexpected fast transient postural disturbances (Diener, Dichgans, Guschlbauer, & Mau, 1984), which are mostly compensated for by long-loop reflexes (Nashner, 1976). Since no major changes in the tendon reflex seem to occur in healthy active elderly as a whole, age-related difficulty in mastering dynamic balance control during locomotion seems unlikely to be due to lower level reflex deficits. Nevertheless, reflex activity may have a perturbing influence on dynamic balance control because of the reappearance of short-latency spinal reflexes in the elderly in postural situations that do not appear to elicit responses of this kind in younger subjects, possibly due to elderly subjects' reduced central control (Magladery, Teasdall, & Norris, 1958). Sporadic reflex activity has been observed in elderly persons who were voluntarily controlling their equilibrium while small, slow perturbations were applied to the platform (Stelmach, Teasdale, DiFabio, & Phillips, 1989). Woollacott, Shumway-Cook, and

Nashner (1982) reported that elderly people subjected to rotational perturbations frequently (82% of the population) exhibited monosynaptic stretch reflexes, whereas the younger subjects exhibited these reflexes only occasionally. These authors suggested that the cortical impairment that is potentially responsible for the release of lower centers in the hierarchy is likely to be variable, which may partly explain the inter- and intraindividual variability of the responses produced by the ageing postural system. Depending on the degree of neural impairment and on the strength of the sensory excitation, either higher centers may generate a long-latency response, or if the threshold level of activity triggering central processing cannot be reached, a myotatic reflex may be released.

SUPRASPINAL POSTURAL RESPONSES AND SENSORY INTEGRATION MECHANISMS

Long-latency automatic postural responses, which are also elicited by proprioceptive input, are polysynaptic, involve controversial supraspinal pathways, and are functionally much more closely involved in dynamic balance control than the monosynaptic stretch reflex is. The question as to the extent of their contribution to the decrease in postural control in the elderly has often been raised in connection with the deficits affecting higher level integrative mechanisms. A study in which proprioception was excluded by ischemia showed that the contribution of proprioceptive input to the regulation of dynamic balance disturbed by a shifting support depends on the frequency of the disturbance (Diener et al., 1984). A reflex-like regulation of postural balance sufficed to counteract the high frequency, transient postural disturbances, based on information from muscle spindles originating from above the ischemia (ankle), whereas this did not suffice during low frequency sinusoidal stimulation. The latter involves slower and more continuous regulatory mechanisms where proprioceptive information has to be complete and well integrated with the other sensory cues. Since these higher level integrative mechanisms coordinate redundant visual, vestibular, proprioceptive, and plantar tactile information about postural events, peripheral damage to any one of these sensory channels, resulting in the well known increase in the elderly's sensory thresholds, will reduce this redundancy and restrict the safety margin of the postural system. Impaired processing of these messages by the central nervous system might diminish the quality of the information collected about the body and the environment, and degrade the triggering and regulation of postural responses.

The idea that the deterioration of regulatory responses to disequilibrium frequently observed in the elderly might be mainly due to deficits at a higher level of postural control is a widely-held opinion. In the above study showing that elderly subjects' postural stability, which was lower in forward leaning than in an upright position, could improve with repeated trials, the authors ascribed the practice effects to cortical involvement in the control of posture in this situation (Hasselkus & Shambes, 1975). Defective central control of posture has been thought to explain not only age-related balance problems (McLennan et al., 1980), but also the declining postural control which begins long before senescence and probably results from physiological causes and diseases of the central nervous system (Overstall et al., 1977).

Stelmach et al. (1989) investigated the age-related changes affecting the reflex vs. high level integrative mechanisms of postural control, by comparing young and elderly people's postural reactions to large-fast and small-slow ankle rotation postural disturbances. Large-fast rotations, which activated the long-loop reflexes, were found to be equally disturbing to both age groups, whereas the small-slow ones, which involved the higher integration level of postural control, induced greater sway in the elderly subjects, who, unlike the younger group, showed no ability to adapt to these conditions. The authors concluded that elderly people are at a disadvantage when their posture is controlled by slower, higher level sensory integrative mechanisms. The specific difficulty which elderly people have with these mechanisms has been demonstrated by either reducing the information from the various sensory systems involved in posture control (Straube, Bötzel, Hawken, Paulus, & Brandt, 1988; Teasdale, Stelmach, & Breunig, 1991) or by generating a conflict between information of different kinds (Woollacott et al., 1986; Mirka, Peterka, Horak, & Black, 1988). When confronted with the need for sensory reweighting, the older subjects tended to loose balance, whereas the younger ones did not; however the older subjects were able to adapt to the situation during subsequent trials (Woollacott et al., 1986).

It is not quite clear from these experiments whether the lack of sensory reweighting and the subsequent imbalance resulted from a decrease in the sensitivity of the remaining peripheral system or from the deterioration of central processing operations. Teasdale and his colleagues (Teasdale, Bard, Dadouchi, Fleury, Larue, & Stelmach, 1992) addressed this question using an experimental procedure based on augmented rather than reduced sensory conditions (which rules out an exclusive peripheral interpretation), and focusing on the subject's behavior during the transient period subsequent to the reduced sensory condition.

They showed that unlike the younger subjects, the elderly exhibited greater sway dispersion during transitions from no vision to vision. This inability to adapt to an augmented sensory condition was interpreted by the authors as specifically indicating that the central integrative mechanisms responsible for reconfiguring the postural set were impaired. Facing these difficulties in central integration, older persons tend to allocate more of their attentional resources to the postural task than younger subjects, as indicated by increased reaction times (Teasdale, Bard, Larue, & Fleury, 1993).

In these higher level mechanisms, the role of sensory input does not consist merely of detecting a stimulus and triggering a postural response. As pointed out by Horak, Shupert, and Mirka (1989), these messages are also involved in establishing an internal representation of the position of the center of gravity and the characteristics of the environment, which is used by the central nervous system to activate the appropriate motor synergy to restore balance. In the sensory information about balance control, there is surprisingly little data on age-related deficits due to the deterioration of the proprioceptive system. Proprioceptive loss in the elderly leads to higher movement detection thresholds and to a decrease in joint angle copying or matching accuracy (Kenshalo, 1979; Skinner, Barrack, & Cook, 1984; Kaplan, Nixon, Reitz, Rindfleish, & Tucker, 1985; Stelmach & Sirica, 1987; Ferrell, Crighton, & Sturrock, 1992). This loss has generally been found to be greater in the lower than in the upper extremities (Laidlow & Hamilton, 1937; Kokmen, Bossemeyer, Barney, & Williams, 1978; Schiano, Marchetti, Bardot, Sambuc, Bardot, & Serratrice, 1988). It appears fairly important to determine the effects of propriomuscular loss on age-related deficits in balance control, since this information channel seems to be critical for sway perception around the ankle, as suggested by Fitzpatrick and McCloskey (1994), who found that the proprioceptive threshold in young adults was lower than the visual and vestibular thresholds for perceiving body sway. According to these authors, the spindle sensitivity to muscle stretch at ankle level is quite sufficient to account for the proprioceptive threshold. But changes in muscle force with body sway, such as reflex contractions evoked by subliminal movements (Fitzpatrick, Taylor, & McCloskey, 1992) may be an additional and indirect source of sensory input through their effect on Golgi tendon organs firing.

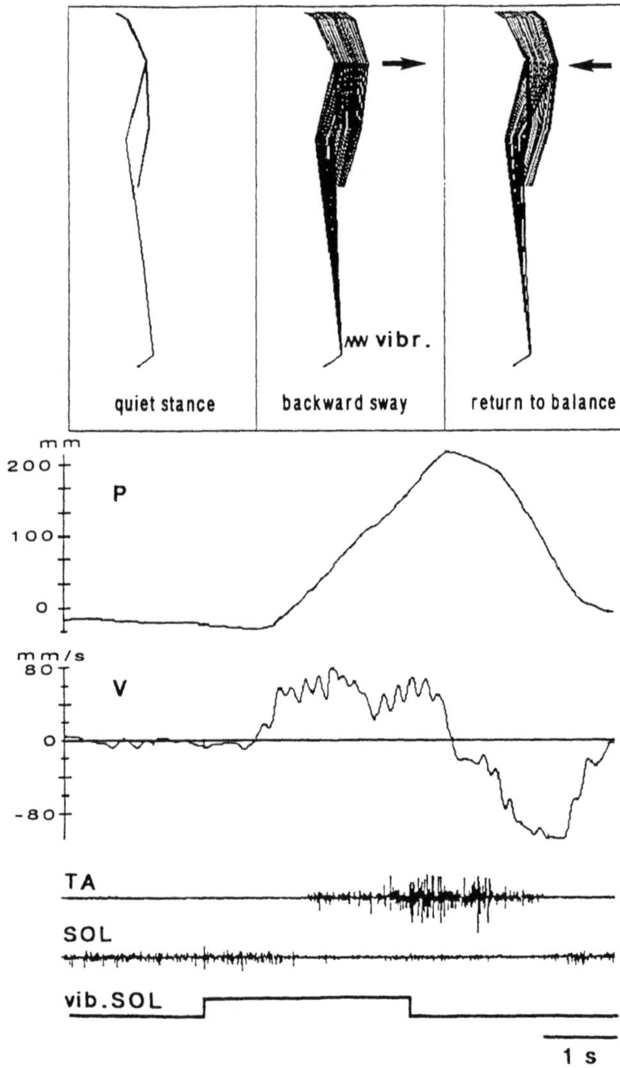

FIGURE 2. Illustration of Vibration-Induced Falling: the vibratory stimulation applied to the soleus muscles tendons during quiet stance induced a backward sway (VIF), as shown by the anterior-posterior movement of a shoulder marker (first phase of the position and velocity curves). When the stability limit was reached, a movement in the opposite direction was performed to restore balance (second phase of the curves). The EMG recordings show the antagonist (tibialis anterior) activity triggered in order to stop the VIF and restore balance. Note that the onset of the VIF was not accompanied by a marked increase in soleus muscle activity.

Vibration-induced falling

The tendon vibration technique has been used to test the specific role of propriomuscular input in postural responses to disequilibrium. Besides the TVR response described above, vibratory stimulation applied to the ankle muscle distal tendons of a standing subject induces an involuntary sway of the whole body, the direction of which depends on whether the tibialis anterior (forward sway) or the soleus (backward sway) muscles are vibrated (Figure 2).

This postural response has been described as "vibration-induced falling" (VIF) by Eklund (1972a, 1973), who pointed out that it could not be confused with an elicited TVR because of its long latency. The exact levels of the nervous system responsible for this reaction have not yet been determined, but it is generally agreed that its origin is supraspinal (Eklund, 1972a; Hagbarth, 1973), and VIF can be assumed to result from an integrative mechanism involved in postural regulation based on propriomuscular afferents (Roll, Vedel, & Roll, 1989b). VIF may reflect a corrective response to artificially generated muscle spindle input, taken by the nervous system to mean that a body tilt has occurred. This response can be used in the elderly to assess the specific contribution of propriomuscular information to the control of upright posture, since the propriomuscular message triggering the postural reaction is emitted in the absence of any movement.

This method was applied to the same young and elderly populations as described above. The subjects were standing quietly with a comfortable posture. Two vibrators were fixed to the distal tendons of either both soleus or both tibialis anterior muscles. The electromyographic activity of both ankle muscles and body movement kinematics were recorded during each experimental sequence, which consisted of a preliminary 2-second phase of normal stance, followed by a 3-second vibratory stimulation, and a return to the normal stance. The subjects were tested under eyes-open and eyes-closed conditions. Applying vibration to the ankle muscles induced body sways in all subjects, whatever their age, the direction of which depended on which muscles were vibrated, as previously described. The main age-related trends are illustrated in Figure 3, which gives individual tracings of the shoulder marker velocity on the anterior-posterior axis.

The latency of the sway (time between the vibration onset and the start of the sway) did not appear to change with age, the visual conditions, or the muscles stimulated, as shown in Table 1.

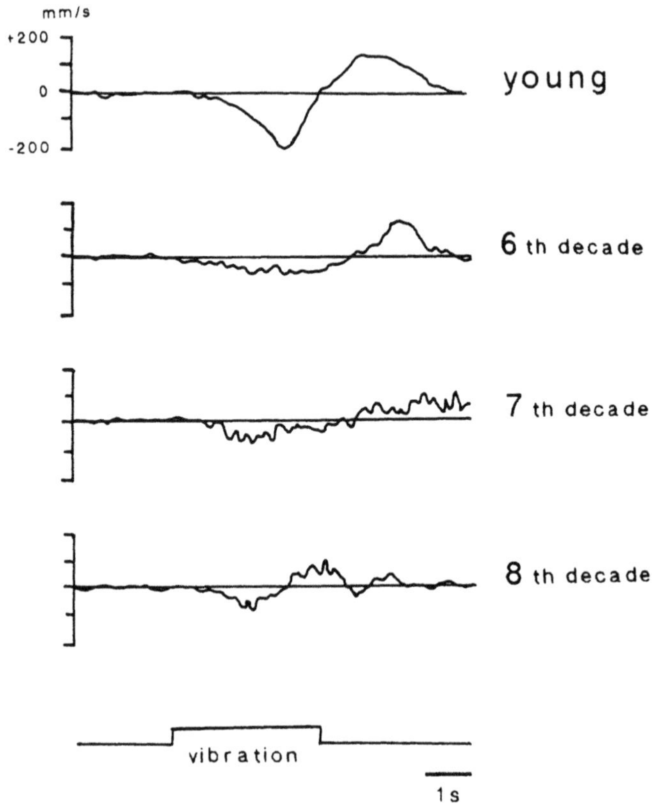

FIGURE 3. Individual tracings of shoulder marker velocity during backward sways induced by soleus vibration, in four age groups.

The fact that the response latency did not increase with age is fairly consistent with the results mentioned above (Woollacott, 1986, 1988; Manchester et al., 1989; Panzer et al., 1992), which were obtained with long latency postural responses to platform shifts. Even at a higher level of postural regulation, the "alerting" function of propriomuscular messages does not seem to be markedly affected by ageing. Stelmach and Worringham (1985) stressed the need for a heuristic device to organize our knowledge and hypotheses on motor control deficits, and proposed a scheme of processing stages dealing with responses to disequilibrium. When a potential fall is imminent, any sensory input generated by the perilous situation will "alert" those centers involved in selecting the response. This "alerting process" must therefore be correctly achieved, irrespective of the appropriateness of the subsequent postural response, for recovery of balance to be possible. Assuming that the VIF reflects a postural response to a purely propriomuscular message signalling a disequilibrium, it seems possible that within this sensory system, deficits in the alerting process might not account for the main balance problems of the elderly, which can be more satisfactorily explained by investigating the later response selection or execution stages.

TABLE 1.
Latency (ms) of the backward and forward sway induced by tendon vibration of the soleus or tibialis anterior muscles, under eye-open (EO) and eye-closed (EC) conditions, as a function of age (mean and *standard deviation*). As shown by an analysis of variance, no effect of Age, $F(3, 31) = .84$, Vision, $F(1, 31) = .78$, or Muscle vibrated, $F(1, 31) = 2.88$), could be found (Quoniam et al., 1995).

	Young	60-69	70-79	80-89
Soleus EO	780 *250*	600 *210*	680 *250*	740 *240*
Soleus EC	710 *210*	680 *200*	660 *190*	600 *90*
Tibialis EO	560 *110*	650 *120*	710 *240*	550 *110*
Tibialis EC	650 *120*	610 *110*	680 *130*	620 *250*

As mentioned above, the direction (backward or forward) of the responses could be correctly selected at all ages in accordance with the muscle being stimulated. The magnitude of these postural responses, as measured by the amplitude, mean velocity, and mean acceleration of the VIF, was found to vary with the subjects' age. It decreased between the young adult age and the sixth decade, and levelled off between the sixth and eighth decades. This trend can be seen in Figure 4, which shows the mean acceleration of the sway (ratio between the peak velocity and the time to reach that value) by age group, eye condition, and muscles vibrated.

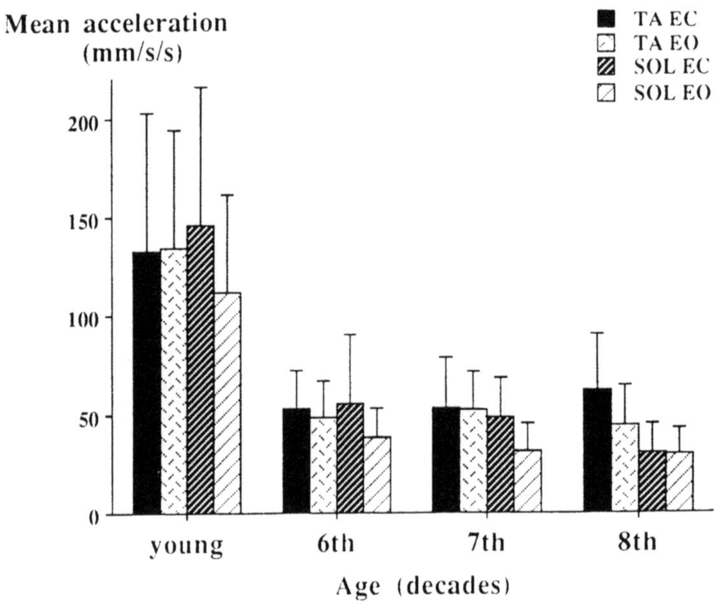

FIGURE 4. Mean acceleration of the sway (ratio between peak velocity and time-to-peak) induced by soleus and tibialis anterior vibration, with the eyes closed and the eyes open, in four age groups. An analysis of variance showed a significant effect of Age, $F(3, 31) = 9.51$, $p < .001$, Vision, $F(1, 31) = 6.19$, $p < .025$, and Muscle vibrated, $F(1, 31) = 5.22$, $p < .05$; the effect of Age was entirely determined by the difference between the group of young adults and the three older groups, $F(1, 31) = 28.53$, $p < .001$, who did not differ from each other, $F(2, 23) = .29$.

The observed age-related changes in VIF suggest that the amplitude, velocity, and strength of the compensatory responses are lower in older than in younger adult subjects confronted with imminent disequilibrium signalled by the same propriomuscular input. This difference cannot be explained solely by the fear of falling in the elderly (Maki, Holliday, & Topper, 1990) which would lead them to refrain more from swaying, since the decrease in the response strength was apparent even before the onset of the antagonist activity (Quoniam et al., 1995). Horak, Shupert, and Mirka (1989) observed that older subjects exhibited smaller forward and backward sways than young adults when asked to voluntarily lean until reaching their stability limit. The authors concluded that the perceived stability limit may decrease with age and become smaller than it actually is, since the older subjects were able to control their center of gravity much farther during quiet stance. This suggests that the decrease in the corrective postural response reflected in the VIF might be partly due to the underestimation of the proprioceptive message signalling a disequilibrium. Otherwise, smaller perceived stability limits would have led to stronger postural responses in the elderly than in the young.

When subjects were stimulated under the eyes-open condition, they could realize from the available visual information that they were not swaying, since the spindle message was induced by vibration and not by an actual body sway. Nevertheless, as shown in Figure 4, this visual condition did not eliminate the postural response but only reduced it. The forward sway induced by tibialis anterior muscle stimulation was stronger than the backward sway induced by soleus muscle stimulation, in all of the age groups tested. Since the peak velocity was the same no matter which muscles were stimulated, the difference was due to a lag in the time-to-peak velocity, which was shorter with tibialis anterior than with soleus muscle stimulation. This means that an afferent message equivalent to that generated by a backward sway stretching the tibialis anterior muscles induced stronger compensatory responses than a message generated by a forward sway. This might be attributed to the fact that backward sways are normally less readily compensated for than forward sways (Horak et al., 1989), due to the biomechanical constraints involved. The tolerance limits for backward imbalance may therefore be lower, so that the compensatory response to backward sway information is more rapidly recruited. An underestimation of the response to be produced in order to recover balance might be more disastrous in the case of backward than forward sway, and might partially account for the fact that backward falls occur more frequently in the elderly than in younger people (Sabin, 1982).

The age-related decrease of the vibration effect on postural sway was even found to be more severe in a study where a 100-Hz vibration was applied during 30 seconds to the Achilles tendons of young and elderly subjects (Nakagawa, 1992), since it completely disappeared in the older subjects. The peripheral proprioceptive degeneration put forward by this author to explain his results might not be the only cause of this information loss, as suggested by our previously mentioned data. Moreover, as illustrated in Figure 2, in agreement with other findings (Hayashi, Miyake, & Watanabe, 1981), the onset of the VIF was not directly correlated with the activity of the distal vibrated (soleus) muscles; vibration-induced activity can be observed in muscles other than the vibrated ones, such as the antagonist muscles (Latash & Gurfinkel, 1976; Roll et al., 1980) and even in muscles which are quite distant from the vibration point (Quoniam, Roll, Deat, & Massion, 1990; Redon, 1992). This suggests that complex synergies and a high integrative level of processing are involved in these postural responses.

Nevertheless, the decreasing ability to produce this particular kind of postural response may also be due to several other age-related factors, such as nerve conduction (Dorfman & Bosley, 1979), muscle strength and joint flexibility deficits, and increasing muscle coactivation (Woollacott, 1990), all of which might contribute to deteriorating the execution of the selected response. As noted by Horak, Shupert, and Mirka (1989), in addition to requiring the ability to select an appropriate corrective response, effective postural control involves the capacity "to match the magnitude of the response to the magnitude of the disturbance". Underresponding may not only result from the underestimation of the disturbance, but also from the inability to perform the response properly because of motor constraints. One such possible impediment to the correct execution of a compensatory response might be elderly people's tonic postural activity. The electromyographic activity of the ankle muscles of young and elderly subjects was recorded during quiet stance. As shown in Figure 5, the tonic activity, which was not surprisingly found to be much higher in the soleus than in the tibialis anterior muscles, increased with age.

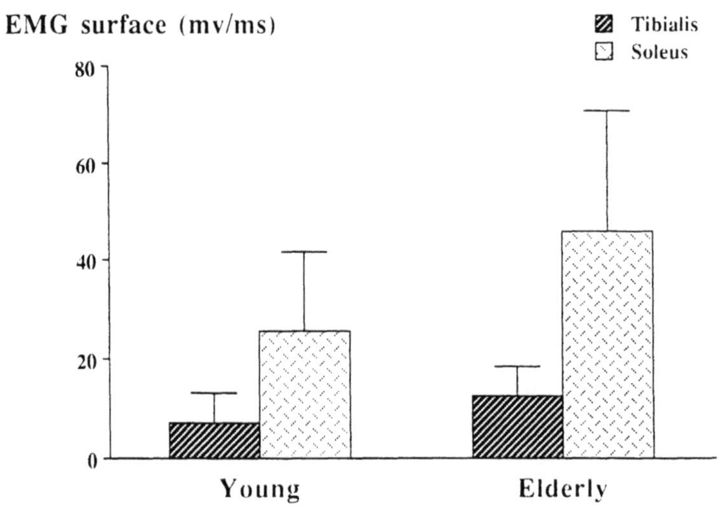

FIGURE 5. Tonic postural activity (EMG integrals) of the soleus and tibialis anterior muscles during quiet stance, in the young and elderly groups. An analysis of variance showed a significant effect of Muscle, $F(1, 31) = 50.29$, $p < .001$, and Age, $F(1, 31) = 8.16$, $p < .01$.

This increased tonic postural activity possibly anchors the body to its support and makes it more difficult to produce a postural response which is correctly scaled to the amplitude and velocity of the disturbances. This higher tonic activity may also be a perturbing factor at the sensory level; the occurrence of a stronger and possibly more variable tonic activity in both antagonist muscles is likely to prevent the muscle being stretched during postural sway from correctly fulfilling its sensory function, since more noise would then be associated with the relevant signal. This notion of neural noise, which was first used by Welford (1981) in connection with signal detection problems in the elderly, might be a useful means of accounting for age-related deficits in balance control at both the motor and sensory levels.

Vibration-induced postural illusions

As with postural responses to ankle vibration, age-related underestimation was found to occur at the perceptual level in a study which evaluated the postural illusions (perception of sway) induced by ankle muscle vibration in stabilized subjects (Quoniam, Hay, Roll, & Harlay, 1993). In addition to the motor responses described above (TVR and VIF), vibratory stimulation applied to a muscle tendon is known to induce sensations of illusory movement in a motionless limb (Eklund, 1972b; Goodwin, McCloskey, & Mattews, 1972; Roll & Vedel, 1982). This kinesthetic effect, which is equivalent to that potentially elicited by pulling on the tendon (Gandevia & McCloskey, 1976; Mattews & Simmonds, 1974), has been regarded as showing that movement perception relies strongly on muscle proprioception, in agreement with the existence of cortical projections of muscular afferents. When subjects are restrained in order to prevent swaying, they experience an illusion of body sway which corresponds to the stretching of the muscle being vibrated, in the direction opposite to the VIF (Goodwin et al., 1972; Gurfinkel, Lipshits, & Popov, 1977; Gurfinkel, Levick, Popov, Smetanin, & Shlikov, 1988).

The data reported below were collected from the same samples as described above. The subjects were standing in an upright but relaxed position with their eyes closed. Two vibrators were fastened to the distal tendons of either both soleus or both tibialis anterior muscles. An 80-Hz vibration was applied for eight seconds while the subjects were held at the shoulders to prevent any postural sway. They were required to indicate any perceived body sway by moving a vertical lever with their hand, and to attempt to accurately copy the onset, the backward or forward direction, and the velocity of the illusory body sway they perceived. The lever rotations, which were transmitted to a potentiometer and converted into angular displacements, reflected the subjects' postural sensations.

In line with the results of the above studies, the subjects experienced kinesthetic sensations of body sway when vibratory stimulation was applied to their ankle muscles. This postural movement was an illusory one since the subjects' shoulders were fixed, and it involved a body tilt in the direction that would have stretched the stimulated muscle: backward sway with tibialis anterior vibration, and forward sway with soleus vibration. Although vibration-induced postural illusions were elicited in subjects belonging to all of the age groups tested, age-related differences were observed in the timing and intensity of the kinesthetic sensations, as determined from the subjects' imitative responses. The latency of the perceived sway (time between the onset of the vibration and the onset of the response) increased progressively with age, and became nearly twice as long in the eighth decade as in the young control group, as shown in Figure 6.

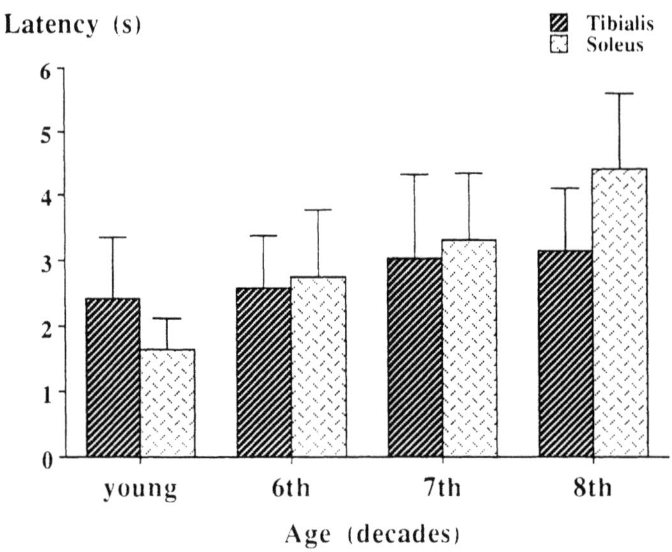

FIGURE 6. Latency of the perceived postural sway (time between the onset of the vibration and the onset of the imitative response) induced by soleus and tibialis anterior vibration, in four age groups. An analysis of variance showed a significant effect of Age, $F(3, 33) = 6.18$, $p<.01$, and an Age × Muscle interaction, $F(3, 33) = 2.97$, $p<.05$.

This increase was stronger with the forward sway sensations induced by vibrating the soleus muscles than with the backward ones induced by vibrating the tibialis anterior muscles. The perceived velocity of the illusory body sway decreased considerably (by more than 50%) with age, whichever of the two muscles was stimulated, as shown in Figure 7. This decrease occurred as early as the sixth decade, and the perceived velocity subsequently levelled off between the sixth and the eighth decades.

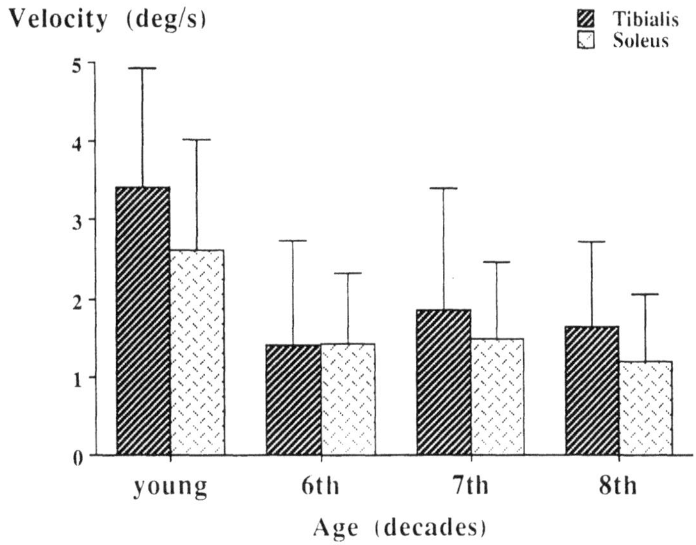

FIGURE 7. Mean velocity of the perceived postural sway induced by soleus and tibialis anterior vibration, in four age groups. An analysis of variance showed a significant effect of Age, $F(3, 33) = 4.05$, $p < .025$, which was determined by the difference between the group of young adults and the three older groups, $F(1, 33) = 11.91$, $p < .01$, who did not differ from each other, $F(2, 25) = .17$.

It therefore seems likely that vibration-induced kinesthetic illusions, which can be taken to reflect the perceptual processing of propriomuscular messages, still occur in the elderly but deteriorate with age. The information loss demonstrated by the drop in the perceived velocity resembles the decrease in the postural responses to ankle vibration (VIF), and suggests that the decrease in VIF may involve sensory deficits instead, although the possible presence of deficits preventing the proper execution of an appropriately selected response cannot be ruled

Eklund, G. (1972a). General features of vibration-induced effects on balance. *Upsala Journal of Medical Sciences, 77,* 112-124.

Eklund, G. (1972b). Position sense and state of contraction; the effects of vibration. *Journal of Neurology, Neurosurgery and Psychiatry, 35,* 606-611.

Eklund, G. (1973). Further studies of vibration-induced effects on balance. *Upsala Journal of Medical Sciences, 78,* 65-72.

Eklund, G., & Hagbarth, K. E. (1966). Normal variability of tonic vibration reflexes in man. *Experimental Neurology, 16,* 80-92.

Era, P., & Heikkinen, E. (1985). Postural sway during standing and unexpected disturbance of balance in random samples of men of different ages. *Journal of Gerontology, 40,* 287-295.

Ferrell, W. R., Crighton, A., & Sturrock, R. D. (1992). Age-dependent changes in position sense in human proximal interphalangeal joints. *NeuroReport, 3,* 259-261.

Fitzpatrick, R. C., & McCloskey, D. I. (1994). Proprioceptive, visual, and vestibular thresholds for the perception of sway during standing in humans. *Journal of Physiology, 478,* 173-186.

Fitzpatrick, R. C., Taylor, J. L., & McCloskey, D. I. (1992). Ankle stiffness of standing humans in response to imperceptible perturbations: reflex and task-dependent components. *Journal of Physiology, 454,* 533-547.

Gandevia, S. C., & McCloskey, D. I. (1976). Joint sense, muscle sense, and their combination as position sense, measured at the distal interphalangeal joint of the middle finger. *Journal of Physiology, 260,* 387-407.

Goodwin, G. M., McCloskey, D. I., & Mattews, P. B. C. (1972). The contribution of muscle afferents to kinaesthesia shown by vibration-induced illusions of movement and by the effects of paralysing joint afferents. *Brain, 95,* 705-748.

Gottsdanker, R. (1982). Age and simple reaction time. *Journal of Gerontology, 37,* 342-348.

Guieu, R., Ribot-Ciscar, E., Tardy-Gervet, M. F., Dano, P., & Roll, J. P. (1990). Les vibrations tendineuses comme moyen clinique d'exploration de la boucle myotatique. *Revue Neurologique, 146,* 699-702.

Gurfinkel, V. S., Lipshits, M. I., Mori, S., & Popov, K. E. (1976). The state of the stretch reflex during quiet standing in man. In S. Homma (Ed.), *Progress in brain research* (Volume 44, pp. 473-486). Amsterdam: Elsevier.

Gurfinkel, V. S., Lipshits, M. I., & Popov, K. E. (1977). Investigation of the system for the vertical posture regulation using vibratory stimulation of the muscle spindles. Human Physiology, Translated by Plenum Publ. C° from *Fiziologiva Cheloveka, 3,* 635-643.

Gurfinkel, V. S., Levick, Y. S., Popov, K. E., Smetanin, B. N., & Shlikov, V. Y. (1988). Body scheme in the control of postural activity. In V. S. Gurfinkel, M. E. Ioffe, J. Massion, & J. P. Roll (Eds.), *Stance and motion, facts and concepts* (pp. 185-193). New York, London: Plenum Press.

Hasselkus, B. R., & Shambes, G. M. (1975). Aging and postural sway in women. *Journal of Gerontology, 30,* 661-667.

Hagbarth, K. E. (1973). The effects of muscle vibration in normal man and in patients with motor disorders. In J. E. Desmedt (Ed.), *New developments in electromyography and clinical neurophysiology, 3* (pp. 428-443). Basel: Karger.

Hayashi, R., Miyake, H., & Watanabe, S. (1981). Postural readjustment to body sway induced by vibration in man. *Experimental Brain Research, 43,* 217-225.

Hinchcliffe, R. (1983). Epidemiology of balance disorders in the elderly. In R. Hinchcliffe (Ed.), *Hearing and balance in the elderly* (pp. 295-353). Singapore: Churchill Livingstone.

Horak, F. B., Shupert, C. L., & Mirka, A. (1989). Components of postural dyscontrol in the elderly: a review. *Neurobiology of Aging, 10,* 727-738.

Juntunen, J., Matikainen, E., Ylikoski, J., Ylikoski, M., Ojala, M., & Vaheri, E. (1987). Postural body sway and exposure to high-energy impulse noise. *The lancet, ii,* 261-264.

Kaplan, F. S., Nixon, J. E., Reitz, M., Rindfleish, L., & Tucker, J. (1985). Age-related changes in proprioception and sensation of joint position. *Acta Orthopaedica Scandinavica, 56,* 72-74.

Kenshalo, D. (1979). Changes in the vestibular and somesthetic systems as a function of age. In J. Ordy & K. Brizzee (Eds.), *Sensory systems and communication in the elderly.* New York: Raven Press.

Kokmen, E., Bossemeyer, R. W., Barney, J., & Williams, W. J. (1977). Neurological manifestations of aging. *Journal of Gerontology, 32,* 411-419.

Kokmen, E., Bossemeyer, R. W., & Williams, W. J. (1978). Quantitative evaluation of joint motion sensation in an aging population. *Journal of Gerontology, 33,* 62-67.

Laidlaw, R. W., & Hamilton, M. A. (1937). A study of thresholds in apperception of passive movement among normal control subjects. *Bulletin of Neurological Institute, 6,* 268-273.

Latash, M. L., & Gurfinkel V. S. (1976). Tonic vibration reflex and position of the body. *Fiziologiya Cheloveka, 2,* 593-598.

Magladery, J. W., Teasdall, R. D., & Norris, A. H. (1958). Effect of aging on plantar fle[illegible] superficial abdominal reflexes in man. A clinical and electr[illegible]ic study. *Journal of Gerontology, 13,* 282-288.

Maki, B. E., Holliday, P. J., & Topper, A. K. (1990). Does fear of falling influence postural performance in the elderly? In T. Brandt, W. Paulus, W. Bles, M. Dieterich, S. Krafczyk, & A. Straube (Eds.), *Disorders of posture and gait 1990* (pp. 316-320). Stuttgart, New York: Georg Thieme Verlag.

Manchester, D., Woollacott, M., Zederbauer-Hylton, N., & Marin, O. (1989). Visual, vestibular and somatosensory contributions to balance control in the older adults. *Journal of Gerontology: Medical Sciences, 44,* M118-M127.

Mattews, P. B. C., & Simmonds, A. (1974). Sensations of finger movement elicited by pulling upon flexor tendons in man. *Journal of Physiology, 239,* 27-28.

McLennan, W. J., Timothy, J. I., & Hall, M. R. P. (1980). Vibration sense, proprioception and ankle reflexes in old age. *Journal of Clinical and Experimental Gerontology, 2,* 159-171.

Milne, J. S., & Williamson, J. (1972). The ankle jerk in older people. *Gerontologia Clinica, 14,* 86-88.

Mirka, A., Peterka, R. J., Horak, F. B., & Black, F. O. (1988). Comparison of postural control in elderly with and without subjective dizziness versus normal young subjects. Presented at Association for Research in Otolaryngology.

Nakagawa, H. M. D. (1992). Postural control in the elderly. *Journal of Otolaryngology Japan, 95,* 1042-1052.

Nashner, L. M. (1976). Adapting reflexes controlling human posture. *Experimental Brain Research, 26,* 59-72.

Ochs, A. L., Newberry, J., Lenhardt, M. L., & Harkins, S.W. (1985). Neural and vestibular aging associated with falls. In J. E. Birren & K. W. Schaie (Eds.), *Handbook of psychology of aging* (2nd ed., pp. 378-399). New York: Van Nostrand Reinhold Company.

Overstall, P. W., Exton-Smith, A. N., Imms, F. J., & Johnson, A. L. (1977). Falls in the elderly related to postural imbalance. *British Medical Journal, 1,* 261-264.

Overstall, P. W. (1983). Rehabilitation of elderly patients with disorders in balance. In R. Hinchcliffe (Ed.), *Hearing and balance in the elderly* (pp. 468-488). Singapore: Churchill Livingstone.

Pailhous, J., & Clarac, F. (1984). Approche comportementale de la locomotion: Eléments d'analyse chez l'homme et l'animal. In J. Paillard (Ed.), *Comportements, 1. La lecture sensori-motrice et cognitive de l'expérience spatiale. Direction et distances* (pp. 135-158). Paris: Editions du C.N.R.S.

Paillard, J. (1960). The patterning of skilled movements. In *Handbook of physiology, section I, Neurophysiology* (Volume III, pp. 1679-1708). Bethesda, MD: American Physiological Society.

Panzer, V., Kaye, J., Edner, A., & Holme, L. (1992). Standing postural control in the elderly and very elderly. In M. Woollacott & F. Horak (Eds.), *Posture and gait: control mechanisms* (pp. 220-231). Eugene, OR: University of Oregon Books.

Pyykkö, I., Aalto, H., Hytönen, M., Starck, J., Jäntti, P., & Ramsay, H. (1988). Effect of age on postural control. In B. Amblard, A. Berthoz, & F. Clarac (Eds.), *Posture and gait: development, adaptation and modulation* (pp. 95-104). Amsterdam: Elsevier.

Pyykkö, I., Jäntti, P., & Aalto, H. (1990). Postural control in elderly subjects. *Age and Ageing, 19,* 215-221.

Quoniam, C., Roll, J.-P., Deat, A., & Massion, J. (1990). Proprioceptive induced interactions between segmental and whole body posture. In T. Brandt, W. Paulus, W. Bles, M. Dieterich, S. Krafczyk, & A. Straube (Eds.), *Disorders of posture and gait* (pp. 194-197). Stuttgart, New York: Thieme Press.

Quoniam, C., Hay, L., Roll, J.-P., & Harlay, F. (1993). Perception and regulation of postural balance by the propriomuscular channel, in the elderly. Symposium: Deafferentation and the Role of Sensory Afferent in Human Motor Control. Quebec: University Laval.

Quoniam, C., Hay, L., Roll, J.-P., & Harlay, F. (1995) Age effects on reflex and postural responses to propriomuscular inputs generated by tendon vibration. *Journal of Gerontology: Biological Sciences, 50A,* B155-B165.

Redon, C. (1992). Reprogrammation posturale d'origine proprioceptive chez l'enfant infirme moteur cérébral. Actes des Cinquièmes Entretiens de l'Institut Garches (pp. 271).

Roll, J.-P., Gilhodes, J.-C., & Tardy-Gervet, M.-F. (1980). Effets perceptifs et moteurs des vibrations musculaires chez l'Homme normal: mise en évidence d'une rΘponse des muscles antagonistes. *Archives Italiennes de Biologie, 118,* 51-71.

Roll, J.-P., & Vedel, J.-P. (1982). Kinaesthetic role of muscle afferents in man, studied by tendon vibration and microneurography. *Experimental Brain Research, 47,* 177-190.

Roll, J.-P., Vedel, J.-P., & Ribot, E. (1989a). Alteration of proprioceptive messages induced by tendon vibration in man: a microneurographic study. *Experimental Brain Research, 76,* 213-222.

Roll, J.-P., Vedel, J.-P., & Roll, R. (1989b). Eye, head and skeletal muscle spindle feedback in the elaboration of body references. *Progress in Brain Research, 80,* 113-123.

Rosenberg, G. (1958). Effect of age on peripheral vibratory perception. *Journal of the American Geriatrics Society, 6,* 471-481.

Sabin, T. D. (1982). Biologic aspects of falls and mobility limitations in the elderly. *Journal of American Geriatrics Society, 30,* 51-58.

Schiano, A., Marchetti, S., Bardot, P., Sambuc, R., Bardot, A., & Serratrice, G. (1988). L'évolution de la sensibilité avec l'âge: étude par des tests cliniques quantitatifs. *Annales de Réadaptation et de Médecine physique, 31,* 309-318.

Sheldon, J. H. (1963). The effect of age on the control of sway. *Gerontologia Clinica, 5,* 129-138.

Sherrington, C. S. (1906). *Integrative action of the nervous system.* New York: Scribner.

Sherrington, C. S. (1910). Flexion-reflex of the limb, crossed extension reflex, and reflex stepping and standing. *Journal of Physiology (London), 40,* 28-121.

Skinner, H. B., Barrack, R. L., & Cook, S. D. (1984). Age-related decline in proprioception. *Clinical Orthopaedics and Related Research, 184,* 208-211.

Stelmach, G. E., & Sirica, A. (1987). Aging and proprioception. *Age, 9,* 99-103.

Stelmach, G. E., & Worringham, C. J. (1985). Sensorimotor deficits related to postural stability: Implications for falling in the elderly. *Clinics in Geriatric Medicine, 1,* 679-694.

Stelmach, G. E., Teasdale, N., DiFabio, R. P., & Phillips, J. (1989). Age related decline in postural control mechanisms. *International Journal of Aging and Human Development, 29,* 205-223.

Straube, A., Bötzel, K., Hawken, M., Paulus, W., & Brandt, T. (1988). Postural control in elderly: differential effects of visual, vestibular and somatosensory input. In B. Amblard, A. Berthoz, & F. Clarac (Eds.), *Posture and gait: development, adaptation and modulation* (pp. 105-114). Amsterdam: Elsevier.

Teasdale, N., Stelmach, G. E., & Breunig, A. (1991). Age differences in visual sensory integration. *Experimental Brain Research, 85,* 691-696.

Teasdale, N., Bard, C., Dadouchi, F., Fleury, M., Larue, J., & Stelmach, G.E. (1992). Posture and elderly persons: evidence for deficits in the central integrative mechanisms. In G. E. Stelmach & J. Requin (Eds.), *Tutorials in motor behavior II* (pp. 917-931). Amsterdam: Elsevier.

Teasdale, N., Bard, C., Larue, J., & Fleury, M. (1993). On the cognitive penetrability of posture control. *Experimental Aging Research, 19,* 1-13.

Welford, A. T. (1981). Signal, noise, performance and age. *Human Factors, 23,* 97-109.

Woollacott, M. H. (1986). Gait and postural control in the aging adult. In W. Bles & T. Brandt (Eds.), *Postural control in the aging adult. Disorders of posture and gait* (pp. 325-336). Amsterdam: Elsevier.

Woollacott, M. H. (1988). Posture and gait from newborn to elderly. In B. Amblard, A. Berthoz, & F. Clarac (Eds.), *Posture and gait: development, adaptation and modulation* (pp. 3-12). Amsterdam: Elsevier.

Woollacott, M. H. (1990). Changes in postural control and the integration of postural responses into voluntary movements with aging: is borderline pathology a contributor? In T. Brandt, W. Paulus, W. Bles, M. Dieterich, S. Krafczyk, & A. Straube (Eds.), *Disorders of posture and gait* (pp. 221-228). Stuttgart, New York: Thieme Press.

Woollacott, M. H., Shumway-Cook A., & Nashner L. M. (1982). Postural reflexes and aging. In J. A. Martiner, F. J. Pirozzolo, & G. Maletta (Eds.), *Advances in neurogerontology, The motor system* (pp. 98-119). New York: Praeger Publishers.

Woollacott, M. H., Shumway-Cook, A., & Nashner, L. M. (1986). Aging and posture control: changes in sensory organization and muscular coordination. *International Journal of Aging and Human Development, 23,* 97-114.

Changes in sensory motor behavior in aging
A.-M. Ferrandez and N. Teasdale (Editors)

POSTURE AND GAIT IN HEALTHY ELDERLY INDIVIDUALS AND SURVIVORS OF STROKE

Karen M. HILL and Anthony A. VANDERVOORT

University of Western Ontario

Abstract

Differences between young and older adults in postural control of equilibrium during upright standing have been detailed. Given the documented changes in sensory receptors, central integration of their input, motor planning and the deficits in the muscular system, it is not surprising that older people exhibit more postural sway and handle perturbations to balance less efficiently than young adults. The survivor of a stroke is at an even greater disadvantage due to the asymmetrical nature of their postural control system, making falls an even greater problem. However, there remains in cognitively intact people considerable capacity for learning to adapt to deficits in the motor system. The patterns of gait exhibited by older groups are also of interest due to the dynamic nature of this challenging motor task. The most obvious of age-related change is a slower walking speed. Whether this is a primary factor, or the result of contributing mechanisms, is not clear. Elderly individuals walk with a shorter step length and a less forceful push-off with the ankle. A decreased swing to stance ratio and increased duration of the double support phase may be compensatory mechanisms to increase stability. Aging is not necessarily the lone determinant of walking speed

---

Correspondence should be sent to Anthony A. Vandervoort, Department of Physical Therapy, University of Western Ontario, London, Ontario N6G 1H1, Canada (e-mail: vandervo@julian.uwo.ca).

and pattern, because of the increasing incidence of disuse and disabilities associated with getting older. The adaptive capacity of the locomotor system of older people is currently of much research interest as we begin to explore how best to optimize neuromuscular function at all ages.

Key words: Aging, neuromuscular function, balance, walking, mobility impairment.

INTRODUCTION

The paramount function of the postural control system in humans is to maintain a state of equilibrium in an upright position. During standing, this is achieved when the centre of body mass is located over the base of support with a minimal amount of sway (Shumway-Cook, Anion, & Haller, 1988). In order to maintain a state of equilibrium, this sway has to be counteracted by neuromuscular forces. Therefore, the essential task of balance is to produce moments of force about the lower extremity joints and trunk which counteract perturbing forces and maintain the centre of gravity within a relatively small support base. Humans, however, are not restricted to a stationary base and balance must be maintained during many changing and destabilizing conditions, for example: walking, running, reaching, lifting an object from the floor. It will be seen that in addition to providing stability, the human neuromuscular system has to be adaptable to allow for smooth movement and activity.

This adaptability is demonstrated by the response of the human body to aging, which itself has a strong influence on neuromuscular performance. In addition to aging, the incidence of disease processes increases with age, thus creating additional stresses on the neuromuscular system which may be overwhelming. Stroke, for example, can have disabling effects on balance and ambulation, and yet a large number of survivors of stroke learn to walk again. Therefore, an examination of posture and gait following stroke helps illustrate the adaptability of the human postural control system. This chapter first reviews aspects of neuromuscular function which are affected by aging and in turn have an influence on standing and ambulation. In addition, the effect of hemiplegia following stroke is given some attention in relation to alterations of posture and gait.

STANDING BALANCE IN THE HEALTHY ELDERLY

Standing balance has been shown in several investigations to decline with aging. Postural sway, which is the most frequently reported laboratory measure of balance, increases with aging (Era & Heikkinen, 1985; Alexander, 1994). Both the amplitude and frequency of sway have been found to increase with age (Sheldon, 1963; Hasselkus & Shambes, 1975; Overstall, Exton-Smith, Imms, & Johnson, 1977; Brocklehurst, Robertson, & James-Groom, 1982b; Era & Heikkinen, 1985; Lucy & Hayes, 1985). For example, Lucy and Hayes (1985) reported 52% greater postural sway in the anterior-posterior (AP) direction for a sample of healthy elderly subjects aged 70-80 years, versus younger subjects aged 30-39 years. Speed of postural sway has also been found to be related to age, with the exception of elderly people living in nursing homes where the reasons for institutionalization may be stronger contributors to poor balance than age itself (Dornan, Fernie, & Holliday, 1978; Fernie, Gryfe, Holliday, & Llewellyn, 1982). Maximum voluntary excursion of the centre of gravity within the support base, or the greatest amount of leaning without taking a step, is reduced in elderly subjects, particularly in the posterior direction (Blaszczyk, Lowe, & Hansen, 1994). In addition, elderly subjects demonstrate greater variability of sway paths (Blaszczyk et al., 1994).

Overstall et al. (1977) observed that postural sway is greater in women than men in both young and elderly age groups from the community, however, Fernie et al. (1982) did not find sex to be a predictive factor in their group of institutionalized elderly subjects. Postural sway on average is greater in older people who have a history of falling (Fernie et al., 1982; Crilly, Richardson, Roth, Vandervoort, Hayes, & Mackenzie, 1987) and may be the best test in predicting falls (Topper, Maki, & Holliday, 1993).

Although postural sway has been the focus of many studies concerned with balance in the elderly, measures commonly used in clinical evaluation of balance performance also show a decline in function with aging. For example, Bohannon et al. (1984) found that subjects older than 60 years were unable to stand on one supporting leg (the other foot was held off the floor) for as long a period as younger subjects. Era and

Heikkinen (1985) reported that only 41% of subjects aged 71-75 were able to perform the single limb balance test for at least 8 seconds. Potvin et al. (1980) found the single limb balance test with eyes closed to have the highest correlation with chronological age among several standard clinical tests. However, other clinical investigators have argued that many elderly people cannot perform such a difficult task and have emphasized the need for performance of activity based tests (Tinetti, 1986; Berg, Wood-Dauphinee, Williams, & Gayton, 1989). Mathias, Nayak, and Isaacs (1986) for example, devised an assessment of balance based on the functional task of "Get Up and Go", in which scores correlated significantly with postural sway measures.

Although the stereotypical view of elderly posture is characterized by a stooped and forward leaning position, elderly subjects as a group do demonstrate minor postural differences when compared with younger individuals. A photographic study of 41 elderly females, aged 65 and older, demonstrated increased kyphosis (as measured by degrees in a sagittal view), a more posterior position of the hip in relation to the ankle joint (but with the hip joint still anterior to the ankle joint), and greater forward leaning of the trunk (more anterior position of the average centre of gravity above the hips) (Woodhull-McNeal, 1992). The centre of gravity of the whole body remained a mean of 4.5 ($\pm$ 1.6) cm anterior to the ankle joint and was not statistically different from young subjects. Therefore, on the average, these older subjects tended to lean forward from the hips but at the same time maintain similar anterior-posterior balance by compensating with a further posterior hip position. Individual variation in these subjects was reported to be great and no particular pattern of posture was observed, i.e., all of the above changes did not necessarily appear together in all subjects. This report concurs with the findings of Brocklehurst, Robertson, and James-Groom (1982b) who also found postural changes in the elderly to be extremely variable.

Subjects who reported greater activity levels tended to have less forward lean (Woodhull-McNeal, 1992). The delicately balanced positions of the centres of gravity of the portions of the body above the hip and knee in younger subjects tend to hyperextend these joints and therefore, increase the stability of each joint (Woodhull, Maltrud, & Mello, 1985). A change in this balanced alignment in elderly subjects would suggest that postural muscles need to exert more than minimal force in order to maintain an upright position. Therefore, the compensations of further forward leaning and further posterior hip position likely result in less mechanical advantage and greater muscle use. Further description of muscle activity and posture in the elderly is needed to establish these connections.

The processes which lead to kyphosis in elderly subjects are likely different from the processes which lead to forward leaning. Kyphosis is related to a decrease in height which is consistent with the anterior "wedging" of vertebrae found as a result of progressive mechanical factors or compression fractures as a result of osteoporosis (Neumann, 1993). Forward leaning however, is believed to be related to lower activity levels which is secondary to weakness or fear of falling (Woodhull-McNeal, 1992).

In discussing the changes in balance function in the elderly, two points should be kept in mind. First, there is a large amount of variability in the performance of elderly subjects (Hasselkus & Shambes, 1975). Secondly, an individual's perception of activity may vary. For example, Woodhull-McNeal (1992) made an effort to recruit less active subjects from senior centres. However, the self-reported activity level of the 41 subjects was fairly active (walked 5 - 10 blocks a week). Therefore, the representativeness of these observations should be considered with caution.

Sensory motor changes with aging

In addition to the musculoskeletal dynamics of an upright posture considered above, neural processes are necessary for the regulation and control of posture. The control of balance requires the coordinated effort of the neuromuscular system to obtain accurate sensory input, organize motor programs, and generate effective motor output responses.

Information about the position of the body in relation to the environment is received from three primary sources: proprioception (muscle, joint, and cutaneous receptors), vision, and the vestibular system (Nashner, 1976). Each of these will be reviewed briefly in relation to the aging process.

Proprioceptive sensation is mediated via three principal types of receptors: 1) those originating in muscles and tendons (neuromuscular spindles and Golgi tendon organs), 2) articular mechanoreceptors (transient, velocity and position-velocity receptors, and 3) deep pressure receptors in the plantar aspects of the feet (Brooks, 1986). The muscle receptors are believed to have a more important role than the joint receptors. A discussion of the role of stretch reflexes in control of posture will follow.

The threshold for perception of vibratory stimuli generally increases with age, particularly in the lower extremities (Brocklehurst et al., 1982a; Potvin et al., 1980; MacLennan, Timothy, & Hall, 1980;

Olney, 1985). Proprioceptive sensation, as tested by asking subjects to identify the position of their toes, is also reported to show impairment with increasing age (MacLennan et al., 1980). Pyykko, Jantti, and Aalto (1990) found greater loss of proprioceptive sensitivity of calf muscles in very old subjects, aged 85 and older, than control subjects aged 50 to 60 years.

Although normal adults rely heavily on proprioceptive input from joint and muscle receptors to maintain posture, when conditions in which this information is reduced or less useful visual information becomes more important. Dornan et al. (1978) demonstrated this by measuring postural sway with eyes open and eyes closed and comparing a group of above-knee amputees with a nonamputee group. The amputee group showed a significantly greater increase in sway with the eyes closed than did the nonamputee control group.

Specific age-related changes in vision are: reduced acuity, restriction of the visual field, and decreased depth perception (Stelmach & Worringham 1985), which leads to faulty and/or increased time for visual detection of a change in the environment. However, elderly subjects rely heavily on the visual sense to maintain balance and have great difficulty standing on one leg with eyes closed for more than a few seconds (cf. Potvin et al., 1980; Bohannon et al., 1984; Stones & Kozma, 1987). Deprivation of visual information, particularly loss of peripheral vision, results in a significant loss of stability in elderly subjects (Manchester, Woollacott, Zederbauer-Hylton, & Marin, 1989; Pyykko et al., 1990).

The vestibular system assists in maintaining head stability and upright posture. The vestibular apparatus, the otoliths and semicircular canals, detect angular and linear accelerations of the head and influence antigravity tone via the lateral vestibulospinal tract (Markham, 1987). The role of vestibular signals in maintaining upright stance is not as easy to evaluate as other sensory modalities, partly because it is difficult to selectively stimulate the vestibular receptors (Woollacott, Shumway-Cook, & Nashner, 1982; Horak, Shupert, Dietz, & Hortsmann, 1994). Therefore, most of our information about the role of the vestibular system in the control of balance has been gained from studies of patients with peripheral vestibular lesions. Although the vestibular system is probably the least sensitive regulator of posture, when environmental conditions create conflicts among the different sensory modalities, the vestibular system becomes critically important. It has been suggested that vestibular input is essential as an absolute orientation reference against which proprioceptive and visual inputs are compared (Nashner, 1976).

Age-related changes reported in the vestibular system include a decrease in the number of sensory cells in the semicircular canals and utricular and saccular otoliths, a loss of fibres in the vestibular nerve and up to a 40% reduction of hair cells (Rosenhall, 1973). Brocklehurst et al. (1982b) assessed vestibular function in elderly subjects by slowly tilting subjects seated on a table and measuring their ability to maintain an appropriate angle of the neck. Six percent of the subjects showed impaired responses; the majority of these subjects were aged 85 and over. No correlation was found between vestibular function and sway. The authors do state that the tilt test could be affected by proprioception from the buttock area and therefore raise questions as to the validity of the test in assessing vestibular function (Brocklehurst et al., 1982b).

In healthy young adults these three sensory systems appear to provide more than adequate information for postural control. The system has an inherent adaptability which allows incorrect information to be suppressed while the other sensory systems, for example, the visual and vestibular systems, are able to compensate and maintain an upright position (Nashner, 1976). This adaptability may be illustrated by the driver of a car who is stopped at a set of traffic lights, feels that he is moving backwards as a large bus passes beside him, but correctly interprets his position and does not change his posture. However, elderly individuals appear to have less redundancy of sensory information due to changes in the proprioceptive, visual and vestibular systems which occur with age (Woollacott et al., 1982). Therefore, the elderly postural control system has less ability to detect and alert the central nervous system regarding changes in posture, particularly in cases where more than one sensory system is challenged or sensory information is conflicting (Woollacott, Shumway-Cook, & Nashner, 1986).

Peripheral postural reflexes

At one time it was believed that monosynaptic reflexes played a major role in maintaining upright posture (Gurfinkel, 1973). Then, Melvill Jones and Watt (1971) described a second reflex response. A sudden dorsiflexion force applied at the ankle, while subjects voluntarily resisted the force, produced both an early and a late EMG (electromyographic) response in the gastrocnemius muscle. The first response had a short latency of 37 ms and was believed to be monosynaptic. Following a period of silence, a second burst of EMG activity occurred at a latency of 110 ms with a much greater amplitude. Because of the longer latency, this reflex was believed to be polysynaptic and mediated via a supraspinal loop, rather than a monosynaptic reflex arc (Nashner, 1976; Diener, Dichgans, Bootz, & Bacher, 1984; Dietz, Quintern, Berger, & Schenck, 1985). The response is known as the long latency or functional stretch reflex and is more functionally effective than the monosynaptic reflex (Chan, 1984). The subsequent quest to understand the role of these functional stretch reflexes in the control of upright posture has been instrumental in discovering much information about the integrative functions of the neural and musculoskeletal systems.

Two distinct postural response systems are suggested by current neurophysiological research: anticipatory postural adjustments which prepare for an expected loss of balance, and compensatory postural responses which restore balance following an unexpected disturbance or perturbation. Both of these systems are important to consider when discussing the role of postural reflexes in the control of balance.

When a healthy individual stands on a platform and the surface is suddenly and unexpectedly displaced either backwards or toes up, short, medium and long latency stretch responses are evoked in the plantarflexor muscles as detected by EMG electrodes placed on the lower limb muscles. Depending on the velocity, amplitude and of course, direction of the displacement, these reflexes vary in amplitude of muscle response and in frequency of occurrence (Diener et al., 1984). However, continued contraction of the plantarflexors in response to a sudden forward perturbation would be functionally destabilizing and would result in the body being displaced too far backwards. Therefore, the stretch reflexes are normally suppressed prior to reaching a critical balance situation (Nashner, 1976). This modulation of reflexes is also dependent on the destabilizing situation, so that muscle responses may be facilitated as well as suppressed. The time of onset of different muscles may also be modulated (Nashner, 1977).

Following an unexpected disturbance of balance, the order of activation of postural muscles, as detected by surface EMG electrodes placed on the lower extremity muscles, occurs in specific patterns (Nashner, 1977; Nashner, Woollacott, & Tuma, 1979; Badke & Duncan, 1983; Badke & DiFabio, 1985; DiFabio, Badke, & Duncan, 1986; Stelmach, Phillips, DiFabio, & Teasdale, 1989). If the perturbation is applied at the ankles, such as movement of the support surface forward or backward, the recruitment pattern occurs in a distal to proximal order with gastrocnemius and tibialis anterior being recruited before the thigh and trunk muscles. If sway is initiated voluntarily by the subject, such as voluntary forward flexion of the trunk or elevation of the arm, the order of activation is reversed and muscles around the proximal joints are recruited first (Nashner, 1977; Nashner & Cordo, 1981; Oddsson &

Thorstensson, 1987). Thus, the sequencing of postural responses following an anticipated disturbance of balance differ from those in response to unexpected disturbances, but the patterns of response are similar. The difference in onset time between distal and proximal muscles is brief (approximately 12 ms) but the pattern of response remains consistent with different velocities and directions of sway (Nashner, 1977; Thorstensson et al., 1985).

When an individual can predict an upcoming disturbance of balance, postural muscles are activated before the unstable situation occurs (Marsden et al., 1981). These anticipatory postural adjustments may be evaluated during self initiated rapid arm movements performed while standing. The amplitude and latency of EMG responses of muscles in the trunk and lower extremities, during single or bilateral voluntary arm movements, have been recorded by a number of investigators (Cordo & Nashner, 1982; Woollacott, Bonnet, & Yabe, 1984; Friedli, Hallett, & Simon, 1984; Horak, 1987). The results demonstrate that postural adjustments prior to arm movement occur in a consistent order of proximal to distal activation and are specific to the planned voluntary movement.

In relation to aging, elderly individuals take longer to activate postural responses when compared with young control subjects. Woollacott et al. (1986) found significant increases in latency of the gastrocnemius and tibialis anterior in a sample of 12 subjects aged 61 to 78 years when tested by platform perturbation of a static standing position. Five of these subjects also showed a breakdown of the normal distal to proximal sequence of muscle activation. Stelmach et al. (1989) supported these observations and also observed slower postural reflexes and an alteration of the normal stereotypical and synchronous organization of muscle onset patterns in elderly subjects. This was demonstrated in response to perturbation of standing while performing a voluntary sway movement, in contrast to testing of static stance by Woollacott et al. (1986), and suggests that the elderly have difficulty with the functional organization of posture during movement as well as during static positions (Stelmach et al., 1989).

Normal age-related changes in the peripheral nervous system include accumulation of pigment in neurons and Schwann cells, chronic demyelination and remyelination, neuronal loss, and axonal degeneration (Schaumburg, Spencer, & Ochoa, 1983). Although difficult to separate from pathological changes which are due to disease processes, these aging changes are progressive, universal and irreversible (Olney, 1985).

Electrophysiological studies of peripheral nerves demonstrate progressive slowing of both sensory and motor conduction velocity at a rate of up to 1 m/s per decade, plus a decrease in amplitude of sensory nerve action potentials (Schaumburg et al., 1983; Olney, 1985; Hakansson, Vandervoort, & Kramer, 1991; Doherty, Vandervoort, & Brown, 1993). Diminished H reflex responses, an increased proportion of absent ankle reflexes, and weakened ankle muscles are reported in the elderly (MacLennan et al., 1980; Vandervoort & Hayes, 1989; Porter, Vandervoort, & Lexell, 1995). These changes in peripheral neuromuscular function are even more pronounced in elderly fallers (Pack, Wolfson, Amerman, Whipple, & Kaplan, 1985; Whipple, Wolfson, & Amerman, 1987; Vandervoort, Hill, Sandrin, & Vyse, 1990; Studenski, Duncan, & Chandler, 1991). Elderly patients who suffer from peripheral neuropathy, a condition which causes a gradual loss of distal sensation and strength, are apparently at greater risk for falling (Sabin, 1982; Richardson, Ching, & Hurvitz, 1992). Diminished briskness of the peripheral reflexes are probably the result of a combination of chronic changes in the elderly neuromuscular system. Reduced activity along the motor and sensory components of the reflex arc, decreased sensitivity of muscle spindles, a slowing of muscle contraction and increased stiffness of connective tissues are likely contributing factors (Schaumburg et al., 1983; Vandervoort & McComas, 1986; Vandervoort, Chesworth, Cunningham, Paterson, Rechnitzer, & Koval, 1992).

Central coordination of muscular output

Central integration of motor and sensory function plays a major role in balance control and is another important area to consider regarding impairment of balance and aging (Wolfson, Whipple, Amerman, Kaplan, & Kleinberg, 1985). The role of central integrative mechanisms in configuring the postural control system may be summarized by the following three processing stages, as outlined by Stelmach and Worringham (1985); first, sensory input alerts or triggers the response selection centre, secondly, corrective or protective responses are selected which are appropriate to the situation, and finally, the actual response, which is likely to be a specific series of muscular responses, is organized and initiated. There is evidence to suggest that slowing of these central integrative mechanisms, due to a defect or to age, may be another cause for decreased postural control in the elderly (Stelmach & Worringham, 1985, Stelmach et al., 1989). For example, atypical patterns of muscle recruitment found in elderly subjects when subjected to perturbations of static balance and voluntary sway, suggest that there are problems with the central coordination of the postural reflexes (Woollacott et al.,

1986, Manchester et al., 1989; Stelmach et al., 1989). Slowness of postural reflexes and atypical muscle activation patterns cannot be attributed to reduced muscle strength or to delayed peripheral input alone.

Teasdale, Bard, LaRue, and Fleury (1993) tested this central integrative hypothesis further by measuring reaction time to an auditory stimulus as a means of measuring cognitive processing ability for a functional postural task. Young and elderly subjects attempted to maintain an upright posture while standing on a force platform and being subjected to a variety of visual and surface conditions. The results showed that the elderly subjects had greater difficulty maintaining upright balance as the postural task became more challenging due to displacement of the platform surface and reduction of vision. As peripheral sensory information was reduced, it was believed that the complexity of the central integrative mechanisms was increased. Because of these conditions, the subjects had to devote more attention to the task, which was measured as reaction time to the auditory stimulus. These data suggest that greater attentional demands are required of the elderly to maintain upright posture, particularly as sensory information via the visual and proprioceptive systems is decreased. The authors argue that these results do not downplay the importance of peripheral sensation, but rather suggest that both the central integrative mechanisms and the peripheral sensory systems play interdependent roles in the control of posture (Teasdale, Stelmach, & Breunig, 1991; Teasdale et al., 1993). Other authors confer with this idea (Wolfson et al., 1992). However, slowing of peripheral sensory information may challenge the central integrative mechanisms by increasing the demand for attention on the postural task and could thereby, directly influence balance problems.

Faults in central motor programming and discrepancies in timing of motor responses undoubtedly lead to an increased risk for falling. This is a major concern which accompanies a decline in the control of balance in the elderly (Cape, 1978). There does appear to be a relationship between impaired balance responses and the incidence of falls in the elderly. Postural sway is greater in subjects with a history of falls (Overstall et al., 1977; Fernie et al., 1982; Kirshen, Cape, Hayes, & Spencer, 1984), and responses to perturbations of balance are slower (Maki, Holliday, & Topper, 1994). Other clinical tests of balance also demonstrate significantly impaired scores in fallers (Studenski et al., 1991). It should be clarified, however, that falls due to environmental causes (i.e., accidental falls and trips), account for a large percentage (45% to 50%) of falls reported in the elderly (Sheldon, 1960; Overstall et al., 1977; Fleming & Pendergast, 1993) and Overstall et al. (1977)

reported that there was no difference in postural sway measures between subjects who fell because of a trip and those who did not have a history of falls. Therefore, other factors which may influence postural control in the elderly, including postural hypotension and the side effects of medications, should not be neglected in an evaluation of elderly balance (Tinetti, 1986).

Certain neurohistologic changes in the elderly brain may relate to a gradual decline in motor control. For example, loss of dendritic spines in the Betz cells of the cerebellum may account for timing problems and slower responses in the elderly. Similar changes in the striatum and decreased dopamine levels may also contribute to decreased coordination (Scheigel, 1985). Whether these changes are due to pathology or simply a part of the aging process is not clear. However, a marked lesion of the central nervous system caused by an atypical aging process, such as a stroke, may lead to pronounced deficits in balance performance.

In summary, there is evidence for deterioration of the postural control mechanism, during both the information processing and the execution stages of control, in association with the aging process. Although specific changes may be small and insignificant, the combination of deficits may summate to increase the risk of incorrect or inefficient movement and a subsequent loss of balance or a fall, particularly when a functional activity is attempted. Slower postural reflexes alone may not be the cause of a fall, but a delay in the detection of changes in posture, disorganization of central processing, in addition to slow and disorganized muscle activation may in combination result in poor balance. An increase in postural sway may increase the risk for an individual's centre of gravity to be located further from the centre of the base of support at the time that a perturbation occurs and therefore, increases the risk of losing balance and falling. The resulting injury, loss of mobility and death following a fall, are significant problems for the elderly.

The numerous approaches to balance evaluation reflect the complex process of posture control (Horak, Shupert, & Mirka, 1989). Earlier studies were limited to a single measure of balance, for example, postural sway, which was believed to provide a global measure of postural abilities (Sheldon, 1963; Era & Heikkinen, 1985). Although sway does demonstrate age related changes and has been useful in guiding researchers into asking further questions about balance, it has now become obvious that sway alone does not provide a full picture as to how an individual performs in the community or why an individual falls. The present trend is to combine several measures in what may be termed a systems approach to balance evaluation (Woollacott, 1993). From a clinical point of view, this approach is applicable as it emphasizes the interaction between balance ability and adaptability during functional activities. In a systems approach, balance is analyzed from multiple neural and musculoskeletal factors which are believed to play a relevant role in posture control. Considering the present knowledge of posture control, these factors should include: 1) the accuracy of proprioceptive, visual and vestibular systems in receiving information about the environment, 2) the ability of the above sensory systems to adapt to conflicting and redundant information, 3) the ability to activate muscle synergies in appropriate patterns of response, and 4) the ability to activate muscles with adequate force (Woollacott, 1993). This systems approach also proves useful when attempting to examine individuals with pathophysiology such as the effect of a stroke.

BALANCE FOLLOWING STROKE

Stroke, or cerebral vascular accident (CVA), is the third leading cause of death in the United States (Duncan, 1994). The primary conditions which are responsible for stroke, such as arteriosclerosis and atherosclerosis, are chronic and progressive in nature, and thus the elderly are more likely to suffer from stroke than other age groups. Although the incidence of stroke is decreasing, the prevalence appears to be actually increasing due to a growing elderly population as well as an improvement in survival rates following stroke (Heart and Stroke Foundation, 1992).

Depending on the location of the cerebral vascular lesion, stroke survivors experience sensory, motor or cognitive deficits, or any combination of these. Motor deficits, primarily hemiparesis or hemiplegia, may be seen in up to 88% of stroke survivors (Duncan, 1994). Although the incidence of impairment is high, the majority of patients are expected to recover to a mild level of deficit. For example, follow up of long-term stroke survivors in the Framingham, Massachusetts population study revealed that 47% had residual hemiparesis (Gresham, Phillips, Wolf, McNamara, Kannel, & Dawber, 1979). Eighty five percent of these survivors were living at home and 78% were independent in mobility. Even so, a significant number of stroke survivors in this study (mean ages between 64 and 65 years) demonstrated one or more comorbid disease processes, compared with their age-matched controls. This fact highlights the complex interdependence of aging and disease processes associated with advanced years.

TABLE 1.
A sample clinical evaluation of postural control in an elderly individual with examples of measurement tests.

Component	Test
A. History	
1. Onset of functional changes	
2. Medical conditions	
3. Medications	
4. Environmental conditions	
B. Systems assessment	
1. Sensory	
a) Vision	Romberg Test
visual acuity	(Black et al., 1982)
visual field	
contrast sensitivity	
depth perception	
b) Somatosensory	Sensory Integration Test
proprioception	(Shumway-Cook & Horak, 1986)
vibration	
vestibular	
2. Central Processing	
a) Feedforward	Functional Reach
	(Duncan et al., 1990)
b) Feedback	Postural Stress Test
c) Adaptability	(Wolfson et al., 1986)
3. Motor	
a) Strength	
b) Range of Motion	
c) Endurance	6-Minute Walk (Cole et al., 1994)
C. Functional assessment	
1. Mobility skills	Timed Up and Go Test
	(Mathias et al., 1986)
	Berg Balance Scale
	(Berg et al., 1989)
	Tinetti Balance/Gait Scale
	(Tinetti, 1986)
2. Environment	Functional Home Assessment
	(Chandler & Duncan 1993)

[Adapted from Chandler and Duncan, 1993]

Many of the same research techniques used to assess balance and gait in the healthy elderly are now being applied to the disabled, such as survivors of stroke. Although stroke is a sudden disease process, it should be kept in mind that certain chronic, progressive neurological disorders, such as Parkinson's disease, also have significant disabling effects on posture and gait. In addition, disorders of an orthopaedic and musculoskeletal nature, such as osteoarthritis, rheumatoid arthritis and osteoporosis, also may have disabling effects on balance and gait. Of interest is that the motor and sensory impairments following stroke are manifested primarily on one side, with a resulting asymmetry of posture, weight distribution, and muscle activation. However, it should be noted that some motor abnormalities may also be manifested on the nonparetic side (Halaney & Carey, 1989; Thilmann, Fellows, & Garms, 1990).

This section of the review will concentrate on stroke, since the resulting combination of sensory, motor and cognitive deficits may have quite a devastating effect on postural control. A study of how survivors of stroke learn to adapt to altered perceptual and sensory information, delayed muscle responses and altered biomechanical properties may shed some light on the adaptability of the postural system. From a clinical point of view, a study of this population highlights the complex interrelationship of sensory, central integrative and motor systems. Also, evaluation of the control of balance and gait has undergone a shift to focus on functional outcomes, i.e., quantitative evaluation of an activity or components of an activity.

Centre of pressure measures of subjects with stroke

Measurements of postural sway in stroke patients consistently demonstrate a shift of the centre of pressure towards the non hemiplegic side (Shumway-Cook et al., 1988; Winstein, Gardner, McNeal, Barto, & Nicholson, 1989; Dettmann, Linder, & Sepic, 1987). This lateral shift is observed during quiet standing, as well as during weight shifting activities (Dettmann et al., 1987). The shift in the centre of pressure is related to asymmetrical weight bearing through the lower limbs. Healthy subjects typically distribute weight evenly between the two limbs in a 50/50 distribution. However, in one study, hemiplegic subjects bore an average of 36% (± 15%) of body weight on the paretic limb during quiet standing (Dettmann et al., 1987). This is consistent with the findings of others (Shumway-Cook et al., 1988; Sackley, 1991) in which up to 70% of total body weight was carried by the nonparetic limb. These findings confirm clinical observations that hemiplegic patients tend to distribute weight asymmetrically, with weight shifted abnormally towards the nonparetic side and reduced weight bearing through the paretic limb. This fact has formed the basis of rehabilitation treatment approaches for the last 15 or more years (Bobath, 1978; Carr & Shepherd, 1987; Daleiden, 1990).

In addition to an unequal distribution of weight, hemiplegic subjects demonstrate an increase in the total sway area recorded while standing on a static force plate (Shumway-Cook et al., 1988). Although the total movement of the centre of pressure is greater, perhaps due to increased variability of the path of sway, it seems that actual movement of sway is restricted in subjects with stroke. During voluntary movement on a static force plate, Dettmann et al. (1987) found that stroke patients were restricted in the movement of the centre of pressure in both the anterior-posterior and lateral directions. However, during a visually cued task, DiFabio and Badke (1990b) found similar amounts of sway in stroke and healthy subjects in the AP direction, but restriction of lateral sway in the subjects with stroke. The area of stability, defined by Dettmann et al. (1987) as the average position of the centre of pressure during a weight shifting activity, is much smaller in subjects with stroke and is shifted towards the non paretic side. These authors discussed the idea of a "small safe zone" which hemiplegic subjects use during weight shifting movements. It appears that these subjects compensate for lack of steadiness by restricting the distance that the centre of pressure moves.

Lateral weight transfer following stroke

If stroke patients voluntarily move less to the side, does this indicate greater lateral stability and perhaps reduced risk of falling towards the hemiplegic side? This does not seem to be true. Lee, Deming, and Sahgal (1988) found that hemiplegic subjects were able to withstand significantly less maximum static loads than healthy young and elderly adults in all directions, including loading which tended to pull the subjects towards or away from the hemiplegic side.

Pai, Rogers, Hedman, and Hanke (1994) evaluated weight transfer ability in people with a hemiparesis. Subjects were ambulatory and had sustained a left CVA. They were asked to perform single leg flexion movements with the paretic and nonparetic limbs. Assessment of the ability to transfer weight to the opposite limb and hold the position long enough to flex the non weight bearing leg (so that the foot cleared the floor for the duration of the 5 second trial), was classified into four categories: successful transfer and able to hold single limb stance, successful transfer but unable to hold single limb stance, unsuccessful transfer with undershooting, unsuccessful transfer with overshooting.

Both position and displacement of the centre of mass in the frontal plane were recorded. Subjects performed the task successfully to the nonparetic side for 48% of the trials and to the paretic side 20% of the trials. It is no surprise that the hemiplegic subjects had greater difficulty performing the task when standing on the paretic limb rather than the non paretic limb. However, it is interesting that transitions to the non-paretic limb were unsuccessful 52% of the time. This indicates that individuals with stroke have difficulty transferring weight adequately to the non-paretic as well as the paretic limbs. Unsuccessful trials to the nonparetic side were due to undershooting, i.e., lack of adequate displacement of the centre of mass, 26% of the time, and failure to hold the single limb position after the completion of the centre of mass transfer, also 26% of the time. Failure to transfer to the paretic side was due primarily to an inability to hold the position (63% of the time) and secondly, to insufficient displacement of the centre of mass (17%).

Deficits in control of medial-lateral stability have also been demonstrated in elderly subjects who are at risk for falling (Crosbie, Nimmo, Banks, Brownlee, & Meldrum, 1989; Maki et al., 1994). Although postural sway is generally greater in elderly subjects with a history of falls than nonfallers, there is evidence to suggest that control of weight transfer from side to side in the medial-lateral direction may have greater impact on balance than control in the anterior-posterior direction (Maki et al., 1994). This is an area in which there has been little discussion.

Control of balance following a perturbation

In addition to measurement of postural sway, or the centre of pressure, much of the data reported about balance mechanisms in stroke have been obtained from perturbation types of protocols, i.e., unexpected and anticipated displacements to balance, similar to those used to evaluate the healthy elderly and fallers. Stroke survivors with hemiparesis show alterations in both the timing and sequencing of postural muscles when standing balance is displaced unexpectedly by platform movement (Badke & Duncan, 1983; Hocherman, Dickstein, Hirschbiene, & Pillar, 1988). The response time of distal postural muscles is increased in hemiplegic subjects, particularly on the paretic side (Badke & Duncan, 1983; Diener, Ackerman, Dichgans, & Guschlbauer, 1985; DiFabio et al., 1986; Badke, Duncan, & DiFabio, 1987). In addition, the pattern of activation is altered (Badke & Duncan, 1983; DiFabio et al., 1986; Badke et al., 1987). Instead of the normal sequence of distal to proximal muscle activation found in healthy subjects, Badke et al. (1987) found that proximal muscles tend to be activated first in the paretic limb. DiFabio et al. (1986) found a similar pattern and suggested that the longer latency of the distal muscles in the paretic limb is counterbalanced by an earlier than normal response in the proximal muscles in the nonparetic limb. Others (Badke & Duncan, 1983; Hocherman et al., 1988) have found no clear activation pattern in hemiplegic subjects and have observed co-contraction of all four muscles (medial gastrocnemius, tibialis anterior, medial hamstrings, vastus medialis) of the paretic limb.

Hemiplegic subjects may have difficulty with adaptation of postural responses. Hocherman et al. (1988) observed that healthy elderly subjects standing on a continuously moving platform depressed certain EMG responses as they became familiar with the continuous movement of the support surface. However, half of the hemiplegic subjects in this study had major difficulty with modulation of the postural muscles. The typical responses seen in the lower limbs of the hemiplegic subjects included co-contraction of tibialis anterior and gastrocnemius, plus tonic contractions of either tibialis anterior or gastrocnemius for long periods of time, rather than short alternating bursts of activity. It is interesting that these abnormal responses were observed on both the paretic and the nonparetic limbs (Hocherman et al., 1988). DiFabio et al., (1986) also noted some problems with adaptation of postural responses in hemiplegic subjects, but concluded that hemiplegic subjects do retain, or perhaps recover, some ability to modulate postural activity for balance. Certainly, greater variability of muscle response synergies have been found in many hemiplegic groups (Badke & Duncan, 1983; DiFabio et al., 1986; Badke et al., 1987). The above findings regarding the timing and sequencing of postural responses in hemiplegic subjects support current theories that the programming of muscle sequencing is centrally controlled and therefore, is impaired in hemiplegic subjects.

The availability of mechanized platforms and force plates is limited in clinical settings and therefore, there is a need to establish other testing methods for evaluation (DiFabio & Badke, 1990a). We evaluated the effectiveness of using a simpler form of perturbation test (Figure 1) with survivors of stroke (Harburn, Hill, Kramer, Noh, Vandervoort, & Teasell, 1995). The test was adapted from the original postural stress test reported by Wolfson, Whipple, Amerman, and Kleinberg (1986), which to our knowledge, had not been used before with subjects with stroke. It involves a dynamic situation in which the individual received an unexpected perturbation in the backwards direction. Several investigators have demonstrated the vulnerability of healthy older people and survivors of stroke in this direction (Sabin, 1982; Lee et al., 1988; Hill, Vandervoort, & Kramer, 1990; Blaszczyk et al., 1994).

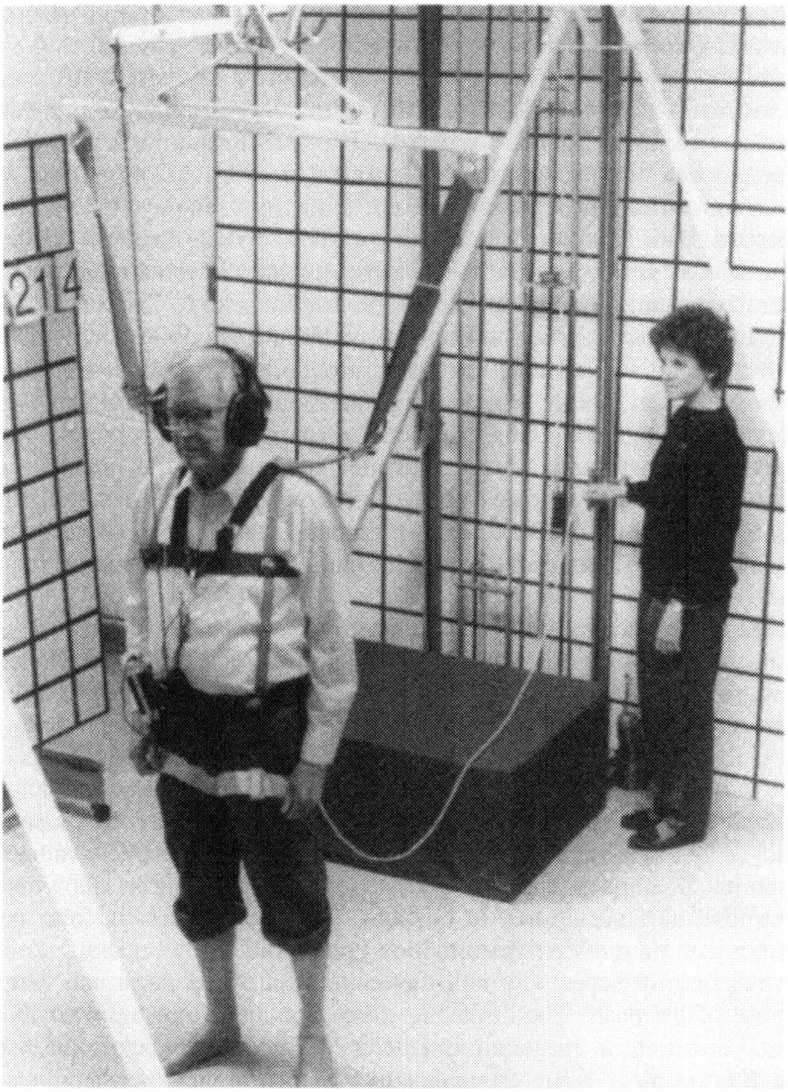

FIGURE 1. Measurement of responses to the Postural Stress Test. Note the safety harness on the subject which is attached to an overhead trolley system. The tester is ready to release the weight for generating a backwards perturbation, via the rope attached at the subject's waistine. (From Hill et al., 1994. Comparison of balance responses to an external perturbation test, with and without an overhead harness safety system. Gait and Posture, 2, 27-31, by permission of the publishers Butterworth-Heinnemann Ltd©).

Videotaped performances of responses to a backward perturbation applied to the waist were graded according to a 10 point scale of adaptive postural responses. The initial perturbation force was reduced to 0.5% of body weight, rather than 1.5% in the original test design, and was increased in 0.5% body weight increments until 4.5% of body weight was reached or a loss of balance occurred. This modification was selected to accommodate for the less stable stance of the subjects with stroke. In addition, subjects were protected from falling by using a safety harness (Harburn, Hill, Kramer, Noh, Vandervoort, & Matheson, 1993). It had been shown previously that the presence of the safety harness did not alter the postural stress test scores in healthy elderly subjects (Hill, Harburn, Kramer, Noh, Vandervoort, & Matheson, 1994). Individuals with stroke in our study demonstrated lower adapted postural stress test total scores than healthy subjects (age and sex matched control subjects) (Harburn et al., 1995). This indicates that more efficient response strategies, i.e., use of ankle or hip strategies, were not used to maintain or restore balance and subjects resorted to stepping backwards or fell. Pai et al. (1994) also observed lack of stepping and hopping movements when hemiplegic subjects lost equilibrium during a weight transferring task.

Since hemiplegic posture is typically asymmetrical in nature, it would be reasonable to expect that responses to a perturbation of balance are different, depending on the direction and site of application. Although this has been shown to be true in an AP direction, there is some disagreement about the lateral direction. Lee et al. (1988) did not find a difference in the maximum load that subjects with stroke could maintain on the paretic or nonparetic sides. This however, was a static load. Wing, Goodrich, Virji-Babul, Jenner, and Clapp (1993) evaluated responses of subjects with stroke to a lateral perturbation by applying a controlled horizontal force to the hips. This evaluation went one step further than the previous perturbation type of studies by breaking down the response of subjects to both the application of the push and to the release of the push. The displacing force was applied at the waist in a lateral direction, as in the postural stress test and as in Lee's static load test (Lee et al., 1988). Compared to a control group, the hemiplegic subjects swayed further to the side and required longer to reach stabilization of the hip position. These differences were noted particularly on the release phase from the push. Also, greater differences between the paretic and nonparetic sides were noted at the release from the lateral push. Thus, on release from a push, hemiplegic subjects swayed less to the nonparetic side, when the force had been directed towards the paretic side, than compared to a push towards the nonparetic side (Wing et al., 1993). This study has implications for clinical evaluation of balance and demonstrates the need to be specific in balance evaluation, i.e., consideration of both the application of the force and the stabilization responses on release of the destabilizing force.

The timing and recruitment of anticipatory postural adjustments are also altered following stroke. Horak, Esselman, Anderson, and Lynch (1984) demonstrated that hemiplegic subjects showed similar sequences of muscle activation compared with healthy subjects prior to performing fast paced, unilateral upper extremity movements. However, the onset of postural responses on the paretic side was delayed. The same response was found when weights were added to the arm (Horak et al., 1984).

Badke et al. (1987) have shown that if hemiplegic subjects are given prior knowledge of a displacement of balance (in their experiment this was knowledge of the time and direction of platform movement), the responses of the paretic limb are quicker and become similar to those on the nonparetic limb. Incorrect instructions regarding the direction of movement resulted in prolonged, and quite variable, muscle responses on the paretic limb. Prior knowledge appeared to improve the onset times of muscle responses during anterior sway only; no difference was found during posterior sway. Badke and DiFabio (1985) have also shown that preloading of one lower limb facilitates EMG activity at a shorter latency on the contralateral side.

This research suggests that some of the motor control problems frequently seen in stroke may be due to impaired sensory function as well as motor system involvement. Not only is sensory information reduced, partly due to unequal weight bearing through the paretic limb, but there is evidence to suggest that the ability to integrate sensory information, particularly proprioceptive information from the lower limbs, is impaired following stroke (DiFabio & Badke, 1991). DiFabio and Badke (1991) found that hemiplegic subjects had significant balance problems when attempting to stand on a foam surface. Balance performance was not nearly as affected by altering visual information.

Individuals with stroke may be dealing with a complex set of mechanical and neurological problems; an asymmetrical body posture which provides unequal sensory input, altered sensory function, and asymmetrical activation of muscles in terms of the amount and timing of muscle activity (Badke & Duncan, 1983; Wing et al., 1993). The formation of an internal reference of correctness is essential to the learning of a new task (Schmidt, 1982). Therefore, the use of feedback, auditory, visual and proprioceptive feedback at least, may be particularly necessary during retraining of balance responses of stroke patients. This

(Finley et al., 1969; Murray et al., 1969, 1970). Toe clearance is not significantly different from younger subjects (Winter et al., 1990), although peak knee extension throughout the gait cycle may be less (Ostrosky et al., 1994). In terms of kinetics, differences have been noted in the force of push-off and foot landing (Winter et al., 1990).

TABLE 2.
A comparison of mean walking speeds reported in healthy young and elderly subjects and individuals with stroke.

Source	Sex	n	Age (decade)	Speed* (m/s)
Healthy young subjects				
Guimaraes & Isaacs, 1980	M+F	24	<7	1.20 (±0.16)
Hageman & Blanke, 1986	F	13	3-4	1.60 (±0.16)
Blanke & Hageman, 1989	M	12	3-4	1.31 (±0.18)
Ostrosky et al., 1994	M+F	30	3-4	1.38 (±0.14)
Healthy elderly subjects				
Guimaraes & Isaacs, 1980	M+F	23	>7	0.71 (±0.22)
Imms & Edholm, 1981	M+F	71	7-10	0.74 (±0.28)
Hageman & Blanke, 1986	F	13	7-9	1.32 (±0.24)
Blanke & Hageman, 1989	M	12	7-8	1.39 (±0.23)
Ostrosky et al., 1994	M+F	30	7-8	1.27 (±0.16)
Stroke subjects				
Wall & Ashburn, 1979	M+F	5	6-8	0.46 (±0.16)
Brandstater et al., 1983	M+F	23	5-8	0.31 (±0.21)
Bohannon, 1987	M+F	37	4-9	0.39 (±0.26)
Dettmann et al., 1987	M	15	5-9	0.47 (±0.30)

* Speeds in meters/second compiled by authors from published data of selected studies.

Outcome measures, such as the above, provide descriptive details of changes in gait patterns associated with aging but still leave questions as to the mechanisms responsible for these changes. From a biomechanical point of view, Winter et al. (1990) offer primary adaptations which could account for these changes. First, a slower velocity, or a shorter step length, may be mechanisms used to increase the time of the double support phase. This would reduce the time spent in the relatively unstable single support phase and thereby improve balance. Alternately, a deficiency of muscle power in some older people may reduce the power of push-off, reducing the step length, increasing the double support time and reducing the speed. However, the question still remains, is the decrease in walking velocity responsible for the changes in gait variables seen in the elderly, or is velocity slowed secondary to other mechanisms?

Jansen, Vittas, Hellbert, and Hansen (1982) assessed the hypothesis that velocity is the primary factor by studying the gait patterns of young and elderly subjects walking at the same speed on a treadmill. At a velocity of 1.11 m/s, the elderly subjects demonstrated a similar gait pattern to the younger subjects. However, when asked to walk at very slow speeds, changes in the gait pattern of elderly subjects may be evident. Gillis et al. (1986) found very little difference in the gait patterns of young and elderly subjects when asked to walk at self-selected slow, normal and fast speeds, except that the speed of the elderly subjects was consistently slower than that of the young subjects. When both groups walked at very slow speeds (less than 0.3 statures/s), stride time was significantly reduced in the elderly group, and the duration of total support was increased, although not significantly. The authors observed that at very slow speeds the elderly subjects walked with a more cautious gait pattern. The fitness level of subjects is a factor to consider in describing the gait patterns of the elderly. For example, elderly women who exercise regularly may be able to successfully walk at a slower pace than elderly women who do not exercise (Leiper & Craik, 1991).

GAIT FOLLOWING STROKE

In addition to the normal aging changes of motor and sensory function, disorders of the central nervous system, such as hemiplegia following stroke, may cause alterations of motor control and therefore deviations of gait. As noted earlier, hemiplegia may be characterized by loss of selective movements, abnormal sensory function of the affected limbs and deficit of postural righting and equilibrium responses (Bobath, 1978).

The gait pattern of hemiplegic subjects is asymmetrical in appearance due to the unilateral involvement and a reluctance to transfer and support full body weight on the affected limb (Bobath, 1978). This asymmetry, with respect to affected versus unaffected side, has been demonstrated by analysis of temporal and distance parameters of hemiplegic gait. The duration of the support phase tends to be decreased on the affected side and increased on the unaffected side (Wall & Ashburn, 1979; Wall & Turnbull, 1986). Poor weight transference to the affected limb and prolonged stance time on the unaffected limb are likely due to impaired balance mechanisms. Hence, it has been shown that intact balance responses are strong predictors of successful gait performance in hemiplegic patients (Bohannon, 1987; Keenan, Perry, & Jordan, 1984).

The duration of swing and support phases account for 40% and 60% of the gait cycle, respectively, in healthy adult subjects (Murray, 1967). Peat, Dubo, Winter, Quanbury, Steinke, and Grahame (1976) reported that hemiplegic subjects demonstrated altered swing and support phases of both lower extremities. Mean swing phase on the affected side was determined to be 33%, and support phase 67%, of the gait cycle. The unaffected limb spent 20% of the cycle in swing and 80% in support phase. The increased support time on the unaffected side, plus the very short swing time, reflects the difficulty hemiplegic subjects have in transferring and supporting weight on the affected limb.

Voluntary movement of the affected hemiplegic limb may be dominated by synergistic patterns of muscle activity (Bobath, 1978). This loss of selective movement is demonstrated by EMG analysis of hemiplegic gait. Peat et al. (1976) synchronized the analysis of lower limb EMG activity with specific temporal components of the gait cycle. Hemiplegic subjects showed peak activity of all muscle groups during the midstance phase and low activity of all muscle groups during the swing phase. Healthy subjects, however, show recruitment of specific muscles of the knee and ankle at the time they are required for either stability or movement of the limb (Dubo, Peat, Winter, Quanbury, Hobson, Steinke, & Reimer, 1976).

The location of the lesion, plus the manner in which the subject may have learned to compensate for the motor loss, will influence the movement pattern. For example, if abnormal tone is strongest in the extensor or antigravity muscle groups, the subject may be able to support weight on the affected limb during the stance phase, but will have difficulty moving the limb forward during the swing phase. The result may be dragging of the toes, or the subject may learn to compensate by circumducting the affected limb at the hip and thereby allowing the extended limb to clear the floor (Bobath, 1978).

Subjects with residual hemiplegia walk slower than urban pedestrians (Finley & Cody, 1970) and even slower than healthy elderly subjects (see Table 2). Walking speed has been shown to be related to isometric strength and the stage of motor recovery of the affected limb, plus the overall balance ability of hemiplegic subjects (Bohannon, 1986a, 1986b, 1987; Brandstater, de Bruin, Gowland, & Clark, 1983). Wall and Ashburn (1979) showed a trend towards an increase in walking speed as patients recovered during the first 9 months post onset of hemiplegia.

REFERENCES

Alexander, N. B. (1994). Postural control in older adults. *Journal of the American Geriatrics Society, 42,* 93-108.

Badke, M. B., & DiFabio, R. P. (1985). Effects of postural bias during support surface displacements and rapid arm movements. *Physical Therapy, 65,* 1490-1495.

Badke, M. B., & Duncan, P. W. (1983). Patterns of rapid motor responses during postural adjustments when standing in healthy subjects and hemiplegic patients. *Physical Therapy, 63,* 13-20.

Badke, M. B., Duncan, P. W., & DiFabio, R. P. (1987). Influence of prior knowledge on automatic and voluntary postural adjustments in healthy and hemiplegic subjects. *Physical Therapy, 67,* 1495-1500.

Berg, K., Wood-Dauphinee, S., Williams, J. I., & Gayton, D. (1989). Measuring balance in the elderly: preliminary development of an instrument. *Physiotherapy Canada, 41,* 304-311.

Black, F. O., Wall, C., Rockette, H., & Kitch, R. (1982). Normal subject postural sway during the Romberg test. *American Journal of Otolaryngology, 3,* 309-318.

Blanke, D. J., & Hageman, P.A. (1989). Comparison of gait of young and elderly men. *Physical Therapy, 69,* 144-148.

Blaszczyk J. W., Lowe, D. L., & Hansen, P. D. (1994). Ranges of postural stability and their changes in the elderly. *Gait and Posture, 2,* 11-17.

Bobath, B. (1978). *Adult hemiplegia: Evaluation and treatment* (2nd edition). London: Heinemann Medical Books.

Bohannon, R. W. (1986a). Relationship between static strength and various other measures in hemiparetic stroke patients. *International Journal of Rehabilitation Medicine, 8,* 125-128.

Bohannon, R. W. (1986b). Strength of lower limb related to gait velocity and cadence in stroke patients. *Physiotherapy Canada, 38,* 204-206.

Bohannon, R. W. (1987). Gait performance of hemiparetic stroke patients: selected variables. *Physical Therapy, 68,* 777-781.

Bohannon, R. W., Larkin, P. A., Cook, A. C., Gear, J., & Singer, J. (1984). Decrease in timed balance scores with aging. *Physical Therapy, 64,* 1067-1070.

Brandstater, M. E., de Bruin, H., Gowland, C., & Clark, B. M. (1983). Hemiplegic gait: Analysis of temporal variables. *Archives of Physical Medicine and Rehabilitation, 64,* 583-587.

Brocklehurst, J. C., Robertson, D., & James-Groom, P. (1982a). Clinical correlates of sway in old age - sensory modalities. *Age and Ageing, 11,* 1-10.

Brocklehurst, J. C., Robertson, D., & James-Groom, P. (1982b). Skeletal deformities in the elderly and their effect on postural sway. *Journal of the American Geriatrics Society, 30,* 534-538.

Brooks, V. B. (1986). *The neural basis of motor control.* New York: Oxford University Press.

Cape, R. D. T. (1978). *Aging: Its complex management.* New York: Harper & Row.

Carr, J. H., & Shepherd, R. B. (1987). *A motor relearning programme for stroke* (2nd ed.). Rockville, MD: Aspen Publishers.

Chan, C. Y. W. (1984). Neurophysiological basis underlying the use of resistance to facilitate movement. *Physiotherapy Canada, 36,* 335-341.

Chandler, J. M., & Duncan, P. W. (1993). Balance and falls in the elderly: Issues in evaluation and treatment. In A. A. Guccione (Ed.), *Geriatric physical therapy.* St Louis, MO: Mosby.

Cole, B., Finch, E., Gowland, C., & Mayo, N. (1994). *Physical rehabilitation outcome measures.* Ottawa: Supply and Services Canada.

Cordo, P. J., & Nashner, L. M. (1982). Properties of postural adjustments associated with rapid arm movements. *Journal of Neurophysiology, 47,* 281-302.

Crilly, R. G., Richardson, L. D., Roth, J. H., Vandervoort, A. A., Hayes, K. C., & Mackenzie, R. A. (1987). Postural stability and Colles fracture. *Age and Ageing, 16,* 133-138.

Crosbie, W. J., Nimmo, M. A., Banks, M. A., Brownlee, M. G., & Meldrum, F. (1989). Standing balance responses in two populations of elderly women: A pilot study. *Archives of Physical Medicine and Rehabilitation, 70,* 751-754.

Cunningham, D. A., Rechnitzer, P. A., Pearce, M. E., & Donner, A. P. (1982). Determinants of self-selected walking pace across ages 19 to 66. *Journal of Gerontology, 37,* 560-564.

Daleiden, S. (1990). Weight shifting as a treatment for balance deficits: a literature review. *Physiotherapy Canada, 42,* 81-87.

Dettmann, M. A., Linder, M. T., & Sepic, S. B. (1987). Relationships among walking performance, postural stability, and functional assessments of the hemiplegic patient. *American Journal of Physical Medicine, 66,* 77-90.

Diener, H. C., Ackerman, H., Dichgans, J., & Guschlbauer, B. (1985). Medium- and long-latency responses to displacements of the ankle joint in patients with spinal and central lesions. *Electroencephalography and Clinical Neurophysiology, 60,* 407-416.

Diener, H. C., Dichgans, J., Bootz, F., & Bacher, M. (1984). Early stabilization of human posture after a sudden disturbance: influence of rate and amplitude of displacement. *Experimental Brain Research, 56,* 126-134.

Dietz, V., Quintern, J., Berger, W., & Schenck, E. (1985). Cerebral potentials and leg muscle EMG responses associated with stance perturbation. *Experimental Brain Research, 57,* 348-354.

DiFabio, R. P., Badke, M. B. (1990a). Relationship of sensory organization to balance function in patients with hemiplegia. *Physical Therapy, 70,* 542-548.

DiFabio, R. P., Badke, M. B. (1990b). Extraneous movement associated with hemiplegic postural sway during dynamic goal-directed weight redistribution. *Archives of Physical Medicine and Rehabilitation, 71,* 365-371.

DiFabio, R. P., Badke, M. B. (1991). Stance duration under sensory conflict conditions in patients with hemiplegia. *Archives of Physical Medicine and Rehabilitation, 72,* 292-295.

DiFabio, R. P., Badke, M. B., & Duncan, P. W. (1986). Adapting human postural reflexes following localized cerebrovascular lesion: analysis of bilateral long latency responses. *Brain Research, 363,* 257-264.

Doherty, T. J., Vandervoort, A. A., & Brown, W. F. (1993). Effects of aging on the motor unit: a brief review. *Canadian Journal of Applied Physiology, 18,* 331-358.

Dornan, J., Fernie, G. R., & Holliday, P. J. (1978). Visual input: Its importance in the control of postural sway. *Archives of Physical Medicine and Rehabilitation, 59,* 586-591.

Dubo, H. I. C., Peat, M., Winter, D. A., Quanbury, A. O., Hobson, D. A., Steinke, T., & Reimer, G (1976). Electromyographic temporal analysis of gait: Normal human locomotion. *Archives of Physical Medicine and Rehabilitation, 57,* 415-420.

Duncan, P. (1994). Stroke disability. *Physical Therapy, 74,* 399-407.

Duncan, P. W., Weiner, D. K., Chandler, J. M., & Studenski, S. (1990). Functional reach: A new clinical measure of balance. *Journal of Gerontology, 45,* M192-M197.

Era, P., & Heikkinen, E. (1985). Postural sway during standing and unexpected disturbance of balance in random samples of men of different ages. *Journal of Gerontology, 40,* 287-295.

Fernie, G. R., Gryfe, C. I., Holliday, P. J., & Llewellyn, A. (1982). The relationship of postural sway in standing to the incidence of falls in geriatric subjects. *Age and Ageing, 11,* 11-16.

Ferrandez, A.-M., Pailhous, J., & Durup, M. (1990). Slowness in elderly gait. *Experimental Aging Research, 16,* 79-89.

Finley, F. R., & Cody, K. A. (1970). Locomotive characteristics of urban pedestrians. *Archives of Physical Medicine and Rehabilitation, 51,* 423-426.

Finley, F. R., Cody, K. A., & Finzie, R. V. (1969). Locomotion patterns in elderly women. *Archives of Physical Medicine and Rehabilitation, 50,* 140-149.

Fleming, B. E., & Pendergast, D.R. (1993). Physical condition, activity pattern, and environment as factors in falls by adult care facility residents. *Archives of Physical Medicine and Rehabilitation, 74,* 627-630.

Friedli, W. G., Hallett, M., & Simon, S. R. (1984). Postural adjustments associated with rapid voluntary arm movements. 1. Electromyographic data. *Journal of Neurology, Neurosurgery and Psychiatry, 47,* 611-622.

Gillis, B., Gilroy, K., Lawley, H., Mott, L., Wall, J. C. (1986). Slow walking speeds in healthy young and elderly females. *Physiotherapy Canada, 38,* 350-352.

Gresham, G. E., Phillips, T. F., Wolf, P. A., McNamara, P. M., Kannel, W. B., & Dawber, T. R. (1979). Epidemiologic profile of long-term stroke disability: the Framingham study. *Archives of Physical Medicine and Rehabilitation, 60,* 487-491.

Guimaraes, R. M., & Isaacs, B. (1980). Characteristics of the gait in old people who fall. *International Rehabilitation Medicine, 2,* 177-180.

Gurfinkel, V. S. (1973). Muscle afferentation and postural control in man. *Agressologie, 14C,* 1-8.

Hageman, P. A., & Blanke, D. J. (1986). Comparison of gait of young and elderly women. *Physical Therapy, 66,* 1382-1397.

Hakansson, D. M., Vandervoort, A. A., & Kramer, J. F. (1991). Effect of cold therapy on nerve conduction in young and elderly women. *Physiotherapy Canada, 43,* Suppl.

Halaney, M. E., & Carey, J. R. (1989). Tracking ability of hemiparetic healthy subjects. *Physical Therapy, 69,* 342-347.

Harburn, K. L., Hill, K. M., Kramer, J. F., Noh, S., Vandervoort, A. A., & Matheson, J. E. (1993). An overhead harness and trolly system for balance and ambulation assessment and training. *Archives of Physical Medicine and Rehabilitation, 74,* 220-223.

Harburn, K. L., Hill, K. M., Kramer, J. F., Noh, S., Vandervoort, A. A., & Teasell, R. (1995). Clinical applicability and test-retest reliability of an external perturbation test of balance using stroke subjects. *Archives of Physical Medicine and Rehabilitation, 76,* 317-323.

Hasselkus, B. R., & Shambes, G. M. (1975). Aging and postural sway in women. *Journal of Gerontology, 30,* 661-667.

Heart and Stroke Foundation of Ontario (1992). *At A glance '93.* Toronto, Canada.

Hill, K. M., Harburn, K. L., Kramer, J. F., Noh, S., Vandervoort, A. A., & Matheson, J. E. (1994). Comparison of balance responses to an external perturbation test, with and without an overhead harness safety system. *Gait and Posture, 2,* 27-31.

Hill, K. M., Vandervoort, A. A., & Kramer, J. F. (1990). Reproductibility of a test of postural responses in healthy elderly females. *Physiotherapy Canada, 42,* 61-67.

Hocherman, S., Dickstein, R., Hirschbiene, A., & Pillar, T. (1988). Postural responses of normal geriatric and hemiplegic patients to a continuing perturbation. *Experimental Neurology, 99,* 388-401.

Horak, F. B. (1987). Clinical measurement of postural control in adults. *Physical Therapy, 67,* 1881-1885.

Horak, F. B., Esselman, P., Anderson, M. E., & Lynch, M. K. (1984). The effects of movement velocity, mass displaced, and task certainty on associated postural adjustments made by normal and hemiplegic individuals. *Journal of Neurology, Neurosurgery, and Psychiatry, 47,* 1020-1028.

Horak, F. B., Shupert, C. L., Dietz, V., & Hortsmann, G. (1994). Vestibular and somatosensory contribution to head and body displacements in stance. *Experimental Brain Research, 100,* 93-106.

Horak, F. B., Shupert, C. L., & Mirka, A. (1989). Components of postural dyscontrol in the elderly: a review. *Neurobiology of Aging, 10,* 727-738.

Imms, F. J., & Edholm, O. G. (1979). The assessment of gait and mobility in the elderly. *Age and Ageing, 8* (suppl), 261-267.

Imms, F. J., & Edholm, O. G. (1981). Studies of gait and mobility in the elderly. *Age and Ageing, 10,* 147-156.

Jansen, E. C., Vittas, D., Hellberg, S., & Hansen, J. (1982). Normal gait of young and old men and women. *Acta Orthopaedica Scandinavica, 53,* 193-198.

Keenan, M. A., Perry, J., & Jordan, C. (1984). Factors affecting balance and ambulation following stroke. *Clinical Orthopaedics and Related Research, 182,* 165-171.

Kirshen, A. J., Cape, R. D. T., Hayes, K. C., & Spencer, J. D. (1984). Postural sway and cardiovascular parameters associated with falls in the elderly. *Journal of Clinical and Experimental Gerontology, 6,* 291-307.

Lee, W. A., Deming, L., & Sahgal, V. (1988). Quantitative and clinical measures of static standing balance in hemiparetic and normal subjects. *Physical Therapy, 68,* 970-976.

Leiper, C. I., & Craik, R. L. (1991). Relationships between physical activity and temporal-distance characteristics of walking in elderly women. *Physical Therapy, 71,* 791-803.

Leonard, E. (1990). Balance tests and balance responses: performance changes following a CVA. A review of the literature. *Physiotherapy Canada, 42,* 68-72.

Lucy, S. D., & Hayes, K. C. (1985). Postural sway profiles: normal subjects and subjects with cerebellar ataxia. *Physiotherapy Canada, 37,* 140-148.

MacLennan, W. J., Timothy, J. I., & Hall, M. R. P. (1980). Vibration sense, proprioception and ankle reflexes in old age. *Journal of Clinical and Experimental Gerontology, 2,* 159-171.

Maki, B. E., Holliday, P.J., & Topper, A.K. (1984). A prospective study of postural balance and risk of falling in an ambulatory and independent elderly population. *Journals of Gerontology: Medical Sciences, 49,* 72-84.

Manchester, D., Woollacott, M., Zederbauer-Hylton, N., & Marin, O. (1989). Visual, vestibular and somatosensory contributions to balance control in the older adult. *Journal of Gerontology, 44,* M118-127.

Markham, C. H. (1987). Vestibular control of muscular tone and posture. *Canadian Journal of Neurological Sciences, 14,* 493-496.

Marsden, C. D., Merton, P. A., & Morton, H. B. (1981). Human postural responses. *Brain, 104,* 513-534.

Mathias, S., Nayak, U. S. L., & Isaacs, B. (1986). Balance in elderly patients: the "get-up and go" test. *Archives of Physical Medicine and Rehabilitation, 67,* 387-389.

Melvill Jones, G., & Watt, D. G. D. (1971). Observations on the control of stepping and hopping movements in man. *Journal of Physiology (London), 219,* 709-727.

Murray, M. P. (1967). Gait as a total pattern of movement. *American Journal of Physical Medicine, 46,* 290-330.
Murray, M. P., Kory, R. C., & Clarkson, B. H. (1969). Walking patterns in healthy old men. *Journal of Gerontology, 24,* 169-178.
Murray, M. P., Kory, R. C., & Sepic, S. B. (1970). Walking patterns of normal women. *Archives of Physical Medicine and Rehabilitation, 51,* 637-650.
Nashner, L. M. (1976). Adapting reflexes controlling the human posture. *Experimental Brain Research, 26,* 59-72.
Nashner, L. M. (1977). Fixed patterns of rapid postural responses among leg muscles during stance. *Experimental Brain Research, 30,* 13-24.
Nashner, L. M., & Cordo, P. J. (1981). Relation of automatic postural responses and reaction-time voluntary movements of human leg muscles. *Experimental Brain Research, 43,* 395-405.
Nashner, L. M., Woollacott, M., & Tuma, G. (1979). Organization of rapid responses to postural and locomotor-like perturbations of standing man. *Experimental Brain Research, 36,* 463-476.
Neumann, D. A. (1993). Arthrokinesiologic considerations in the aged adult. In A. A. Guccione (Ed.), *Geriatric physical therapy*. St Louis, MO: Mosby.
O'Brien, M., Power, K., Sanford, S., Smith, K., & Wall, J. (1983). Temporal gait patterns in healthy young and elderly females. *Physiotherapy Canada, 35,* 323-326.
Oddsson, O., & Thorstensson, A. (1987). Fast voluntary trunk flexion movements in standing: motor patterns. *Acta Physiologica Scandinavica, 129,* 93-106.
Olney, R. K. (1985). Age-related changes in peripheral-nerve function. *Geriatric Medicine Today, 4,* 76-86.
Ostrosky, K. M., VanSwearingen, J. M., Burdett, R. G., & Gee, Z. (1994). A comparison of gait characteristics in young and old subjects. *Physical Therapy, 74,* 637-644.
Overstall, P. W., Exton-Smith, A. N., Imms, F. J., & Johnson, A. L. (1977). Falls in the elderly related to postural imbalance. *British Medical Journal, 1,* 261-264.
Pack, D. R., Wolfson, L. I., Amerman, P., Whipple, R., & Kaplan, J. G. (1985). Peripheral nerve abnormalities and falling in the elderly. *Neurology, 35* (suppl), 79.
Pai, Y. C., Rogers, M. W., Hedman, L. D., & Hanke, T. A. (1994). Alterations in weight transfer capabilities in adults with hemiparesis. *Physical Therapy, 74,* 647-657.

Peat, M., Dubo, H. I. C., Winter, D. A., Quanbury, A. O., Steinke, T., & Grahame, R. (1976). Electromyographic temporal analysis of gait: Hemiplegic locomotion. *Archives of Physical Medicine and Rehabilitation, 57,* 421-425.

Porter, M. M., Vandervoort, A. A., & Lexell, J. (1995). Ageing of human muscle: structure, function and adaptability. *Scandinavian Journal of Medicine Sciene in Sports, 5* (in press).

Potvin, A. R., Syndulko, K., Tourtellote, W. W., Lemon, J. A., & Potvin, J. H. (1980). Human neurologic function and the aging process. *Journal of the American Geriatrics Society, 28,* 1-9.

Pyykko, I., Jantti, P., & Aalto, H. (1990). Postural control in elderly subjects. *Age and Ageing, 19,* 215-221.

Richardson, J. K., Ching, C., & Hurvitz, E. A. (1992). The relationship between electomyographically documented peripheral neuropathy and falls. *Journal of the American Geriatrics Society, 40,* 1008-1012.

Rosenhall, U. (1973). Degenerative patterns in the aging human vestibular neuro-epithelia. *Acta Oto-laryngologica, 76,* 208-220.

Sabin, T. D. (1982). Biologic aspects of falls and mobility in the elderly. *Journal of the American Geriatrics Society, 30,* 51-58.

Sackley, C. M. (1991). Falls, sway, and symmetry of weight-bearing after stroke. *International Disabilities Studies, 13,* 1-4.

Schaumburg, H. H., Spencer, P. S., & Ochoa, J. (1983). The aging human peripheral nervous system. In R. Katzman & R. Terry (Eds.), *The neurology of aging.* Philadelphia, PA: F. A. Davis.

Scheigel, A. B. (1985). Falls, motor dysfunction, and correlative neurohistologic changes in the elderly. *Clinics in Geriatric Medicine, 1,* 671-676.

Schmidt, R. A. (1982). *Motor control and learning: A behavioral emphasis.* Champain, Il: Human Kinetics Publishers Inc.

Sheldon, J. H. (1960). On the natural history of falls in old age. *British Medical Journal, 2,* 1685-1690.

Sheldon, J. H. (1963). The effect of age on the control of sway. *Gerontology Clinics, 5,* 129-136.

Shumway-Cook, A., Anion, D., & Haller, S. (1988). Postural sway biofeedback: its effect on re-establishing stance stability in hemiplegic patients. *Archives of Physical Medicine and Rehabilitation, 69,* 395-400.

Shumway-Cook, A., & Horak, F. B. (1986). Assessing the influence of sensory interaction on balance. Suggestion from the field. *Physical Therapy, 66,* 1548-1550.

Stelmach, G. E., Phillips, J., DiFabio, R. P., & Teasdale, N. (1989). Age, functional postural reflexes, and voluntary sway. *Journal of Gerontology, 44,* B100-106.

Stelmach, G. E., & Worringham, C. J. (1985). Sensorimotor deficits related to postural stability. *Clinics in Geriatric Medicine, 1,* 679-694.

Stones, M. J., & Kozma, A. (1987). Balance and age in the sighted and blind. *Archives of Physical Medicine and Rehabilitation, 68,* 85-89.

Studenski, S., Duncan, P. W., & Chandler, J. (1991). Postural responses and effector factors in persons with unexplained falls: results and methodologic issues. *Journal of the American Geriatrics Society, 39,* 229-234.

Teasdale, N., Bard, C., LaRue, J., & Fleury, M. (1993). On the cognitive penetrability of posture control. *Experimental Aging Research, 19,* 1-13.

Teasdale, N., Stelmach, G. E., & Breunig, A. (1991). Postural sway characteristics of the elderly under normal and altered visual and support surface conditions. *Journals of Gerontology: Biological Sciences, 46,* 238-244.

Thilmann, A. F., Fellows, S. J., & Garms, E. (1990). Pathological stretch reflexes on the "good" side of hemiparetic patients. *Journal of Neurology, Neurosurgery, and Psychiatry, 53,* 208-214.

Thorstensson, A., Oddsson, L., & Carlson, H. (1985). Motor control of voluntary trunk movements in standing. *Acta Physiologica Scandinavica, 125,* 309-321.

Tinetti, M. E. (1986). Performance - oriented assessment of mobility problems in elderly patients. *Journal of the American Geriatrics Society, 34,* 119-126.

Topper, A. K., Maki, B. E., & Holliday, P. J. (1993). Are activity-based assessments of balance and gait in the elderly predictive of risk of falling and/or type of fall? *Journal of the American Geriatrics Society, 41,* 480-487.

Vandervoort, A. A., Chesworth, B. M, Cunningham, D. A., Paterson, D. H., Rechnitzer, P. A., & Koval, J. J. (1992). Age and sex effects on mobility of the human ankle. *Journals of Gerontology: Medical Sciences, 47,* 17-21.

Vandervoort, A. A., & Hayes, K. C. (1989). Plantarflexor muscle function in young and elderly women. *European Journal of Applied Physiology and Occupational Physiology, 58,* 389-394.

Vandervoort, A. A., Hill, K. M., Sandrin, M. L., & Vyse, V. M. (1990). Mobility impairment and falling in the elderly. *Physiotherapy Canada, 42,* 99-107.

Vandervoort, A. A., & McComas, A. J. (1986). Contractile changes in opposing muscles of the human ankle joint with aging. *Journal of Applied Physiology, 61,* 361-367.

Wall, J. C., & Ashburn, A. (1979). Assessment of gait disability in hemiplegics. Hemiplegic gait. *Scandinavian Journal of Rehabilitation Medicine, 11,* 95-103.

Wall, J. C., & Turnbull, G. I. (1986). Gait asymmetries in residual hemiplegia. *Archives of Physical Medicine and Rehabilitation, 67,* 550-553.

Whipple, R. H., Wolfson, L. I, & Amerman, P. M. (1987). The relationship of knee and ankle weakness to falls in nursing home residents: an isokinetic study. *Journal of the American Geriatrics Society, 35,* 13-20.

Wing, A. M., Goodrich, S., Virji-Babul, N., Jenner, J. R., & Clapp, S. (1993). Balance evaluation in hemiparetic stroke patients using lateral forces applied to the hip. *Archives of Physical Medicine and Rehabilitation, 74,* 292-299.

Winstein, C. J., Gardner, E. R., McNeal, D. R., Barto, P. S., Nicholson, D. E. (1989). Standing balance training: effect on balance and locomotion in hemiparetic adults. *Archives of Physical Medicine and Rehabilitation, 70,* 755-762.

Winter, D. A., Patla, A. E., Frank, J. S., & Wait, S. E. (1990). Biomechanical walking pattern changes in the fit and healthy elderly. *Physical Therapy, 70,* 340-347.

Woodhull, A. P., Maltrud, K., & Mello, B. L. (1985). Alignment of the human body in standing. *European Journal of Applied Physiology, 54,* 109-115.

Woodhull-McNeal, A. P. (1992). Changes in posture and balance with age. *Aging - Clinical and Experimental Research, 4,* 219-225.

Wolfson, L. I., Whipple, R., Amerman, P., Kaplan, J., & Kleinberg, A. (1985). Gait and balance in the elderly. Two functional capacities that link sensory and motor ability to falls. *Clinical and Geriatric Medicine, 1,* 649-655.

Wolfson, L. I., Whipple, R., Amerman, P., & Kleinberg, A. (1986). Stressing the postural response. A quantitative method for testing balance. *Journal of the American Geriatrics Society, 34,* 845-850.

Wolfson, L., Whipple, R., Derby, C. A., Amerman, P., Murphy, T., Tobin, J. N., & Nashner, L. (1992). A dynamic posturography study of balance in healthy elderly. *Neurology, 42,* 2069-2075.

Woollacott, M. H. (1993). Age-related changes in posture and movement. *Journal of Gerontology, 48* (Special Issue), 56-60.

Woollacott, M. H., Bonnet, M., & Yabe, K. (1984). Preparatory process for anticipatory postural adjustments: modulation of leg muscles reflex pathways during preparation for arm movements in standing man. *Experimental Brain Research, 55,* 263-271.

Woollacott, M. H., Shumway-Cook, A., & Nashner, L. (1982). Postural reflexes and aging. In J. A. Mortimer, F. J. Pirozzolo, & G. J. Maletta (Eds.), *The aging motor system.* New York: Praeger Publishers.

Woollacott, M. H., Shumway-Cook, A., & Nashner, L. M. (1986). Aging and posture control: Changes in sensory organization and muscular coordination. *International Journal of Aging and Human Development, 23,* 97-114.

Changes in sensory motor behavior in aging
A.-M. Ferrandez and N. Teasdale (Editors)

TESTS IN RODENTS FOR ASSESSING SENSORIMOTOR PERFORMANCE DURING AGING

Bernhard JÄNICKE and Helmut COPER

Free University of Berlin

Abstract

One characteristic feature of aging in all organisms is the continuous loss of adaptability to environmental perturbations. It is the result of a loss of control over the harmonization of a number of individual reactions which should be in a reciprocally interdependent state ensuring homeostasis. This chapter includes six sections which describe some sensorimotor performance tests that can be used to closely analyze age-related diminishing adaptability. The subareas which regulate and safeguard the functional systems for which the tests are models, are also presented. Procedures for measuring reflexes, muscular strength, motor activity and coordination are described by means of examples, in addition to some methods which are suitable for assessing motor behavior associated with cognitive abilities or the expression of emotional reactions. The concluding chapter discusses the possibilities and limitations of the tests for describing the scheme of stimulus-effect relationships (stimulus perception – central nervous system processing of information – responsiveness of the target organ) and stresses the need for an extensive battery of tests.

Key words: Sensorimotor performance tests, age-related changes, rodents.

Correspondence should be sent to Dr. Bernhard Jänicke, Free University of Berlin, Institute for Neuropsychopharmacology, Ulmenallee 30, 14050 Berlin, Germany.

INTRODUCTION

Much of the credit for the increase in our knowledge of aging in its many façets over the last three decades goes to experimental gerontology (Schneider & Rowe, 1990; Arking, 1991). Because aging is a fundamental process and takes place according to the same principles and via similar mechanisms – albeit expressed in different ways – in all higher creatures, observations concerning age-related changes made in animal experiments can be extrapolated, at least in part, to humans (Bartus et al., 1983; Hazzard, 1991; Johnson, 1978). Some examples are the characteristic features of aging, such as muscle stiffness, flexed posture, tremor, and diminished cognitive performance. A variety of methods have been developed to study age-related changes, although their validity and reliability have not always been confirmed. Regardless of this, aging is in particular also part of the biography of an individual, which is influenced by fewer variables in the laboratory animal than in human beings. Moreover, the differentiation of normal aging phenomena from pathological processes is difficult even in the case of humans. Because of this, many questions can frequently be answered in experimental studies only by simplifying the complexity of multiregulated and controlled functional systems and constructing suitable models.

"The best material model for a cat is another, or preferably the same cat", is how Rosenblüth and Wiener (1945) put it, deliberately trivially, but very aptly. At the heart of this trenchant statement is the fact that models are always positioned between an unavoidable remoteness from an original, and the intended approximation of that original, despite having to contain and reproduce its essential characteristics. What is more, however, models are always also "molded" theories, since the decision regarding which properties are important and which are less important is based on abstract ideas about what distinguishes and characterizes the original. As an example, the assumption that gerontologic questions can be answered by specific manipulations using young animals and that age-related declines in performance can be detected in a few selected study designs has proved to be erroneous. The sensorimotor system itself is a good example: motor performance diminishes by degrees and in different ways during aging – especially when greater demands are made – and findings concerning impaired function are more likely to be due to chance when too few expressions of behavior are studied. It has now been adequately documented for various animal species that ontogenetically early modes of behavior such as spontaneous activity and swimming movements are retained up to an advanced age. In contrast, coordination in rats, for example, has already greatly diminished at the start of the third phase of life (Figure 1; Ingram, 1988; Jänicke et al., 1983; Wallace et al., 1980). The results of pertinent studies, then, depend both on the function measured and the magnitude of the demand made, as well as on the age of the animals. Whether lifespan studies with the possibility of detecting progressive changes and the effects of training can produce basically different results than the conventional cross-sectional studies remains an open question. The former could at the least indicate the extent to which a temporal shift in loss of function occurs at a later time as a result of certain demands on performance and experience gained.

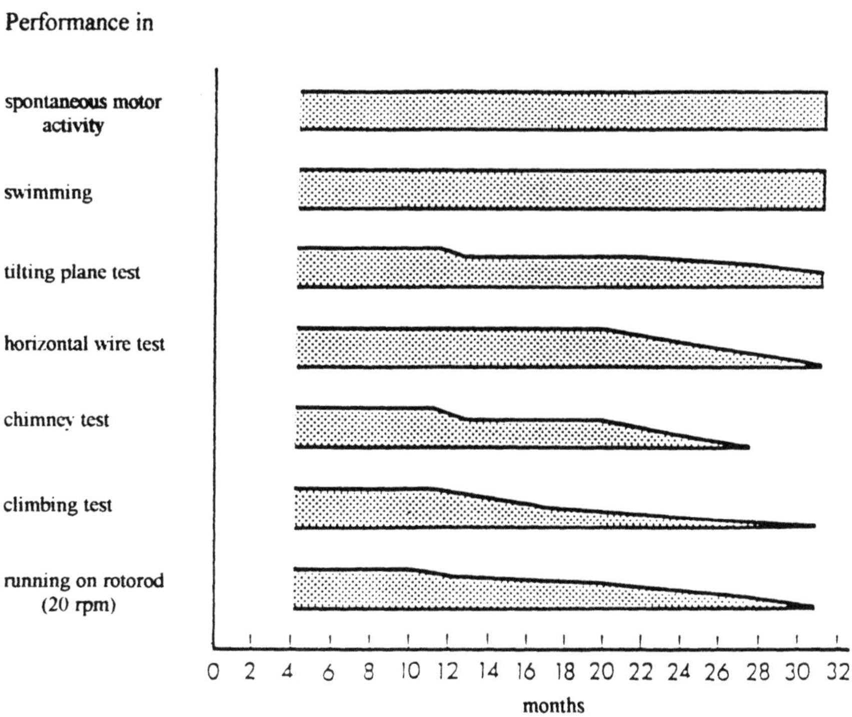

FIGURE 1. Changes in motor performance with increasing age in reference to performance of 4-month-old rats.

Sensorimotor performance is the result of a stimulus-effect relationship. Age-correlated changes due to sensory impairment which may have consequences on the response are possible even at the level of stimulus perception (reduction of visual, auditory, or taste sensitivity, reduction of tactile perception or of temperature oscillation, etc.). Furthermore, in old age the efficiency of the target organ may be diminished in the case of a reduced response to a standardized stimulus – as measured against the maximally possible performance. An example is the impairment of motor functions with a change in bodily posture control and/or with a reduction in the range of movement in the joints following muscular atrophy due to a loss of motoneurons and terminal axons (Davis, 1990; Lexell, 1993). Finally, central nervous system processing of the stimulus may be disturbed, resulting in an altered or even absent response. In this case, a stimulus could lead to a weakened response despite increased sensitivity of the functional system. Such dissociation in information processing is not infrequent in advanced age. It is well-documented that the loss of heat through increased perspiration as a response to thermal stimuli is much reduced in people over 70 years of age compared to younger age groups (Foster et al., 1976). Vascular responses – particularly vasoconstriction in response to a cold stimulus – are also affected (Collins & Exton-Smith, 1983; Wagner & Horvath, 1985). In studies conducted in Great Britain, around 10% of the population above 65 years of age had a slightly reduced body temperature. Remarkably, none of those affected felt any discomfort, and consequently, were not provoked into thermoregulatory behavior (Fox et al., 1973). Parallel to this, however, is the known phenomenon that the elderly can feel as if they are "freezing" at an ambient temperature of more than 20° C and a normal body temperature. Regulation of the water and electrolyte balance also becomes more unstable in old age. The central recognition of nominal value deviations of the fluid volume between the cells and the content of salts is apparently impaired in this phase of life, so that the "thirst" symptom – which normally develops as a result of insufficient fluid intake – fails to occur (Hess, 1987). Dissociation of the stimulus, its processing and appraisal, and its response can also occur in old animals – especially with regard to sensorimotor performance – and can be determined by means of suitable methods. Examples are tests of classical and instrumental conditioning, which can be constructed with varying complexity and multistage stimulus-response relationships (Bartus et al., 1983; Burwell & Gallagher, 1993; Coper et al., 1986; Ingram, 1988; Wallace et al., 1980).

The purpose of this overview is not only to list some tests for assessing sensorimotor performance during aging, but also to provide a critical assessment of the possibilities and limitations of their use as well as their validity. Study conditions and influencing factors such as different strains of rats or mice, housing conditions, pecking order, stress factors, circadian rhythm, handling, etc. – some of which can considerably alter the results – have been ignored, but they must be considered in experiment planning and interpreted separately (Campbell, 1982; Cheal et al., 1987; Ingram et al., 1981; Jänicke et al., 1983; Spangler et al., 1994; Wallace et al., 1980). Also ignored are the causes of age-specific differences found in the various tests, i.e., interindividual variability due to individual aging courses. Because it is relatively rare for biochemical, morphological, and behavior-biology studies to be conducted in parallel, and for the results to be related with each other, only plausible assumptions and not clearly-founded statements can usually be made about the disturbance underlying the loss of function. With these limitations in mind, a selection of tests used in experimental gerontology are described, together with the functional system for which they are models. It should be considered, however, that the methods used in the tests presented have been more or less modified by individual working groups, which often complicates comparability. Using a battery of tests could limit this restriction and enable an overall assessment. The criteria used to select the tests were published data and evidence of the validity and reliability of the main results. One potential disadvantage of a test battery is the risk of interference between individual test performance. However, false interpretations can be corrected by multiple biostatistics. Moreover, some of the tests referred to have as yet only been employed during the adult and juvenile life phases, although we assume that they are also suitable for use during the senile phase.

REFLEXES

Motor reflexes are stereotypical responses of the central nervous system to sensory stimuli, enabling correct voluntary movements. In the case of polysynaptic motor reflexes, several neurons are linked in series, the motoneuron being the last link in the chain. The behavioral response is determined by the electromyographically measurable activity of the corresponding muscle, the muscular strength, and the neuromuscular innervation – variables which must be studied separately when examining even simple reflex responses (Wecker & Ison, 1986). Relevant findings complemented by biochemical and morphological data have shown age-related changes both in the reflex arc and muscular tone (Larsson & Ansved, 1988; Ossowska et al., 1992).

1. Foot reflex

The functional ability of the hind-leg reflexes and any limitation in old age can readily be determined without any technical measures. The examiner places a rat backwards at the edge of a table, and with his/her index finger, pulls the right and then left hind paw from under the animal. Whether the animal pulls the paw back immediately or reacts after a delay is recorded. Marshall (1982) observed a reduced reflex response of the hind legs in male Fischer 344 rats after the age of 24 months.

2. Orienting reactions and placing reflexes

2.1. Righting reflex. In order to study the righting reflex, a rat is placed on its back on a tabletop and the time taken for the animal to right itself through 180° is measured. To study the air-righting reflex, the rat is dropped in the supine position from approximately 40 – 50 cm above a padded tabletop and the righting reaction is observed. This method, described by Altman and Sudarshan (1975), has been employed almost exclusively in the juvenile life phase of rats and mice. To date, further findings of cross-sectional analyses are apparently not available.

2.2. Visual and contact placing. In these tests, the response to different stimulus modalities is determined by means of the behavioral reaction. Directional responses are triggered by touching the vibrissae or by means of olfactory stimuli (Marshall, 1982). No more than marginal age-related changes have consistently been found in different animal species for both visual and contact placing reflexes (Campbell, 1982; Levine et al., 1987; Markowska et al., 1989).

2.3. Negative geotaxis. For the negative geotaxis test, a rat is placed on a platform (41 × 27 cm) with a slightly roughened surface. The platform is then swung through 90° in 3 to 4 seconds and the angle at which the animal loses its balance is measured. This tilting plane test examines the sensory system, and above all, the intactness of the placing reflexes. The results of various cross-sectional and life-span studies have shown that animals display adequate responses in this test up to an advanced age (Ingram et al., 1982; Jänicke et al., 1983; Jänicke & Coper, 1993).

2.4. Startle reflex. The startle reflex examines the response of a laboratory animal to a defined acoustic, optical, or painful stimulus

(Campbell & Gaddy, 1987; Hoffman & Wible, 1970; Wecker & Ison, 1986). The tests can distinguish between an altered sensitivity threshold and responsiveness. In general, this response in rats decreases with advancing age irrespective of the trigger stimulus. The slower response to an acoustic stimulus, for example, can be explained by the change in the sensory perception threshold. In the case of painful stimulation, the target organ also responds more slowly (Campbell, 1982; Krauter et al., 1981).

3. Blinking reflex

The classical conditioning test, the blinking reflex, must also be mentioned. An age-dependent reduction in the blinking reflex response has been observed in rats, rabbits, and cats (Harrison & Buchwald, 1983; Powell et al., 1984; Weiss & Thompson, 1992). The authors have attributed the reduction to diminished neuronal processing in the hippocampus and cerebellum and not to any loss of efficiency of the peripheral sensory system. Studies in humans have produced very similar results (Kimble & Pennypacker, 1963). Generally speaking, an advantage offered by reflex response testing is that the somatomotor response elicited is relatively independent of non-specific influences such as spontaneous activity and higher learning and memory performance. Motivation and emotion towards unconditioned stimuli also tend not to be counted among the causes of the observed age-dependent reduction in classical conditioning. Changes in central "associative" processing are more likely to be responsible for the aging process (Powell et al., 1984).

MUSCULAR STRENGTH

1. Traction test

The traction test measures the length of time that rats can keep their front paws on a horizontally suspended bar. The rod is usually fixed approximately 40 to 50 cm above the tabletop so that when the animal falls, it has sufficient time and space to land on its feet by means of the righting reflex. The motivation to hang onto the bar stems from the aversion to falling. Pellis et al. (1987) described a modified version of this test in which two bars (2 mm) are fixed one above the other with a distance of 8 cm between them. The lower bar is fixed approximately 20 cm above the tabletop. Again, the time during which a laboratory animal clings to the bars is measured. A distinct age-related trend can be observed in this standard test which determines muscular strength. Performance falls steeply after the twentieth month of life and then continues to fall, albeit more slowly (Campbell, 1982; Jänicke et al., 1983; Joseph et al., 1983). The assumption that higher body weight impairs the performance of older animals to a relatively great extent has not been confirmed (Gallagher & Burwell, 1989). If, in the traction test, the animal is allowed to use its tail as well as its front paws to hang onto the bar, old rats and mice rarely make use of this aid (cf. Altman & Sudarshan, 1975; Miquel & Blasco, 1978). The latter authors attributed the age-related difference to reduced elasticity of the fibers of the tail muscles. A disturbance in sensorimotor regulation by the CNS and a reduced spinal stimulus response have also been suggested, however (Welford, 1977). Neither apparently leads to a life-threatening impairment.

The finding of an age-related reduction in muscular strength is well-documented for various animal species as well as for humans (Shock & Norris, 1970; Welford, 1977). The reduction in performance is not uniform, however. For instance, fast-responding muscles lose their strength more quickly (Kovanen et al., 1987; Newton & Yemm, 1986). Several functional levels are involved in the differentiated reduction of muscular strength in old age. Both muscle mass and the number of motor end-plates decrease (Rosenheimer & Smith, 1985), as do glucose utilization by the cells, phosphocreatine content, some enzyme activities, and the number of muscle proteins (Steinhagen-Thiessen et al., 1981). However, this view is opposed by findings which show that deficits in performance should be interpreted as poor spinal and supraspinal regulation of neuromuscular coordination rather than as peripheral myopathological events (Rogers, 1988). Some light has been shed on this subject by the results of recent animal experiments involving the combined use of mechanomyographic and electromyographic methods. Muscular strength can be determined directly by measuring the resistance to consecutive passive bending and stretching in the hind ankle joint (mechanomyogram, MMG) (Ossowska et al., 1992). Recordings of the electromyographic (EMG) activity of the gastrocnemius muscle and the anterior tibial muscle in the old rat sometimes show not only normal values as in young animals, but also two types of EMG activity: (1) a dense, low-amplitude tonic activity (70 – 200 μV), and (2) a rare activity with high-amplitude spikes (880 - 2000 μV) (Ossowska et al., 1992). However, there is considerable interindividual heterogeneity. In parallel with this, striking neuromuscular changes were observed at the neuromorphological level in rats for which the second EMG pattern was typical.

This finding is consistent with Flood and Coleman's (1988) results, which showed an age-related, characteristic loss in the plasticity of denervation and innervation processes. Other studies in old rats have shown that muscular weakness and stiffness are possibly of different origins. Muscular weakness is probably the consequence of neurogenic muscular atrophy (damage to motoneurons), while muscular stiffness is caused by lesions in the cerebral and/or spinal sections of the motor system. Even the simple models so far described clearly illustrate that age-related changes can take place in different subareas of a regulated functional system, the functional ability of which, with its many safeguards, can nevertheless remain more or less intact even when individual control elements are no longer fully active or go faulty. However, the capacity for compensation is no longer sufficient under increased demands. The organism becomes more susceptible to disturbances. This phenomenon is also the determining feature in all of the mentioned experiments for assessing sensorimotor performance during aging.

MOTOR ACTIVITY

1. Spontaneous activity

Investigation of spontaneous locomotor activity is a prerequisite for any study examining sensorimotor performance. Low motor activity, for example, would hardly lead to normal exploration of the new environment in the open field test, while it would simulate good retrieval performance in the passive avoidance test, etc. It is therefore vital to determine whether spontaneous motor activity decreases as a whole with increasing age, in accordance with the general slowing down of movements. There are findings which suggest that the reduction in spontaneous activity is caused by disturbances in the locomotor system and its innervation (degeneration and loss of muscle fibers, decreased neuromuscular transmission, altered metabolic efficiency, etc.) (Brown, 1987; Larsson & Ansved, 1988; Steinhagen-Thiessen et al., 1981; Suominen & Kovanen, 1988). The peripheral structure changes are not uniform, however, and they are not so serious as to affect spontaneous activity significantly. Just how important central regulation is in the slowing down of locomotor processes remains to be clarified in detail. The test situation in which an internal motivation to act can affect the result must also be considered.

The spontaneous activity of an individual animal is measured by recording electrical capacity fluctuations caused by the movements of the laboratory animal. Another method is the three-dimensional control of movement using light barriers. The measurements are usually performed after an adaptation period of at least five days. The variables measured include not only the resting and activity phases, but also – depending on available equipment – righting, comfort behavior, and the spatial and temporal distribution pattern of motor activities. Since, regardless of the measuring technique, the animal strain used, the sex, and the test condition variables (body weight, cage size, adaptation period of laboratory animals, duration of test phase, external stimuli during the test, etc.) varied across the experiments reported, it is not surprising that the data published on age-related changes of spontaneous motor activity do not agree on every detail. It is generally undisputed, however, that spontaneous circadian activity as a whole hardly changes up to the onset of aging (beyond the 50% survival limit) and that it decreases only slowly even then, although this does not mean that some expressions of behavior do not display deviations. Sustained stability of the basic activity required to fulfill vital needs is biologically useful and desirable (Figure 2; Jänicke et al. 1983; Willig et al., 1987).

2. Reactive motor activity

The open field test is suitable for assessing motor activity as a reaction to an unknown environment, i.e., locomotion motivated by exploration (Archer, 1973; Roth & Katz, 1979). The field (1 m^2) is divided into 16 squares and surrounded by walls approximately 18 cm high. Scientists have varied the test period as well as the illumination of the field in their experiments. The illumination ranges from white light of varying intensity during the light phase, to red light – to which the rats are insensitive – during the night phase. Studying rats under red light during the dark phase has the advantage that the rats' activity phase is also taken into account (Jänicke & Coper, 1993). The test animal is placed in a corner square with its head in the corner. The main variables recorded during the next 5 or 10 minutes are: (a) latency (the time taken to leave the starting square), (b) crossings (number of times the line of a square is crossed with all 4 legs), (c) rearings (number of times the animal stands on its hind legs), (d) grooming (frequency of grooming activity), and (e) number of defecation boli.

In agreement with the results of other authors (Ammassari-Teule et al., 1994; Nabeshima et al., 1993), we found (Jänicke & Coper, 1991) a decrease in the rearing frequency, an expression of directed exploration in the adult phase of life, although locomotion only decreases nearing the senile phase; accordingly, a dissociation exists between exploration and general motor activity (locomotion). Grooming – which is often interpreted as displacement behavior – and latency in the starting square – likewise associated with emotionality – remain relatively unchanged (Ammassari-Teule et al., 1994). Defecation as an expression of emotional reactivity increases distinctly in older rats (cf. Dean et al., 1981; Ingram & Reynolds, 1986; Ingram, 1988; Lhotellier et al., 1993; Nabeshima et al., 1993).

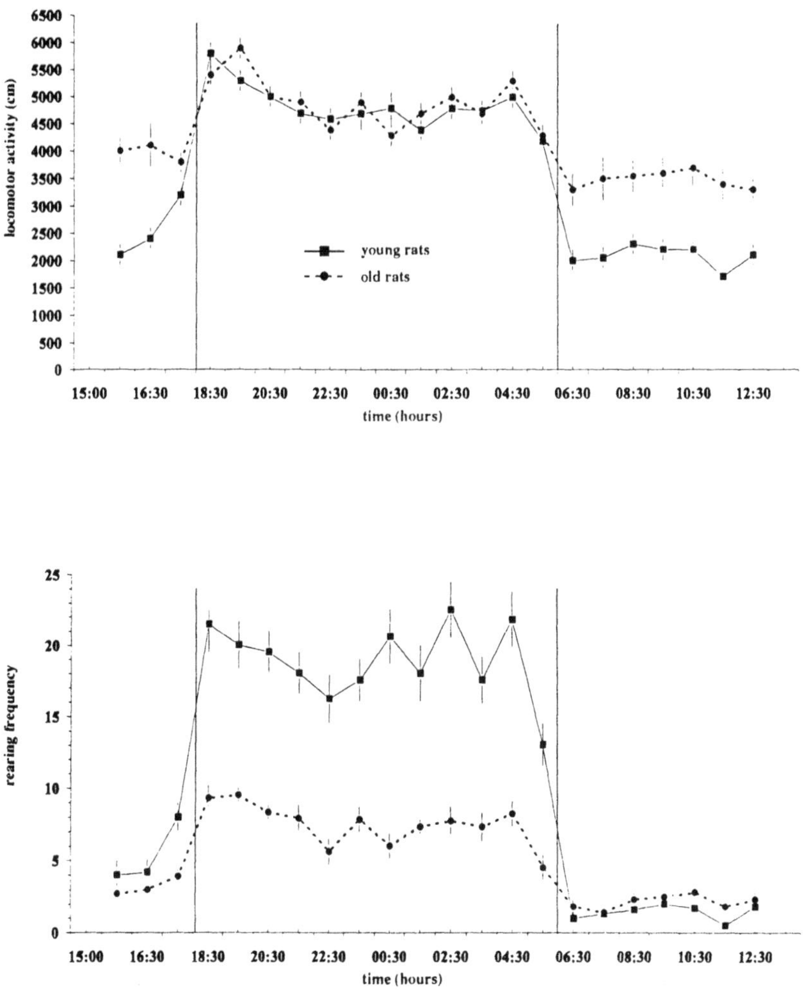

FIGURE 2. Spontaneous motor activity in young and old rats. Mean + sd, n = 15 per group.

Further knowledge can be gained if the conditions under which the rats are kept are designed to stimulate activity, for example, by introducing different objects or spatial structures. The studies by Willig et al. (1987) showed that the exploration activity of old rats is reduced to a lesser extent in a diversified than in a monotonous environment. Structured spaces are created by making use of height, so that vertical movements (climbing, hanging, jumping) are also possible. This can provide additional information about reactions to stimulating situations, which again presuppose an "internal motivation to act" in the animal on the one hand, and demand "cognitive knowledge" with respect to the appropriate qualitative behavioral procedure, the "how", on the other. In this second-floor test, as it is called, the laboratory animal is offered more possibilities to explore than in a simple open field. The design consists of two floors above each other connected via a staircase, which makes access to the two areas difficult. On the other hand, the staircase is both attractive and frightening to the rats. If an animal is placed in the middle of the upper floor, it often spends time in the vicinity of the staircase, but rarely on it (Wolffgramm & Heyne, in preparation). No comparative studies between young and old animals have been conducted so far, and the test is mentioned only to indicate the direction in which the tools for measuring motor activity have developed.

3. Swimming

Marshall and Berrios (1979) determined the swimming movements (vigor) and head position (success) of old rats during a 15-minute test. The test calls for a water temperature of approximately 23° C and a tank so deep that the animal is unable to touch the floor. The investigators have found a significant difference in the swimming behavior of 24- and 27-month-old male Fischer 344 rats. However, this result may be specifically influenced by the strain since in several studies it was demonstrated that especially the speed of swimming movements diminished with age; nevertheless, elementary sequential movements remain largely effective up to an advanced age (Gower & Lamberty, 1993; Jänicke et al., 1983; Kay & Birren, 1958; Lamberty & Gower, 1990).

4. Water avoidance test

This test is suitable not only for observing basic locomotor behavior, but also and above all for studying appropriate reactions. According to Wilcock (1972), the situation of falling into water by accident and without previous experience is aversive to the rat. The author used the situation-related behavioral reaction of the animal – that of leaving the water as quickly as possible – as a measure of its "vigilance". The test involves dropping an unaware animal into the middle of a cylindrical water-filled basin from a closed box. The animal then has an equal chance of reaching any part of the edge of the basin. This possibility of moving in any direction avoids the occurrence of inappropriate reactions by the animal, e.g., despair behavior. The test does not, however, take endurance or swimming postures into account. No results are available as yet from studies employing this method in rats or mice of different ages. One advantage it offers over the Morris water maze (Morris, 1981), in which reactions are similarly motivated by escape behavior, is that it does not require spatial learning with proximal or distal cues.

To summarize, it can be concluded that the test situations provoke simple coordinated movements in all experiments. Exploration and locomotion can be differentiated and emotional reactivity can be assessed in the open field test. The situation in the swimming test is basically aversive; this test allows for observation not only of the animals' general ability to swim, but also of the effectiveness of the corresponding performance during the test period in a cross-sectional study. The animal's reaction in the water avoidance test reflects the ability to overcome an aversive situation appropriately, quickly, and effectively.

MOTOR COORDINATION

Greater demands on the motor system, i.e., ones involving complex coordinated movements, require the use of several variables such as muscle strength, endurance, coordination, and balance, and therefore also involve different sensory modalities including tactile, proprioceptive, optic, acoustic, olfactory, and vestibulo-motor perception. Even if it is not always possible to make a satisfactory, differentiated assessment of the individual actions which form a part of a complex movement, or of performance deficits, a comparison of the results of these tests for individuals of different ages shows that the complexity of the action demanded is an important factor in the motor performance decline. It is well-documented especially in humans that with advancing age, increased demands on the motor system uncover deficits in performance (Light & Spirduso, 1990; Potvin et al., 1980; Sabin, 1982; Tinetti & Ginter, 1988). Coordination of movements and balance regulation are controlled primarily by the integrative functions of central nervous system. Above all, the basal ganglions and the cerebellum are responsible for these motor functions. Morphometric and electrophysiological investigations have discovered age-dependent changes in Purkinje cells, climbing fibers, parallel fibers, or interneurons, although these results can only be partially correlated to impaired behavior (Bickford-Wimer et al., 1991; Parfitt & Bickford-Wimer, 1990; Rogers et al., 1980; Rogers, 1988).

1. Climbing test

The task for the rat is to climb vertically up a 60-cm long and 7-cm wide metal ladder (distance between rungs 4 to 6 cm). The goal at the upper end of the ladder is a dark chamber in which the animal finds a resting place. To motivate the animals to climb, a box containing chaff into which they would otherwise fall, is placed beneath it. The animals are allowed to make between 2 and 3 dummy runs before the actual test run. The variables recorded are the climbing time and the number of times the rungs are grasped with the front paws. The stepping speed and stepping width can be calculated from these data (Jänicke et al., 1983). Old animals require significantly more time (statistically speaking) than younger animals to reach the goal. Fluid, coordinated movements of front and hind feet are hardly present anymore in old animals, they miss the rungs relatively often, and their stepping width and speed are lower. They also take rests more frequently (Jänicke et al., 1983).

2. Chimney test

This test was originally introduced by Boissier et al. (1960): the animal must climb backwards out of a 30-cm long vertical tube. The tube is grooved inside to provide grip. The variable measured is the time taken by the rat to complete the climb, i.e., until its head appears outside the opening. Even 12-month-old animals need more time to complete the exercise than 4-month-old ones. The time taken increases with advancing age (Jänicke et al., 1983). As in the previous test, the results indicate that the movements of old rats do not become more precise despite the lower speed. Brinley observed such a dissociation of speed and coordination in elderly humans as early as 1965. The mistakes made even on slow movements were explained by the focusing of attention on the goal, to the exclusion of everything else. The different age-related time courses of the speed factor – which becomes increasingly slower in old age – and the power factor – which remains stable for a long time – have also discussed in this connection (Welford, 1988). According to this interpretation, the CNS integration of stored and current information is impaired, i.e., the envisaged goal is reached, but psychomotor speed is reduced (Buckles, 1993; Welford, 1984).

3. Running belt test

When testing running performance on a running belt, the result is determined not only by endurance and coordinating ability, but also by motivation and stress. A 30 × 10 cm running belt is placed in a cage (60 × 40 × 40 cm). To discourage the animal from leaving the running belt, the rest of the floor area is fitted with electrified rods. The speed is increased to 12 m/min after a 3-minute adaptation phase (3.6 m/min). The test is discontinued when the rat stops running or when the set time limit of 10 minutes has been achieved. Although different variations of the method have been used, it is evident that running performance diminishes continuously with age (Hofecker et al., 1981; Jänicke & Coper, 1994). Despite the fact that this test can certainly reveal more than just age-typical weaknesses in physiological dynamic muscle activity and in the ability to coordinate movements, its main limitation is the difficulty distinguishing between animals which are unwilling to run or learn and those with diminished performance. In addition, the ability to perform the task successfully in adult rats differs considerably from individual to individual.

4. Rotorod test

This test examines the ability of rats to run on a rod against the direction of rotation. Avoiding a fall is again the motivation for keeping step and balance with the speed of rotation. If the number of rotations is increased by steps, e.g., from 10 rotations per minute to 20 and then to 40, the overtaxation limit can also be determined (principle of "testing the limits"). The number of attempts required by the animal to remain on the rod for 1 or 2 minutes can be used as the criterion of success. However, the number of animals which meet the demands made on them after three or five attempts can also be used. Whatever criteria are used in the test, performance decreases steadily with increasing age. As in the climbing test, older animals achieve the goal, but require far more attempts to do so. Also striking is the fact that old rats display uncoordinated running movements much more frequently, sometimes running too fast and sometimes too slow. The consequence of such "mistakes" is a fall from the rod. On the other hand, they can largely make up for initial deficits in performance by means of practice. Without continuous practice, however, the training effect is lost again much more quickly than in younger animals (Campbell, 1982; Cheal et al., 1987; Ingram et al., 1981; Jänicke et al., 1983; Wallace et al., 1980).

In general, tests involving motor coordination tasks are suitable for studying individual motor activities of varying complexity. In the climbing test, it is a directed, negative geotactic forwards movement which requires strength, but which also takes individual speed into account. In the chimney test, the directed, negative geotactic backwards movement is facilitated by the fact that the entire body can be used as a prop. The running belt test measures endurance more than anything else, speed being determined by the task. In the rotorod test also, the speed of movement is dictated by the rotation rate of the rod, and the variables recorded are the balancing reactions, continuity of coordinated running actions, and endurance. Attention, concentration, motivation, and emotional reactivity also play a role in these tests. However, these factors can be controlled relatively well by accustoming the animals to the test situation. Comparable studies in humans have produced very similar results (Ferrandez et al., 1988; Woollacott et al., 1993).

MOTOR BEHAVIOR IN ASSOCIATION WITH COGNITIVE DEMANDS

In the absence of other means of communication, conclusions about the cognitive abilities of an animal can be drawn from directed, species-specific motor behavior. However, decreases in mental performance can also manifest themselves as altered motor reactions. The influence of aging on learning and memory performance has been investigated in a wide variety of study designs. A plethora of reproducible age-typical learning deficits have been demonstrated in animals, e.g., impaired long-term/short-term memory, reduced ability to adapt, deal simultaneously with multipart tasks, ignore disturbing stimuli, and associate stimulus-response relations on delayed presentation, and a poor ability to respond appropriately to a certain temporal sequence of stimulus-response patterns without additional external cues or impaired learning in spatial tasks (Gallagher & Pelleymounter, 1988; Kubanis & Zornetzer, 1981; Ordy et al., 1988; Winocur, 1988). In order for the results obtained in these learning tests to be interpreted properly, it is important to determine the extent to which sensorimotor abilities are responsible for the deficits in performance, as the interactions between cognitive and motor functions are well-documented in both humans and animals (Salthouse, 1988). The experiments which include a training program merit particular mention here, since impaired performance of subfunctions can often be compensated for by training. It is worth noting that no agreement between sensorimotor weakness and learning deficits was found in several studies (Gallagher & Burwell, 1989; Gower & Lamberty, 1993; Markowska et al., 1989).

A number of published studies have demonstrated that, although peripheral perception of stimuli decreases with age, diminishing cognitive performance is more attributable to subsequent functions of central nervous system processing such as disturbed selection, assessment, conversion to a response, or storage and reduced retrieval of the stored information, i.e., impaired memory performance. Emotional reactivity and locomotor activity are also regarded as secondary factors in age-related learning deficits (Campbell et al., 1980; Gower & Lamberty, 1993; Kubanis & Zornetzer, 1981). The same holds true for reaction time latency. It has been shown that the increase in latency is age-dependent. However, the start and rate of progress are individually quite different. At least concerning performance in classical conditioning, it is evident that accuracy increases with training, even in old age, although it does not seem to be correlated to changes in reaction time (Burwell & Gallagher, 1993). A selection of learning experiments is presented below in an attempt to show which tests can serve as models and which factors can modify the results.

1. Passive avoidance test

In the passive avoidance test, the laboratory animal (rat or mouse) learns to avoid an unpleasant stimulus by inhibiting locomotion and exploration, which is why it is also called the inhibitory avoidance test. At the start of the test, a rat, for example, is placed in a brightly lit compartment of the test box. As is the habit with rats, the animal very quickly moves into an adjacent, dark compartment where it receives an unavoidable shock. In the test phase – usually 24 hours later – one notes whether the animal reenters the dark compartment. Several authors have consistently demonstrated an age-related decrease in response in this test in different animal species. In general, cognitive performance (learning and memory) declines with increasing latency between acquisition and retrieval. This relationship is worse in old animals (Bartus et al., 1983; Markowska et al. 1989). The assumption, for example, that old rats are at an advantage concerning the avoidance reaction because of diverse lesions of the hind feet proved to be unfounded. This is supported by the fact that, at the start of the test, old animals enter the dark compartment just as quickly as younger animals. Nor is there any great difference in the pain threshold between younger and old animals of the same species (Lippa et al., 1980). It seems instead that information-handling processes are disturbed in old animals, although hypotheses diverge as to which subprocess is affected. One hypothesis is that old animals are less effective at consolidating what has been learned (Weiskrantz & Warrington, 1975), and another is that the disturbing factor is great perseverance, i.e., excessive resistance to extinction of what has been learned and to the integration of new information (Lapsley & Enright, 1983). Greater agreement exists, however, with respect to neuroanatomic and neurofunctional, age-typical changes in the hippocampus, which are directly associated with learning deficits (Landfield, 1988; Winocur, 1988).

2. Active avoidance test

In the active avoidance test, the test animal learns to avoid an aversive stimulus by changing locations. At the start of the test, a rat is placed in one of two compartments. After habituation, a stimulus (light or tone) is presented for a fixed period of time and followed by electrical stimulation of the paws. The rat learns to avoid the shock by moving into the adjacent compartment upon the appearance of the conditioned stimulus. Even if it is justifiable to assume that sensorimotor abilities play a major role in this learning test and that older animals with gradually diminishing performance are therefore non-specifically disadvantaged, the results still show an increase in the avoidance rate during the test, which is, however, lower in older than in younger animals. It appears that the factors responsible for the poorer performance on the test are not lower sensitivity to the painful stimulus, slower movement or reduced attentiveness, a lower arousal level, or faster habituation, but more limited acquisition and retention of experience gained (Kubanis & Zornetzer, 1981; Lamberty & Gower, 1990).

3. FR/DRL relearning test

One of the typical learning weaknesses in old age is the inability to integrate new events into existing memory. This inability can be studied by means of a conditioning scheme in which two tasks succeed each other, i.e., the test animal must relearn its behavioral response. To increase motivation, the rat's weight is reduced by around 20% before the test. In the test, a rat must first learn a fixed ratio of 10 (FR 10; every tenth lever pressing is rewarded). Subsequently, it learns a DRL (differential reinforcement of low rates) program: after pressing a lever and being rewarded, the rat must wait for at least 10 seconds, which is indicated by a discriminatory stimulus (light) before pressing the lever again, otherwise the interval is extended by another 10 seconds. The test criterion is to obtain an average of 90 pellets/30 minutes on two consecutive days. The variables measured are acquisition time, response rate, and reinforcement frequency within the test period. Results of cross-sectional studies of rats have shown that the FR 10 is learned in the same amount of time by 30-month-old and 4-month-old female animals (Jänicke & Schulze, 1987). However, to adapt successfully to the DRL, 20-month-old rats already required much more time than young rats, while 30-month-old animals did not fulfill the criterion even after 3 weeks. Apparently, their reactions did not reach the same level of efficiency as that of younger animals (Figure 3). Structural rigidity is held responsible for the inferior performance of older animals (Lapsley & Enright, 1983) in responding appropriately to situational stimuli, and in

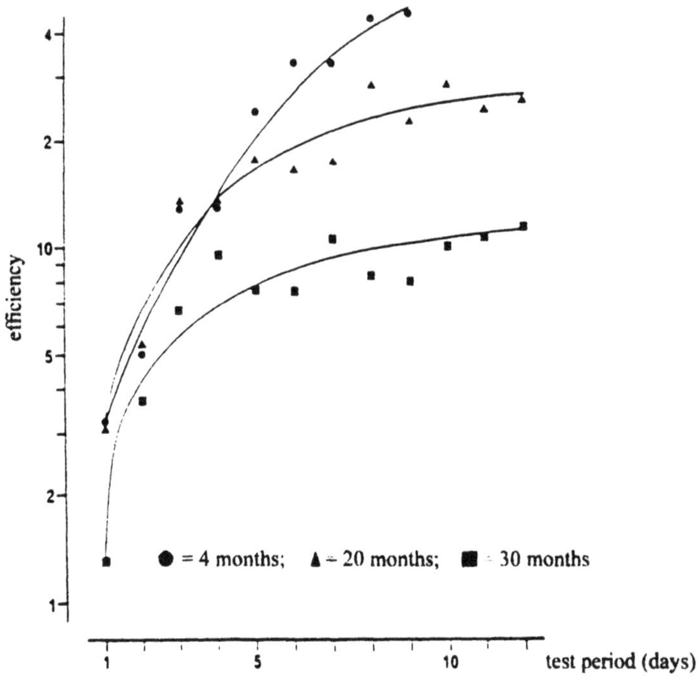

$$\text{efficiency} = \frac{\text{reinforcement} \times \text{reinforcement}}{\text{response} - \text{rate}}$$

FIGURE 3. Influence of age on efficiency of FR/DRL-learning in rats of different ages, n = 10 per group.

modifying previously valid stimulus-response relations. Thompson et al. (1986) speak of response flexibility – the cognitive ability to discard unsuccessful responses and to attempt new responses until an appropriate solution is found.

Another interpretation also includes the state of vigilance. Vigilance is defined as a state of alertness in which an appropriate response is made to a stimulus (Parasuraman, 1984). The light in the DRL phase, indicating the active lever, is such a stimulus, to which the rat must respond. Old rats' poorer efficiency in the DRL test phase may therefore be due to lower vigilance, which hinders them from responding to the discriminatory stimulus. This is further confirmation that diminished cognitive performance in old individuals is not caused by declining sensorimotor performance, but by a susceptibility of neuronal processing to disturbances (Gower & Lamberty, 1993; Kubanis & Zornetzer, 1981).

4. Maze tests

The tests which make greater demands on cognitive abilities in relation to the tests mentioned in sections 6.1 to 6.3 include those which examine the processing of spatial information, e.g., position discrimination tests and different types of mazes (Jucker et al., 1988; Schulze et al., 1988; Stone, 1929). In these tests, a laboratory animal must leave a starting point and find its way through, for example, a system of corridors or tubes to the exit, where it is rewarded with food. Among others, Schulze et al. (1988) conducted an extensive number of experiments of this type. The tube maze they used consisted of six right/left decisions. A computer-assisted recorder was employed to measure, inter alia, the total time taken to reach the exit, the frequency of correct and incorrect decisions, the time taken to make decisions, repetitions of routes, speed of movements, and the relationship between mobile vs. stationary phases. Baseline performance is determined, after which the ability of rats of different ages (5, 20, 28 months) to improve their performance for a given number of training sessions is studied. Even baseline performance indicated substantial age-dependent differences. Over the five-day training period, performance improved asymptotically in young and old rats, although the final performance of old animals was always poorer for the majority of variables measured. While a variety of stimuli (tactile, proprioceptive, acoustic, olfactory) must be recognized to complete the task and the motor system also plays a major role, the results again illustrate that learning and memory deficits, longer decision-making times, and poorer continuity of mobile phases – which are clearly indicative of reduced central processing abilities – are the main reasons for the longer time spent by old rats in the maze (see also Barnes, 1988; Ingram, 1986; Rapp et al., 1987; Wallace et al., 1980).

MOTOR BEHAVIOR AND EMOTIONALITY

A number of models have been employed to document emotional reactions in animal experiments. They make use of pecking orders, social conditions such as isolation, and anxiety/fear situations (Blanchard & Blanchard, 1989; Blanchard et al., 1990; Davis, 1990; Treit, 1985). Their systematic application to gerontologic studies is rare. Numerous studies in humans show that emotionality fluctuates markedly in old age and has non-specific effects on performance (Bäckman & Molander, 1986; Levonson et al., 1991). There are also numerous indications from animal experiments that age-dependent changes in emotional reactivity take place (Ingram, 1988; Lhotellier et al., 1993). Frequently used tests are the open field and the plus maze (Hall & Ballachey, 1932; Lister, 1987). In these tests, the dropping rate of defecation boli and the ratio of either the frequency and/or the duration of stays in the central versus boundary zones of the field (open field) or in exposed versus protected runs (plus maze) are interpreted as an expression of emotional reactivity. The validity of these variables is frequently refuted, with the argument that emotional reactions are difficult to distinguish from the effects of habituation and exploration in these tests (Gentsch et al., 1988). Conflict tests may be a better approach to investigating emotional components.

1. Geller-Seifter test

The Geller-Seifter test procedure consists of a multiple schedule alternating periods of food-maintained behavior with discriminated periods in which this behavior is suppressed with a response-contingent shock. First of all a rat learns to press a lever under a continuous reinforcement (FR; each lever press is reinforced), then a VI (variable interval) of 20, in which the first lever pressing on the average is rewarded with food after 20 seconds. In the actual test phase lasting 30 minutes, these two conditioned tasks are alternated several times, but a shock is now administered to the paws when the lever is pressed for food in FR. The change in the FR is announced by means of a light signal. Young and adult animals quickly learn to distinguish between the phases; old animals, on the other hand, are less successful. The effect of the punishment stimulus is apparently much more pronounced in old rats, with the result that the animals obtain only a few food pellets even during the VI phases – the efficiency of their conditioned responses is clearly reduced (Jänicke & Coper, 1994). If our knowledge of the relatively unimpaired sensorimotor abilities of old individuals and our increasingly detailed knowledge about learning processes are taken into account, this test is suitable for assessing the emotional reactivity of old animals as an inhibiting factor in learning success.

2. Shock probe conflict procedure

The following sections describe two tests which have not been employed in cross-sectional analyses but seem to be appropriate for assessing emotional reactivity.

In the shock probe conflict procedure, which was introduced by Meert and Colpaert (1986), an animal is exposed to a conflict between "approach" and "avoidance" by placing a shock electrode at one point in an open field. The animal explores its new surroundings, including the electrode, with which contact imparts a painful stimulus. The subsequent avoidance behavior and willingness to explore are interpreted as an emotional reaction with a certain decision-making process. The variable measured is the spatial distribution of the exploration.

3. Defensive test battery

In the paper "Attack and defense in rodents as ethoexperimental models for the study of emotion", Blanchard and Blanchard (1989) describe a test series in which defensive behavioral responses to a present threatening stimulus are measured in different situations (immobility, escape behavior, defensive threatening and defensive attacking). As mentioned in the previous section, this battery has not yet been employed in experimental gerontology but by all means appears suitable for the acquisition of potentially far-reaching findings.

DISCUSSION AND CONCLUSIONS

In 1984, Welford pointed out the need for an in-depth investigation of the different causes of age-typical, differentiated sensorimotor performance. He called for the development of methods that could be used to analyze certain aspects of behavioral traits such as perceptual sensitivity, discriminatory capacity, orientation, selecting behavior, sequential movements, and time patterns. The study of the sensorimotor system is thus an important prerequisite to understanding complex behavioral reactions. The extent to which this analysis can be achieved with the above tests is discussed below. Impulse, and hence, information processing in the brain takes place via a dense network of nerve cells, most of which are connected to each other via synapses. The different pieces of information are selected with the few possible reactions available at the synapses, namely conduction or retention of the impulses, and the result of this process is subjected to multiple control. Within this regulated system, the central nervous system in every higher creature mediates not only precise information about its environment through the senses of hearing, seeing, touching, smelling and tasting, but also collects, processes, weighs, assesses, and stores it in memory, converts it into motivation and emotions, and transforms it into motor activity. This process is regulated and safeguarded by feedback mechanisms. It is obvious that an impairment in sensory input due to hearing or visual weakness results in deficient central processing of these external stimuli. It is equally self-evident that the above conversion processes runs through a number of stages from stimulus perception to appropriate response, one or more of which may no longer be fully functional in old age. In this connection, as with many other disturbances, a feature or a complex of symptoms often describes only the final stage of preceding damage, which is not necessarily located at the same site. As yet we know little about where and how information is converted and integrated into different abilities within the central nervous system. Use of the tests described in the preceding sections can nevertheless provide us with indications about the nature and extent of impaired performance based on deviations from that of adult animals. However, as Comfort suggested for humans in 1969, this requires a comprehensive battery of tests, as that now employed for the identification of mobility dysfunctions in elderly patients (Tinetti & Ginter, 1988). Comfort also pointed out that periodic repetition of the measurements is essential. These measuring and assessing tools can easily be adapted for use in animal experiments, and various test batteries with which physiological, sensorimotor, or behavior-biological performance can be measured have already been suggested (Coper et al., 1986; Dean et al., 1981; Ingram, 1988; Wallace et al., 1980). However, there are differences between the study groups as to the choice of and value attached to the tests employed and also as to the usefulness of forming an index from the overall results. The systematic implementation of accepted test batteries would not only support basic research on the causes and functional relations of behavioral movements and their respective pathological changes but would also be necessary to standardize devices for testing new pharmacological compounds.

Reflexes, muscular strength, and spontaneous and reactive motor activity must always be included in a study program, since these basic parameters provide information about peripheral impairment of motility even when the reactions are regulated and controlled primarily from the center. Interestingly, relatively simple locomotor activities of old animals are not impaired to any significant extent, although the structure and dynamics of the functional elements are clearly altered. The functional reserves for the maintenance of vital abilities are apparently particularly large. Performance deficits are recognizable only under "testing the limits" conditions. They become manifest much earlier in the case of complex locomotor activities, which place increased demands on information processing despite the fact that morphological and biochemical deviations from the norm are usually minimal in the brain. Appropriate behavioral tests can therefore be highly sensitive indicators of disturbances within stimulus-response relationships. Again, however, such disturbances can only be characterized with a battery of tests. Despite the fact that the results can be modified by a large number of variables, the tests often allow for proper differentiation between the neurobiological mechanisms involved in sensorimotor performance. To this extent, a battery of tests suitable for examining different aspects of motor behavior can be a powerful tool for documenting sensorimotor performance during aging. However, one condition for determining performance or its changes is that the tests be valid and reliable, and the results, reproducible. They must also be able both to measure the current state of performance and to record time patterns. Moreover, the experience gained in many studies that interindividual variability becomes more pronounced with increasing age must also be taken into account. It is a recurring phenomenon that some subpopulations – this is valid for all species studied – display good performance even in old age. Life-span studies could contribute important information in this respect, in that among other things, they determine whether these individuals had more vitality and were better performers in the younger phases of their lives as well. A few indications are available which suggest that such an association exists (Burwell & Gallagher, 1993).

Simple tests conducted on a systematic and routine basis must be extended by increasing the demands to find individual performance limits. This can be done by employing the principle of "testing the limits", which entails step-wise increases in the difficulty of an exercise so that the performance range of an age group can be defined and individual performance can be assessed. In this way, both gradually declining abilities and interindividual differences in the capacity for compensation or even for increasing performance can be ascertained. In particular, repetition of tasks and demands as an important element of a test method is suitable for demonstrating age-dependent changes in the power and speed factors. This is true not only for the behavioral level, but also for almost all functional levels affected by an organ, tissue, cell, or biochemical-physiological process. This methodological approach is becoming all the more important since the individual functions often display changes which are independent of each other. Herein lies the strength of animal studies, where the genetic factors and biography of the experimental animals are known, making it possible to systematically clarify the functional association of a sensorimotor behavior that diminishes with chronological age and the delayed reduction of performance due to physical training (Chodzko-Zajko & Ringel, 1987; Fries & Crapo, 1981).

REFERENCES

Altman, J., & Sudarshan, K. (1975). Postnatal development of locomotion in the laboratory rat. *Animal Behaviour, 23,* 896-920.

Ammassari-Teule, M., Fagioli, S., & Rossi-Arnaud, C. (1994). Radial maze performance and open field behaviours in aged C57BL/6 mice: Further evidence for preserved cognitive abilities during senescence. *Physiology and Behavior, 55,* 341-345.

Archer, J. (1973). Test for emotionality in rats and mice: A review. *Animal Behaviour, 21,* 205-235.

Arking, R. (1991). *Biology of aging.* Englewood Cliffs, NJ: Prentice Hall.

Bäckman, L., & Molander, B. (1986). Adult age differences in the ability to cope with situations of high arousal in a precision sport. *Psychology and Aging, 1,* 133-139.

Barnes, C. A. (1988). Aging and the physiology of spatial memory. *Neurobiology of Aging, 9,* 563-568.

Bartus, R. T., Flicker, C., & Dean, R. L. (1983). Logical principles for the development of animal models of age-related memory impairments. In Th. Crook, S. Ferris, R. Bartus (Eds.), *Assessment in geriatric psychopharmacology* (pp. 263-298). New Canaan, CT: Powely Association Inc.

Blanchard, R. J., & Blanchard, D. C. (1989). Attack and defence in rodents as ethoexperimental models for the study of emotion. *Progress in Neuro-Psychopharmacology and Biological Psychiatry, 33,* 3-14.

Blanchard, R. J., Blanchard, D. C., Rodgers, J., & Weiss, S. M. (1990). The characterization and modelling of antipredator defensive behavior. *Neuroscience and Biobehavioral Reviews, 14,* 463-472.

Boissier, J.-R., Tardy, J., & Diverres, J.-C. (1960). Une nouvelle méthode simple pour explorer l'action "tranquillisante": le test de la cheminée. *Medicina Experimentalis, 3,* 81-84.

Brown, M. (1987). Changes in fibre size, not number, in ageing skeletal muscle. *Age and Ageing, 16,* 244-248.

Brinley, J. F. (1965). Cognitive sets, speed and accuracy of performance in the elderly. In A. T. Welford & J. E. Birren (Eds.), *Behavior aging and the nervous system.* Springfield, IL: Thomas C.C. Publishers.

Buckles, V. D. (1993). Age-related slowing. In G.E. Stelmach & V. Hömberg (Eds.), *Sensorimotor impairment in the elderly. Behavioural and social sciences* (Vol. 75, pp. 73-87). Dordrecht: Kluwer Academic Publishers.

Burwell, R. D., & Gallagher, M. (1993). A longitudinal study of reaction time performance in long-Evans rats. *Neurobiology of Aging, 14,* 75-64.

Campbell, B. A., Krauter, E. E., & Wallace, J. E. (1980). Animal models of aging: Sensory-motor and cognitive function in the aging rat. In D. G. Stein (Ed.), *Psychology of aging: Problems and prospectives* (pp. 201-225). Amsterdam: Elsevier-North Holland Publishers, Co.

Campbell, B. A. (1982). Behavioral markers of aging in the Fischer 344 rat. In NIH Publication No. 82-2221, *Biological markers of aging* (pp. 78-86). Washington, DC.

Campbell, B. A., & Gaddy, J. R. (1987). Rate of aging and dietary restriction: sensory and motor function in the Fischer 344 rat. *Journal of Gerontology, 42,* 154-159.

Cheal, M. L., Foley, K., & Kastenbaum, R. (1987). Brief periods of environmental enrichment facilitate adolescent growth of gerbils. *Physiology and Behavior, 36,* 1047-1051.

Chodzko-Zajko, W. J., & Ringel, R. L. (1987). Physiological fitness measures and sensory and motor performance in aging. *Experimental Gerontology, 22,* 317-328.

Collins, K. J., & Exton-Smith, A. N. (1983). Thermal homeostasis in old age. *Journal of the American Geriatrics Society, 7,* 519-524.

Comfort, A. (1969). Test-battery to measure ageing-rate in man. *Lancet (7635),* 1411-1415.

Coper, H., Jänicke, B., & Schulze, G. (1986). Biophysiological research on adaptivity across the life-span of animals. In P. Baltes, D.

Featherman, & R. Lerner (Eds.), *Life-span development and behavior* (pp. 207-232). London: Lawrence Earlbaum Association.

Davis, M. (1990). Animal models of anxiety based on classical conditioning: The conditioned emotional response (CER) and the fear-potentiated startle effect. *Pharmacology and Therapeutics, 47,* 147-165.

Dean, R. L., III, Scozzafava, J., Goas, J. A., Regan, B., Beer, B., & Bartus, R. T. (1981). Age-related differences in behavior across the life span of the C57BL/6J mouse. *Experimental Aging Research, 7,* 427-451.

Ferrandez, A. M., Pailhous, J., & Serratrice, G. (1988). Locomotion in the elderly. In B. Amblard, A. Berthoz, & F. Clarac (Eds.), *Development, adaptation and modulation of posture and locomotion* (pp. 115-124). Amsterdam: Elsevier.

Flood, D. G., & Coleman, P. D. (1988). Neuron numbers and sizes in aging brain: Comparisons of human, monkey, and rodent data. *Neurobiology of Aging, 9,* 453-463.

Foster, K. G., Ellis, F. P., Dose, & C. (1976). Sweat response in the aged. *Age and Ageing, 5,* 91-101.

Fox, R. H., Woodward, P. M., Exton-Smith, A. N., Green, M. F., Donnison, D. V., & Wicks, M. H. (1973). Body temperature in the elderly: A national study of physiological, social, and environmental conditions. *British Medical Journal, 1,* 200-206.

Fries, J. F., & Crapo, L. M. (1981). *Vitality and aging.* San Francisco, CA: W.H. Freeman Co.

Gallagher, M., & Pelleymounter, M. A. (1988). An age-related spatial learning deficit: choline uptake distinguishes "impaired" and "unimpaired" rats. *Neurobiology of Aging, 9,* 363-369.

Gallagher, M., & Burwell, R. D. (1989). Relationship of age-related decline across several behavioral domains. *Neurobiology of Aging, 10,* 691-708.

Gentsch, C., Lichtsteiner, M., & Feer, H. (1988). Genetic and environmental influences on behavioral and neurochemical aspects of emotionality in rats. *Experientia, 44,* 482-490.

Gower, A. J., Lamberty, Y. (1993). The aged mouse as a model of cognitive decline with special emphasis on studies in NMRI mice. *Behavioral Brain Research, 57,* 163-173.

Hall, C. S., & Ballachey, E. L. (1932). A study of the rats behavior in a field: a contribution to method in comparative psychology. *University of California Publication in Psychology, 6,* 1-12.

Harrison, J., & Buchwald, J. (1983). Eyeblink conditioning deficits in the old cat. *Neurobiology of Aging, 4,* 45-51.

Hazzard, D. G. (1991). Relevance of the rodent model to human aging studies. *Neurobiology of Aging, 12,* 645-649.
Hess, T. (1987). Durst- und Flüssigkeitshaushalt im Alter. *Schweizerische Medizinische Wochenschrift, 117,* 491-495.
Hofecker, G., Niedermüller, H., Skalicky, M. (1981). Der altersbedingte Leistungsabfall und seine Beeinflussung im Tierexperiment. *Acta Gerontologia, 11,* 188-194.
Hoffman, H. S., & Wible, B. L. (1970). Role of weak signals in acoustic startle. *Journal of the Acoustical Society of America, 47,* 489-497.
Ingram, D. K., London, E. D., Reynolds, M. A., Waller, S. B., & Goodrick, C. L. (1981). Differential effects of age on motor performance in two mouse strains. *Neurobiology of Aging, 2,* 221-227.
Ingram, D. K., Archer, J. R., & Harrison, D. E. (1982). Physiological and behavioral correlates of life-span in aged C57BL/6J mice. *Experimental Gerontology, 17,* 295-303.
Ingram, D. K., & Reynolds, M. A. (1986). Assessing the predictive validity of psychomotor tests as measures of biological age in mice. *Experimental Aging Research, 12,* 155-162.
Ingram, D. K. (1986). Complex maze learning in rodents as a model of age-related memory impairment. *Neurobiology of Aging, 9,* 457-485.
Ingram, D. K. (1988). Motor performance variability during aging in rodents. In J. A. Joseph (Ed.), *Central determinants of age-related declines in motor function* (Vol. 515, pp. 70-96). New York: Annals of the New York Academy of Sciences.
Jänicke, B., & Coper, H. (1991). Effects of age and drugs on food and fluid intake. *Growth, Development and Aging, 55,* 139-150.
Jänicke, B., & Coper, H. (1993). The effects of prenatal alcohol exposure on the behavior of rats during their life span. *Journal of Gerontology, 48,* B156-167.
Jänicke, B., & Coper, H. (1994). The effects of prenatal exposure to hypoxia on the behavior of rats during their life span. *Pharmacology, Biochemistry and Behavior, 48,* 863-873.
Jänicke, B., & Schulze, G. (1987). Influence of normobaric hypoxia on learning capacity in different aged rats. *Neurobiology of Aging, 9,* 495-500.
Jänicke, B., Schulze, G., & Coper, H. (1983). Motor performance achievements in rats of different ages. *Experimental Gerontology, 18,* 393-407.
Johnson, H. A. (1978). The relevance of the rodent as a model system of aging in man. In D. C. Gibson (Ed.), *Development of the rodent*

as a model system of aging (Vol. 1, pp. 3-6). Rockville, MD: National Institute of Child Health and Human Development.

Joseph, J. A., Bartus, R. T., Clody, D., Morgan, D., Finch, C., Beer, B., & Sesack, S. (1983). Psychomotor performance in the senescent rodent: reduction of deficits via striatal dopamine receptor up-regulation. *Neurobiology of Aging, 4,* 313-319.

Jucker, M., Oettinger, R., & Bättig, K. (1988). Age-related changes in working and reference memory performance and locomotor activity in the Wistar rat. *Behavioral and Neural Biology, 50,* 24-36.

Kay, H., & Birren, J. E. (1958). Swimming of the albino rat: Fatigue, practice, and drug effects of age and sex differences. *Journal of Gerontology, 13,* 378-385.

Kimble, G. A., & Pennypacker, H. S. (1963). Eyelid conditioning in young and aged subjects. *Journal of Genetics and Psychology 103,* 283-289.

Kovanen, V., Suominen, H., & Peltonen, L. (1987). Effects of aging and life-long physical training on collagen in slow and fast skeletal muscle in rats. *Cell Tissue Research, 248,* 247-255.

Krauter, E. E., Wallace, J. E., & Campbell, B. A. (1981). Sensory-motor function in the aging rat. *Behavioral Neurology Biology, 31,* 367-392.

Kubanis, P., & Zornetzer, S. F. (1981). Age-related behavioral and neurobiological changes: A review with an emphasis on memory. *Behavioral and Neural Biology, 31,* 115-172.

Lamberty, Y., & Gower, A. J. (1990). Age-related changes in spontaneous behavior and learning in NMRI mice from maturity to middle age. *Physiology and Behavior, 47,* 1137-1144.

Landfield, Ph. W. (1988). Hippocampal neurobiological mechanisms of age-related memory dysfunction. *Neurobiology of Aging, 9,* 571-579.

Lapsley, D. K., & Enright, R. D. (1983). A cognitive developmental model of rigidity in senescence. *International Journal of Aging and Human Development, 16,* 81-93.

Larsson, L., & Ansved, T. (1988). Effects of age on the motor unit. In J. A. Joseph (Ed.), *Central determinants of age-related declines in motor function* (Vol. 515, pp. 303-313). New York: Annals of New York Academy of Sciences.

Levine, M. S., Lloyd, R. L., Fisher, R. S., Hul, C. D., & Buchwald, N. A. (1987). Sensory, motor and cognitive alterations in aged cats. *Neurobiology of Aging, 8,* 253-263.

Levonson, R. W., Carstensen, L. L., Friesen, W. V., & Ekman, P. (1991). Emotion, physiology, and expression in old age. *Psychology and Aging, 6,* 28-35.

Lexell, J. (1993). What is the cause of the ageing athrophy ? Assessment of the fiber type composition in whole human muscles. In G. E. Stelmach & V. Hömberg (Eds.), *Sensorimotor impairment in the elderly* (pp. 143-153). Dordrecht: Kluwer Academic Publishers.

Lhotellier, L., Perez-Diaz, F., & Cohen-Salmon, C. (1993). Locomotor and exploratory activity in three inbred strains of mice from young adulthood to senescence. *Experimental Aging Research, 19,* 177-187.

Light, K. E., & Spirduso, W. W. (1990). Effects of adult aging on the movement of complexity factor of response programming. *Journal of Gerontology, 45,* P107-109.

Lippa, A. S., Pelham, R. W., Beer, B., Critchett, D. J., Dean, R. L., & Bartus, R. T. (1980). Brain cholinergic function and memory in rats. *Neurobiology of Aging, 1,* 13-20.

Lister, G. L. (1987). The use of a plus maze to measure anxiety in the mouse. *Psychopharmacology, 92,* 180-185.

Markowska, A. L., Stone, W. S., Ingram, D. K., Reynolds, J., Gold, P. E., Conti, L. H., Pontecorvo, M. J., Wenk, G. L., & Olton, D. S. (1989). Individual differences in aging: behavioral and neurobiological correlates. *Neurobiology of Aging, 10,* 31-43.

Marshall, J. F. (1982). Sensorimotor disturbances in the aging rodent. *Journal of Gerontology, 37,* 548-554.

Marshall, J. F., & Berrios, N. (1979). Movement disorders of aged rats: Reversal by dopamine receptor stimulation. *Science, 206,* 477-479.

Meert, T. F., & Colpaert, F. C. (1986). The shock probe conflict procedure: A new assay responsive to benzodiazepines, barbiturates and related compounds. *Psychopharmacology, 88,* 445-450.

Miquel, J., & Blasco, M. (1978). A simple technique for evaluation of vitality loss in aging mice by testing their muscular coordination and vigor. *Experimental Gerontology, 13,* 389-396.

Morris, R. G. M. (1981). Spatial localisation does not require the presence of local cues. *Learning and Motivation, 12,* 239-260.

Nabeshima, T., Hasegawa, M., Nakayama, S., Kinoshita, H., Amano, M., & Hasegawa, T. (1993). Impairment of learning and memory and the accessory symptom in aged rats as senile dementia model (1) - Emotional behavior. *Japanese Journal of Psychopharmacology, 13,* 65-72.

Newton, J. R., & Yemm, R. (1986). Changes in the contractile properties of the human first dorsal interosseous muscle with age. *Gerontology 32,* 98-104.

Ordy, J. M., Thomas, G. J., Volpe, B. T., Dunlap, W. P., & Colombo, P. M. (1988). An animal model of human-type memory loss based on aging, lesion, forebrain ischemia, and drug studies with the rat. *Neurobiology of Aging, 9,* 667-683.

Ossowska, K., Schulze, G., Wolfarth, S., Coper, H., & Kaminska, A. (1992). Hausmanowa-Petrusewicz, I. Muscle stiffness and continous electromyographic activity in old rats: an animal model for spasticity? *Neuroscience, 51,* 167-175.

Parasuraman, R. (1984). Sustained attention in detection and discrimination. In R. Parasuraman, D.R. Davis (Eds.) *Varieties of attention* (pp. 243-271). Orlando, FL: Academic Press.

Pellis, S. M., Pellis, V. C., O'Brien, D. P., de la Cruz, F., & Teitelbaum, Ph. (1987). Pharmacological subtraction of the sensory controls over grasping in rats. *Physiology and Behavior, 39,* 127-133.

Potwin, A. R., Syndulko, K., Tourtellotte, W. W., Lemmon, J. A., & Potvin, J. H. (1980). Human neurologic function in the aging process. *Journal of the American Geriatrics Society, 28,* 1-9.

Powell, D. A., Buchanan, S. L., & Hernandez, L. L. (1984). Age-related changes in pavlovian conditioning: central nervous system correlates. *Physiology and Behavior, 32,* 609-616.

Rapp, P. R., Rosenberg, R. A., & Gallagher, M. (1987). An evaluation of spatial information processing in aged rats. *Behavioral Neuroscience, 101, 3*-1.

Rogers, J. (1988). The neurobiology of cerebellar senescence. In A. Joseph (Ed.), *Central determinants of age-related declines in motor function* (Vol. 515, pp. 251-268). New York: Annals of the New York Academy of Sciences.

Rosenblueth, A., & Wiener, N. (1945). The role of models in science. *Philosophy of Science, 121,* 316-321.

Rosenheimer, J. L., & Smith, D. O. (1985). Differential changes in the end-plate architecture of functionally diverse muscles during aging. *Journal of Neurophysiology, 53,* 1567-1581.

Roth, K. A., & Katz, R. J. (1979). Stress, behavioral arousal, and open field activity - a reexamination of emotionality in the rat. *Neuroscience and Biobehavioral Reviews, 3,* 247-263.

Sabin, T. D. (1982). Biologic aspects of falls and mobility limitations in the elderly. *Journal of the American Geriatrics Society, 30,* 51-58.

Salthouse, T. A. (1988). Cognitive aspects of motor functioning. In J. A. Joseph (Ed.), *Central determinants of age-related declines in motor* function (Vol. 515, pp. 33-41). New York: Annals of the New York Academy of Sciences.

Schneider, E., & Rowe, J. W. (1990). *Handbook of the biology of aging* (3rd ed., pp. 306-329). New York: Academic Press.

Schulze, G., Coper, H., & Jäicke, B. (1988). Animal models for evaluation of age-related changes in behaviour. *Drug Development Research, 14,* 363-368.

Shock, N. W., & Norris, A. H. (1970). Neuromuscular coordination as a factor in age changes in muscular exercise. In E. Jokl & D. Brunner (Eds.), *Physical activity and aging* (Vol. 4). Basel: S. Karger.

Spangler, E. L., Waggie, K. S., Hengemihle, J., Roberts, D., Hess, B., & Ingram, D. K. (1994). Behavioral assessment of aging in male Fischer 344 and brown Norway rat strains and their F1 hybrid. *Neurobiology of Aging, 15,* 319-328.

Steinhagen-Thiessen, E., Reznick, A. Z., & Hilz, H. (1981). Positive and negative adaptation and muscle enzymes in aging mice subjected to physical exercise. *Mechanisms and Ageing Development, 16,* 363-369.

Steinhagen-Thiessen, E., & Reznick, A. Z. (1987). Effect of short- and long-term endurance training on creatine phosphokinase activity in skeletal and cardiac muscles of CW-1 and C57BL mice. *Gerontology, 33,* 14-18.

Stone, C. (1929). The age factor in animal learning: I. Rats in the problem box and the maze. *Genetics and Psychology Monographs, 5,* 1-130.

Suominen, H., & Kovanen, V. (1988). Effects of physical training on mechanical properties of skeletal muscle in aged rats. In E. Steinhagen-Thiessen & D.L. Knook (Eds.), *Trends in biomedical gerontology* (pp. 87-90). Den Haag: CIP Gegevens Koninklijke Bibliotheek.

Thompson, R., Huestis, P. W., Crinella, F. M., & Yu, J. (1986). The neuroanatomy of mental retardation in the white rat. *Neuroscience and Biobehavioral Reviews, 10,* 317-338.

Tinetti, M. E., & Ginter, S. F. (1988). Identifying mobility dysfunctions in elderly patients. Standard neuromuscular examination or direct assessment? *Journal of the American Medical Association, 259,* 1190-1193.

Treit, D. (1985). Animal models for the study of anti-anxiety agents: A review. *Neuroscience and Biobehavioral Reviews, 9,* 203-222.

Wagner, J. A., & Horvath, S. M. (1985). Influences of age and gender on human thermoregulatory responses to cold exposures. *Journal of Applied Physiology, 58,* 180-186.

Wallace, J. E., Krauter, E. E., & Campbell, B. A. (1980). Animal models of declining memory in the aged rat: Short-term and spatial memory in the old rat. *Journal of Gerontology, 35,* 355-363.

Wecker, J. R., & Ison, J. R. (1986). Effects of motor activity on the elicitation and modification of the startle reflex in rats. *Animal Learning Behavior, 14,* 287-292.

Weiskrantz, L., & Warrington, E. K. (1975). The problem of the amnesic syndrome in man and animals. In R. L. Isaacson & K. H. Pribram (Eds.), *The hippocampus* (Vol. 2, pp. 411-428). New York: Plenum Press.

Weiss, C., & Thompson, R. F. (1992). Delayed acquisition of eyeblink conditioning in aged F1 hybrid (Fischer-344 x brown norway) rats. *Neurobiology of Aging, 13,* 319-323.

Welford, A. T. (1977). Motor performance. In J. E. Birren & K. W. Schaie (Eds.), *Handbook of the psychology of aging.* New York: Van Nostrand Reinhold.

Welford, A. T. (1984). Between bodily changes and performance: Some possible reasons for slowing with age. *Experimental Aging Research, 10,* 73-88.

Welford, A. T. (1988). Reaction time, speed of performance, and age. In A. Joseph (Ed.), *Central determinants of age-related declines in motor function* (Vol. 515, pp. 1-17). New York: Annals of the New York Academy of Sciences.

Wilcock, J. (1972). Water-escape in weanling rats: A link between behaviour and biological fitness. *Animal Behaviour, 20,* 543-547.

Willig, F., Palacios, A., Monmaur, P., M'Harzi, M., Laurent, J., & Delacour, J. (1987). Short-term memory, exploration and locomotor activity in aged rats. *Neurobiology of Aging, 8,* 393-402.

Winocur, G. (1988). A neuropsychological analysis of memory loss with age. *Neurobiology of Aging, 9,* 487-494.

Woollacott, M. H., Moore, S., & Hu, M. H. (1993). Improvements in balance in the elderly through training in sensory organization abilities. In G. E. Stelmach & V. Hömberg (Eds.), *Sensorimotor impairment in the elderly* (pp. 377-392). Dordrecht: Kluwer Academic Publishers.

Changes in sensory motor behavior in aging
A.-M. Ferrandez and N. Teasdale (Editors)

ATTENTIONAL DEMANDS FOR WALKING: AGE-RELATED CHANGES

Yves LAJOIE, Normand TEASDALE, Chantal BARD, and Michelle FLEURY

Université Laval, Québec

Abstract

The present chapter presents three experiments that examine age-related attentional demands for walking. The cognitive requirements for walking are evaluated: (a) in a constraint-free walking environment, (b) in a goal-directed condition requiring precise foot positioning, and (c) at gait initiation. In all three experiments, a classical dual-task paradigm is used to evaluate if the attentional requirements vary with the walking requirements. Auditory stimuli are given at toe-off and at heel contact to evaluate if the cognitive demands also vary within the walking cycle. Overall, the behavior of the elderly and young persons was very consistent across the three experiments. The reaction times of the elderly persons were systematically slower than those of the young adults and the elderly never showed a significant stance phase effect with the reaction times being slower when stimuli were presented at toe-off than when they were presented at heel contact, suggesting that balance control within the gait cycle is not automatic and loads differentially the higher level cognitive system. For the elderly, modification of the gait pattern (for example, slower walking speed and shorter stride length) may serve to reduce balance requirements.

Key words: Aging, attention, balance, gait, posture.

Correspondence should be sent to Normand Teasdale, Université Laval, Laboratoire de Performance Motrice Humaine, PEPS, Québec, PQ G1K 7P4, Canada (e-mail: Normand.Teasdale@pmh.ulaval.ca).

INTRODUCTION

Walking is a highly practiced activity that is often considered "automated". Nevertheless, a high incidence of falls for the elderly occurs during walking (Lipsitz et al., 1991). Falls represent the failure of the postural control system, usually defined as the maintenance of the body center of mass over its base of support or, more generally, within the limits of stability (Alexander, 1994; Lee, 1989). The present chapter first reports experiments suggesting that a gradual sensory and motor systems degeneration is responsible for the decreased postural stability associated with aging. In a second section, three experiments are presented. Young and elderly subjects were submitted to various postural and walking tasks. A classical dual-task paradigm was used to evaluate if the attentional requirements vary with the task constraints. The cognitive requirements for walking were evaluated: (a) in a constraint-free walking condition, (b) in a goal directed condition requiring precise foot positioning, and (c) at gait initiation. For all three experiments, auditory stimuli were given at toe off and at heel contact to evaluate if the cognitive demands vary within the walking cycle.

Aging and disease of the sensory systems or parts of the motor system

A small and progressive degeneration of sensory input from the lower extremities is often the first and most common manifestation of aging (Calne, 1985). With normal aging, a degeneration of all sensory systems is also observed. For example, diminution of the axon population in the optic nerves (Johnson et al., 1987), loss of visual acuity (particularly in the periphery; Sekuler et al., 1980), reduction of hair cells in the semicircular canals (Rosenhall & Rubin, 1975), higher proprioceptive thresholds and decreased position sense (Lord et al., 1991; Meeuwsen et al., 1993; Skinner et al., 1984; Whanger & Wang, 1974) have all been reported. Decreased muscular strength is also observed (Larsson et al., 1979; Lord et al., 1991). The presence of pathological processes (even in subclinical stages) has been also suggested to accentuate any given elderly persons' disequilibrium (Horak et al., 1989; Manchester et al., 1989). For example, Manchester et al. used dynamic posturography and reported that several falls occurred when ankle proprioception was made incongruent with postural sway and when only foveal vision was available. The elderly persons that were the most affected were those that were diagnosed as having "subclinical" manifestations of neurological disease. Although there are several reports suggesting that these age-related changes affect maintenance of postural control (see Alexander, 1994, for a recent review), attempts to predict balance control using sensory and motor tests have yielded poor results. For example, Lord et al. (1991) have used 11 different tests to measure proprioceptive, visual and vestibular acuity and muscular strength. These tests were not sufficient to explain most of the variance in the measures of sway and balance of elderly persons. Lord et al.'s results agree with other published work (e.g., Brocklehurst et al., 1982; Era & Heikkinen, 1985). The above observations suggest that a degeneration of one of the sensory systems or parts of the motor system do not necessarily lead to functional postural deficits.

Central contribution to postural control

In addition to a gradual degeneration of all peripheral sensory systems, there are several recent proposals that the integrity of higher level systems is of significance for posture control (e.g., Mulder et al., 1993; Shumway-Cook et al., 1993; Teasdale et al., 1992, 1993). In agreement with this, Alexander (1994) mentioned that the importance of higher cortical/central factors has been underemphasized in its potential influence on postural control. Indeed, the higher level mechanisms responsible for posture control are those that deteriorate most with aging (Stelmach et al., 1989; Teasdale et al., 1993). Gait and posture control are often accomplished while performing a secondary task. For example, when walking, one is often confronted with the task of talking and listening. In other circumstances, walking is utilitarian to a goal-directed movement; for example the act of grasping an object located on a shelf is usually programmed while the person is approaching the shelf. According to Kahneman (1973), interference (e.g., a failure of the postural control system) could occur if the demands of competing activities exceed a general limited central capacity. As such, the slower processing capacity that comes with age (Salthouse, 1985; Welford, 1988) and deficits in the ability to divide attention and/or to allocate properly the resources between concurrent tasks (Baron et al., 1988; Craik & Byrd, 1982; Teasdale et al., 1992) may compromise the postural stability of older adults (Stelmach et al., 1990; Teasdale et al., 1993).

Attention is often considered (at least by cognitive theorists) as an extension of sensory processes (Cohen & Sparling-Cohen, 1993), thus providing an interesting link for studying the interdependence between sensorimotor and cognitive systems for the control of posture. Without going into the specific details of cognitive resource models, let us mention that there are three basic assumption to the use of such models: (a), central processing capacity is limited, (b) performing a task requires a given portion of this capacity, and (c), if two tasks performed simultaneously require more than the total capacity, the performance on one or both will be affected negatively (Kahneman, 1973; Parasuraman, 1981; Wickens, 1984).

Using this general approach, several authors have demonstrated that maintaining an upright posture requires attention (Geurts et al., 1992a, b; Kerr et al., 1985; Stelmach et al., 1990; Teasdale et al., 1993). Similarly, walking has been shown to require more cognitive processing than simple sitting or upright standing posture (Bardy & Laurent, 1991; Girouard, 1980; Lajoie et al., 1993). The present paper examines whether there are age-related changes in the attentional requirements for walking. This was evaluated for (a) normal constraint-free walking, (b) while pointing at a target with the dominant foot, and (c) at gait initiation. For all three experiments reported, the subjects were not familiar with the specific hypotheses tested. The elderly were volunteers obtained from aging coalitions and support groups; the young adults were undergraduate and graduate students. Elderly subjects were chosen from a large pool of subjects. They were active residents able to ambulate independently without walking aids. They were selected after an interview if: (a) they were not treated for any musculoskeletal deficits (e.g., arthritis, osteoarthritis, osteoporosis); (b) they were not taking any medication; (c) were not undergoing treatment for central nervous system disorders; and (d) had no history of falling. All subjects gave informed consent to participate.

EXPERIMENT 1: ATTENTIONAL DEMANDS WHEN WALKING AT CONSTANT VELOCITY

In this first experiment, we wanted to examine if the elderly persons need to allocate a greater proportion of their attentional resources to the postural task than young adults when (1) standing upright, and (2) walking at constant velocity. Eight young adults (5 males and 3 females, mean age = 26, range = 22-34, standard deviation = 4.2) and eight elderly persons (4 males and 4 females, mean age = 71, range = 66-79, standard deviation = 4.2) participated. A classical dual-task paradigm was used to test the attentional demands of the postural tasks. In the series of experiments presented in this paper, the postural tasks were always the primary tasks and the secondary task always consisted of responding verbally ('top') as rapidly as possible to a randomly presented auditory probe stimulus. Poorer performance on the secondary task (i.e., increased probe reaction time) is indicative of high concurrent demands being placed on the subject's information-processing capacity by the primary task. In this first experiment, all subjects performed four primary tasks alone: (1) sitting, (2) standing upright with a broad support base, (3) standing upright with a narrow support base, and (4) walking. They also performed the same tasks with the probe reaction time (secondary task). For the upright tasks, subjects adopted their preferred position for the broad-support task and had their feet together for the narrow-support base. For the first three postural tasks performed in a dual-task situation, 20 probe stimuli were randomly presented following a variable foreperiod (3, 3.5, 4, 4.5, or 5 s). When maintaining an upright posture, reducing the base of support increases the amplitude of postural oscillations. For example, decreasing the inner distance between the feet from 20 to 0 cm yields a 30% increase of the sway path for young adults (Kolleger et al., 1989). It was postulated that the narrow base of support would affect more the reaction time of the elderly than that of the young persons. For the walking task, subjects walked at their own preferred pace (10 control trials) to establish their normal walking pattern, unaffected by the addition of a secondary task. For all walking trials, they were asked to maintain a constant speed of progression. When both tasks were performed concurrently, the auditory probe stimuli were presented in the second, third, or fourth walking cycle when the subject's left foot was at toe off (i.e., at the offset of the left foot contact with the ground) or at heel contact (i.e., at the onset of the left heel contact with the ground). When a probe was given, subjects were asked to maintain constant their speed of progression. Eight trials for each condition were given (48 trials). In addition, eight catch trials were randomly presented. The evaluation of all walking events and the presentation of all stimuli were computer-controlled.

Results

An important prerequisite of the double-task methodology consists of verifying that the addition of the probe-RT procedure (i.e., the verbal response to the auditory stimulus) does not affect the performance of the primary task (e.g., to control that subjects did not stop or slow their gait when stimuli were presented). This prerequisite, however, cannot be validated for the postural task. During quiet standing the human body continually moves about in an erratic fashion and the act of maintaining a stable erect posture can be viewed as a stochastic process (Collins & Deluca, 1993, 1994). This precludes the identification of direct causal relationships between postural oscillations and the secondary cognitive task. During walking, however, we have examined the prerequisite by comparing the horizontal heel displacement across the different experimental conditions. Both groups did not modify their gait pattern when stimuli were introduced. Figure 1 presents average cycle length, dura-

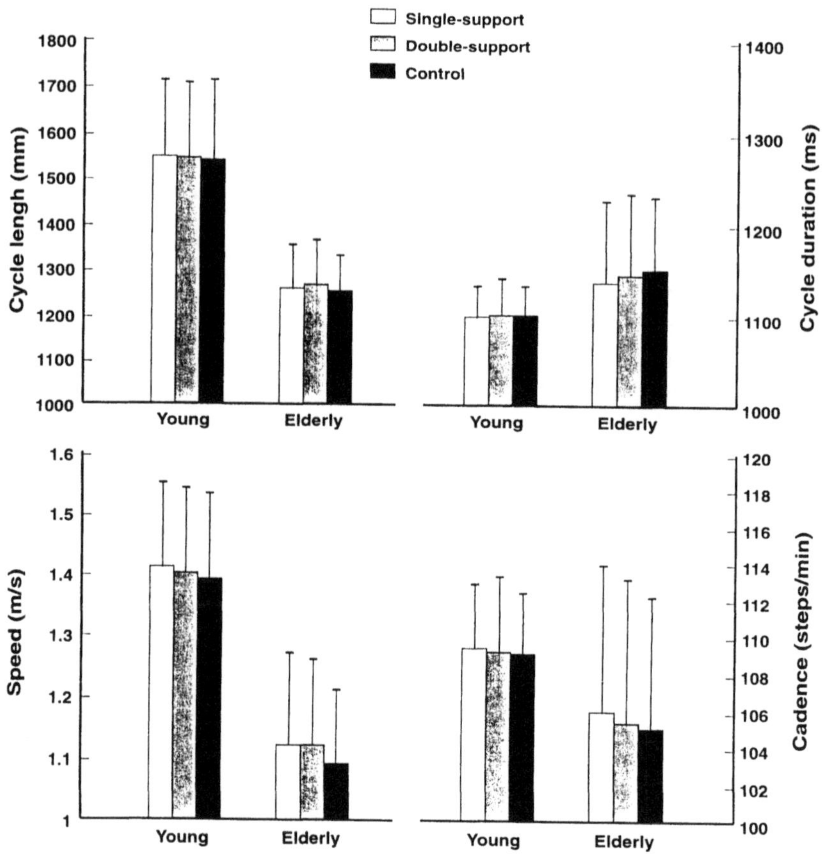

FIGURE 1. Means and between-subject standard deviations for cycle length and duration, walking speed and cadence. Data are presented for the three walking conditions (when stimuli were presented at the onset of the single-support and double-support phases and for the control no-stimulus condition) for both age groups.

tion, speed and cadence obtained for the two age groups. Kinematic results compare well with published data (Winter, 1991) for both young adults (1.50 m for cycle length; 1.08 s for cycle duration; 1.39 m/s for speed; 111.0 steps/min for cadence) and elderly persons (1.25 m for cycle length; 1.15 s for cycle duration; 1.10 m/s for speed; 104.9 steps/min for cadence). Elderly persons adopted a slower walking speed than young persons (1.10 vs. 1.39 m/s, respectively) and also had a shorter stride length (1.25 vs. 1.50 m, respectively), a longer cycle duration (1.15 s vs. 1.10 s), and a slower cadence (105 steps/min vs. 109 steps/min) than young adults.

Attentional requirements

The attentional demands of the sitting, standing, and walking tasks were assessed using the Probe-RT procedure. Before submitting the data to a Group by Tasks ANOVA, a logarithmic transformation was applied to the means because they were positively correlated with their variance (Elliott & Allard, 1985; Sokal & Rohlf, 1981). The ANOVA showed significant main effects of Group and Task, and a significant interaction of Group × Task. For all tasks, elderly were slower to respond to the stimulus than young persons ($p < .005$). A priori comparisons of means showed that for both groups, RTs for the sitting task were faster than those for the postural tasks ($ps < .005$). There has been suggestions that elderly persons, when compared to young adults, need to allocate a greater proportion of their attentional resources to their postural system when it is stressed (Lajoie et al., 1993; Larish et al., 1988; Shumway-Cook et al., 1993; Teasdale et al., 1993). For this reason, the two-way interaction was decomposed into specific contrasts evaluating (1) the effect of stressing the postural system by reducing the base of support during standing, and (2), the effect of a smaller base of support during walking (i.e., comparing RTs obtained for single and double-support phases). The two contrasts are presented in Figures 2 and 3, respectively. The first contrast showed that elderly persons were significantly more affected by the reduction of the base of support than young adults ($p < .05$). The second contrast showed that elderly also behaved differently during walking. Contrary to young adults who responded faster when the stimuli were given during the more stable double-support phase, the elderly showed no phase effect ($p < .01$). These results suggest that, for the standing tasks, elderly allocated more attentional resources than young adults to the postural task. For the walking task, the elderly were not allocating resources the same way

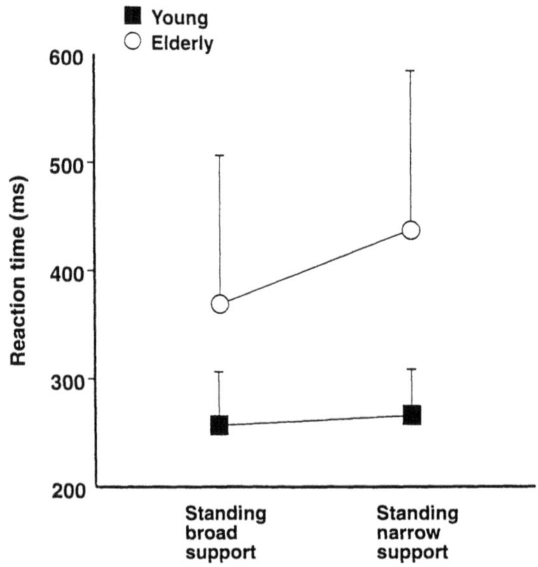

FIGURE 2. Mean reaction times and between-subject standard deviations for the broad-support standing and standing with feet together conditions.

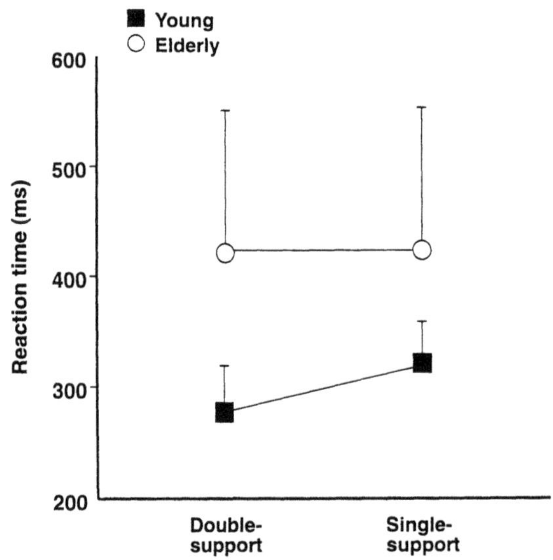

FIGURE 3. Mean reaction times and between-subject standard deviations when stimuli were presented during walking at the onset of the single-support (toe-off) and double-support (heel-contact) phases.

that young persons did. The lack of difference observed between the RTs for the single- and double-support phases suggests that the elderly may use a slower and more conservative speed of progression to reduce the threat to balance during walking.

Aging modifies the attentional demands for posture control and gait

This first experiment demonstrates that, compared to a seated condition, maintaining an upright standing position and walking require attention. The smaller base of support required the elderly persons to allocate more of their attentional resources to the postural task than young adults (as indexed by the slower RTs observed for the smaller base of support). Furthermore, the greater attentional demands observed for the walking task suggest that, from an information processing viewpoint, walking cannot be considered as an automated task requiring no (or hardly any) cognitive processing. The increased reaction time from the sitting condition to the walking condition replicates earlier anecdotal observations (e.g., Kahneman, 1973) and empirical findings by Girouard (1980), Bardy and Laurent (1991), and Lajoie et al. (1993). As shown in a previous experiment (Lajoie et al., 1993), the young adults showed slower RTs for the single than for the more stable double support phase (321 vs. 264 ms, respectively; $p < .05$). The older persons showed slower probe-RTs than the young adults ($p < .01$). Moreover and in contrast to young adults, the probe-RTs for both support phases were nearly identical (413 and 406 ms for the single and double support phases, respectively; $p > .05$). These results agree with the general suggestion that movement is more attention demanding for the elderly than for the young persons (Diggles Buckles, 1993).

EXPERIMENT 2: ATTENTIONAL DEMANDS DURING GOAL-DIRECTED GAIT

Walking often requires modification of the stereotypical forward progression gait pattern. Changes in direction, speed, obstacle avoidance, and precise foot placement are common locomotory tasks. Those adjustments require supraspinal involvement (Armstrong, 1988; Dietz, 1992b; Patla et al., 1991). In the second experiment, the attentional demands were evaluated when young and elderly persons had to walk and stop accurately onto a target embedded in the floor. With this experiment, we wanted to examine whether the step length adjustments necessary to point and stop on a target requires the elderly to allocate a greater proportion of their attentional resources to the walking task than young adults.

Six young adults (3 males and 3 females, mean age = 26.5, range = 22-34, standard deviation = 4.6) and 6 elderly persons (4 males and 2 females, mean age = 72.0, range = 66-78, standard deviation = 4.1) participated. The secondary task again consisted of giving a verbal response to a randomly presented auditory stimulus. The primary tasks were: (a) a seated task, and (b) a goal-directed walking task that consisted of stopping with the right foot on one of three randomly presented targets. The targets were rows of 30 light emitting diodes embedded in the floor at 4.8, 5.1, and 5.4 m from the initial starting position. To establish the normal walking behavior (control condition), subjects first performed 20 walking trials without secondary task. Then, during the dual-task conditions, auditory probe stimuli were presented one, two, or three cycles before target contact when the subject's left foot was at toe off or at heel contact. Subjects performed 96 walking trials including 12 catch trials. All conditions were randomly presented. The target was always turned on before subjects started to walk and represented the signal to initiate a trial.

Attentional requirements

Overall, the mean RTs for the elderly were significantly slower than that observed for the young ($p < .01$) and both groups showed slower RTs when walking than when sitting (410 vs. 375 ms for the elderly; 303 vs. 261 ms for the young adults; $p < .01$). RTs for the walking conditions are presented in Figure 4. For both groups, the attentional demands did not vary within and between cycles ($p > .05$). These results contrast with those of Bardy and Laurent (1991). These authors reported that the attentional demands increased as subjects were approaching the target. In Bardy and Laurent's study, the task consisted of pointing with the nose on targets placed at eye level. A possible interpretation for this discrepancy resides in the greater precision constraints associated with the nose-pointing task of Bardy and Laurent.

Step-length adjustments prior to target contact

In our experiment, a precise foot positioning was requested. To examine whether subjects modified their walking pattern when reaching for a target, the cycle length of trials with and without target presentation were compared. More specifically, the length of the first step and of the next 3 complete cycles were submitted to a Group × Target Condition (with or without target) × Cycle before target contact ANOVA with repeated measures on the last two factors. The analysis showed a Group effect with the elderly showing a shorter cycle length than young adults ($p < .05$) but no effect of Target Condition ($p > .05$). All interactions were not significant ($ps > .05$). The spatial constraints were limited in the present experiment since subjects knew in advance the position of the target and the initial starting position was calibrated for each subject (7 steps from target contact). There are evidence that step length is adjusted several steps before target contact when the spatial constraints are more important (e.g., in the long jumping experiment of Lee, Lishman, and Thomson, 1982; and in the gait experiments of Bardy and Laurent, 1991). There are also evidence of anticipatory and reactive locomotor adjustments within the ongoing cycle for clearing obstacles (McFadyen & Winter, 1991; Patla et al., 1991). Alternatively, it is possible that, in the present experiment, subjects programmed their locomotor behavior at gait initiation according to the spatial constraints. This hypothesis is tested in the next experiment.

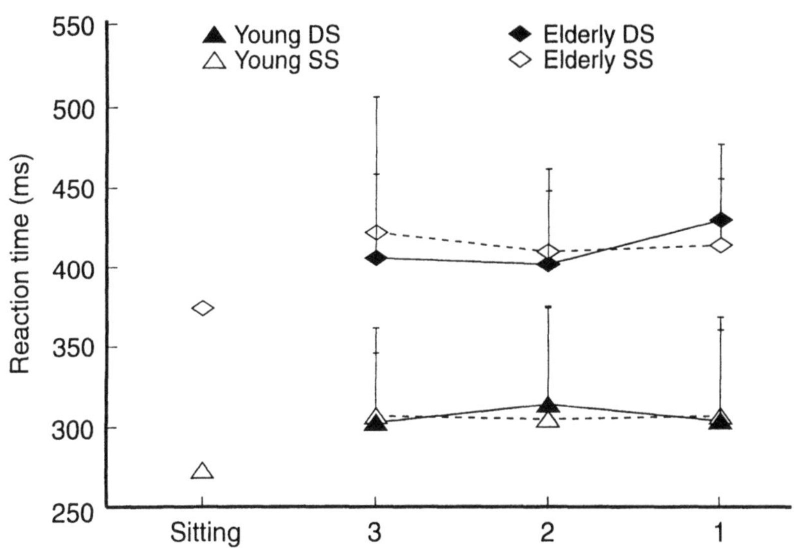

FIGURE 4. Mean reaction times and between-subject standard deviations for the sitting and the walking tasks when stimuli were presented 3 cycles, 2 cycles or 1 cycle prior to target contact. The stimuli were given at the onset of the single-support (SS) or double-support (DS) phases.

EXPERIMENT 3: ATTENTIONAL DEMANDS AT GAIT INITIATION

For arm pointing movements, there are several evidence that the programming and initiation phases are more attention-demanding than the execution phase (Fleury et al., 1993; Posner & Keele, 1969). On the other hand, there has been suggestion that, when crossing a street and stepping over a curb, subjects anticipate the future position of their foot and make adjustments to their step length without being aware of it, suggesting the presence of relatively automatic processes (Pailhous & Clarac, 1984). In this third experiment, we wanted to examine whether gait initiation, because of the spatial constraints imposed by the programming of the locomotor behavior, would require the elderly to allocate more of their attentional resources to the task than young adults. Constraints imposed by the postural destabilization induced by the onset of the walking behavior may also prove more attention demanding than walking without spatial or speed constraints.

Eight young adults (5 males and 3 females, mean age = 25.7, range = 22-34, standard deviation = 3.8) and eight elderly persons (5 males and 3 females, mean age = 71.0, range = 67-72, standard deviation = 1.7) participated. Subjects were first submitted to a dual-task standing upright condition (10 trials with a preferred base of support). Subjects were also asked to walk (8 trials) without stimulus presentation to characterize their normal walking behavior. To examine if spatial constraints affect gait initiation, subjects were instructed to walk toward one of three randomly presented targets (embedded in the floor at 4.8, 5.1, and 5.4 m from the initial starting position) starting with left foot. Stimuli were presented randomly in the first step on left foot toe off or on left foot heel contact (135 walking trials).

Attentional demands

The RTs were submitted to a Group x Task (upright standing, walking single-support and double-support) ANOVA. For both groups, RTs at gait initiation were significantly slower than for the standing task (305 and 330 ms for the double-support and single-support phases vs. 291 ms for the standing condition for the young adults; 406 and 404 ms vs. 331 ms, for the elderly; $p < .01$ for the main effect of Task). These results are similar to that of Experiment 1. Specifically, elderly showed slower RTs than young adults ($p < .01$) but their RTs were not affected by the support phase. On the contrary, young adults showed significantly longer RTs when the stimuli were presented at the onset of the single-support phase ($p < .01$ for the Group by Stance phase interaction).

Across the three experiments, gait initiation did not require more attention than goal-directed walking or normal constraint-free walking. For comparison purposes, RTs for the three experiments are presented in Figure 5. Data for the experiments were submitted to a Group by Walking Stance-phase MANOVA. The MANOVA showed a significant Group × Stance phase effect. The behavior of the elderly persons was very consistent across the three experiments; their RTs were systematically slower than those of the young adults and the elderly never showed any stance phase effects. On the contrary, young adults showed a significant stance phase effect with the RTs being slower when stimuli were presented at the onset of the single-support phase than when they were presented at the onset of the double-support phase.

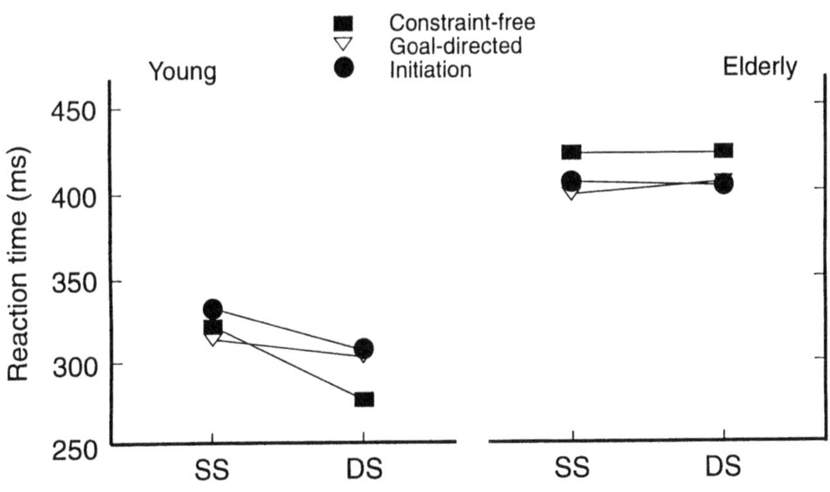

Figure 5. Mean reaction times obtained for the constraint-free (Experiment 1), goal-directed (Experiment 2), and gait initiation experiments (Experiment 3). The reaction times are presented for the young adults and the elderly persons when stimuli were given at the onset of the single-support (SS) and double-support phases (DS).

These results do not confirm our hypotheses that gait initiation requires more attention than normal walking (with or without spatial constraints). The probe-RTs obtained for the elderly persons clearly suggest the use of a secure stereotypical walking behavior. The adoption of such a stable behavior reduces uncertainty and minimizes disequilibrium. Indeed, Patla, Frank, Winter, Rietdyk, Prentice, and Prasad (1993) have shown that elderly have a lower peak force and a slower weight transfer time for gait initiation compared to young adults. Alternatively, subjects were not instructed to initiate their gait as fast as possible in response to the target presentation. More specifically, the primary task was a self-paced task and it is possible that both, young adults and elderly persons, programmed their initial steps before gait initiation (and not at gait initiation as measured in this experiment). This strategy would allow (1) to decrease the attentional demands necessary to regulate the first steps, and (2) provide attentional demands similar to those of Experiments 1 and 2. This interpretation is supported by several studies that have shown that subjects prepare to initiate locomotion with specific anticipatory actions (e.g., Brenière & Do, 1991; Carlsoo, 1966; Herman et al., 1973). Further experiments are necessary to answer this question and more precisely determine the attentional demands for programming gait initiation in reactive (externally triggered) and predictive (self-paced) conditions.

GENERAL DISCUSSION

The three experiments presented support the suggestion that higher cortical central factors are important regulating determinant of postural and locomotor behaviors. With normal aging, the balance demands may require a greater proportion of the attentional resources available. This could also be the case if the decreasing peripheral sensibility associated with normal aging yields an increased cognitive regulation and/or a decreased automaticity in the sensory integration processes. Consistent with this suggestion, Teasdale et al. (1993) reported that, while standing on a force platform, elderly persons' attentional demands increased when sensory information available was reduced (through removing vision and/or adding a foam surface that reduces the quantity and/or quality of the somatosensory information at the ankle). Contrary to young persons, the elderly persons were particularly affected by the combined reduced visual and somatosensory condition. Thus, with aging, the maintenance and the regulation of balance require that a greater amount of the attentional resources be allocated to the postural task. The gait adaptation found in the elderly may also originate from a lost of automaticity resulting in a more attention demanding task. A normal degeneration of the peripheral sensory systems could foster balance problems because of increasing attention demands for the postural task.

Recently, Boucher et al. (1995) demonstrated that the postural stability of diabetic patients suffering from peripheral neuropathy was strongly related with the severity of the neuropathy. Also, when compared to control subjects, the neuropathic patients were less stable when they were submitted to an enriched sensory condition (i.e., no-vision to vision transition requiring a reconfiguration of the postural control system). Alexander (1994) recently suggested that this link between central and peripheral systems has been underemphasized in its potential influence on postural control. Risk avoidance, anxiety, lack of confidence, and fear of falling may also increase task complexity and compromise postural stability.

Indeed, older adults are particularly sensitive to the complexity of the cognitive task (Cerella et al., 1980), and are significantly slower than young persons in divided attention tasks (McDowd & Craik, 1988). It is possible that more complex tasks would not only yield slower RTs but could also compromise the postural stability of the elderly. Several experiments support this possibility. For example, Stelmach et al. (1990) showed that, following a simple arm-swinging task, elderly persons took more time to recover their balance when a simple single-digit mental calculation task was performed concurrently (see also Kerr et al., 1985; McIlroy & Maki, 1993). Recently, Shumway-Cook (1993) also reported that competing demands for attentional resources by a challenging cognitive task (judgment of line orientation and sentence completion) contributed to instability in the elderly. When walking, Mulder et al. (1993) reported that healthy elderly persons, contrary to young persons, significantly reduced their gait speed when asked to perform a concurrent mental calculation task.

When requested to increase their walking speed, elderly are able to do so and there has been suggestion that a decreased speed serves chiefly to produce (1) a more secure and less destabilizing gait and (2) an energy efficient speed of progression (Elble et al., 1991; Ferrandez et al., 1990; Larish et al., 1988; Winter et al., 1990). It is possible that these kinematic gait adaptations also serve to reduce the attention demands necessary to control the continuous disequilibrium intrinsic to walking.

There is also neurophysiological evidence of supraspinal contribution on the peripheral muscular system during the walking cycle. Both, Armstrong and Edgley (1984; see Armstrong, 1988, and Dietz, 1992a, for recent reviews) and Drew (1988) have reported cortical activity varying with the walking cycle. For example, Armstrong and Edgley (1984) observed that the discharge of nucleus interpositus neurons was greater during the swing than during the stance phase. They suggested that nucleus interpositus (through interposito-rubral and rubro-spinal projections) may help the spinal central pattern generators to regulate the levels of flexor muscle contraction that initiate and sustain the swing phase. Although it is hazardous to compare animal and human gait, our observations suggest an alternative hypothesis for the contribution of nucleus interpositus. The slower RTs observed during the single-support phase (swing phase) for the young adults raise the possibility that the programming and dynamic control of balance over the alternating leg movements are cognitively expensive. Ongoing balance regulations may occur at a high level within a walking cycle with the double support phase serving to restabilize balance. Thus, nucleus interpositus activity could also reflect the high level balance evaluation and regulation required by the walking cycle. In line with this suggestion, it is possible that the gait adaptation observed for the elderly persons (e.g., the slower walking speed, shorter step length, slower cadence) serve to minimize the balance requirements of the single-support phase.

Overall, our results emphasize that balance control within the gait cycle is not automatic and loads differentially the higher level cognitive system. Balance control may require a continuous regulation and integration of sensory inputs; the rapidity and efficiency of these high level processes may depend upon the integrity of the peripheral systems and the balance requirements. We have recently tested this possibility by submitting a deafferented patient deprived of sensory information below the neck[1] to conditions similar to that of Experiment 1. While the patient produced RTs faster than the mean RTs obtained for control subjects in the seated condition, the RTs were dramatically slower for the walking task, suggesting that a large portion of the attentional resources available had to be allocated to the control of balance. We believe these observations pose fundamental questions regarding the interdependence between the different levels of organization necessary for an adaptable and optimal gait pattern. Ongoing balance regulations may occur at a high level within a walking cycle with the double support phase serving to restabilize balance. Alternatively, subjects can also modify their walking pattern (e.g., by decreasing their walking speed) to reduce balance requirements.

---

1. The patient suffered from a permanent loss of large myelinated sensory fibers below the neck. A description of the patient is available in Cole et al. (1991). Thanks to Jonathan Cole, Yves Lamarre, and Jacques Paillard for their collaboration on testing this patient.

ACKNOWLEDGEMENTS

Special thanks to Benoit Genest and Gilles Bouchard for programming and technical assistance, respectively and to Richard Courtemanche and Mélanie Hamelin for help in collecting and analyzing data. This project was supported by NSERC and FCAR grants.

REFERENCES

Alexander, N. B. (1994). Postural control in older adults. *Journal of the American Geriatrics Society, 42,* 93-108.

Armstrong, D. M. (1988). The supraspinal control of mammalian locomotion. *Journal of Physiology (London), 405,* 1-37.

Armstrong, D. M., & Edgley, S. A. (1984). Discharges of nucleus interpositus neurones during locomotion in the cat. *Journal of Physiology (London), 351,* 411-432.

Bardy, B. G., & Laurent, M. (1991). Visual cues and attention demands in locomotor positioning. *Perceptual and Motor Skills, 72,* 915-926.

Baron, A., Myerson, J., & Hale, S. (1988). An integrated analysis of the structure and function of behavior: Aging and the cost of dividing attention. In G. Davey & C. Cullen (Eds.), *Human operant conditioning and behavior modification* (pp. 139-166). New York: John Wiley & Sons.

Boucher, P., Teasdale, N., Courtemanche, R., Bard, C., & Fleury, M. (1995). Postural stability in diabetic polyneuropathy. *Diabetes Care* (in press).

Brenière, Y., & Do, M. C. (1991). Control of gait initiation. *Journal of Motor Behavior, 23,* 235-240.

Brocklehurst, J. C., Robertson, D., & James-Groom, P. (1982). Clinical correlates of sway in old age: Sensory modalities. *Age and Ageing, 11,* 1-10.

Calne, D. B. (1985). Normal aging and the nervous system. In R. Andres, L. Bierman, & W. R. Hazard (Eds.), *Principles of geriatric medecine* (pp. 231-235). New York: McGraw-Hill.

Carlsoo, S. (1966). The initiation of walking. *Acta Anatomica, 65,* 1-9.

Cerella, J., Poon, L. W., & Williams, D. M. (1980). Age and the complexity hypothesis. In L. W. Poon (Ed.), *Aging in the 1980s* (pp. 332-339). Washington, DC: American Psychological Association.

Cohen, R. A., & Sparling-Cohen, Y. A. (1993). Response selection and the executive control of attention. In R. A. Cohen (Ed.), *The neuropsychology of attention* (pp. 49-73). New York: Plenum Press.

Collins, J. J., & De Luca, C. J. (1994). Random walking during quiet standing. *Physical Review Letters, 73,* 764-767.

Collins, J. J., & De Luca, C. J. (1993). Open-loop and closed-loop control of posture. A random-walk analysis of center-of-pressure trajectories. *Experimental Brain Research, 95,* 308-318.

Craik, F. I. M., & Byrd, M. (1982). Aging and cognitive deficits: The role of attentional ressources. In F. I. M. Craik & S. E. Trehub (Eds.), *Advances in the study of communication and affect: Aging and cognitive processes* (pp. 191-211). New York: Plenum Press.

Dietz, V. (1992a). Human neuronal control of automatic functional movements: Interaction between central programs and afferent input. *Physiological Reviews, 72,* 33-69.

Dietz, V. (1992b). Neuronal control of stance and gait. In G. E. Stelmach & J. Requin (Eds.), *Tutorials in motor behavior II* (pp. 917-931). Amsterdam: North Holland.

Diggles Buckles, V. (1993). Age-related slowing. In G. E. Stelmach & V. Homberg (Eds.), *Sensorimotor impairment in the elderly* (pp. 73-87). Dordrecht: Kluwer Academic Publishers.

Drew, T. (1988). Motor cortical cell discharge during voluntary gait modification. *Brain Research, 457,* 181-187.

Elble, R. J., Sienko Thomas, S., Higgins, C., & Colliver, J. (1991). Stride-dependent changes in gait of older people. *Journal of Neurology, 238,* 1-5.

Elliott, D., & Allard, F. (1985). The utilization of visual feedback information during rapid pointing movements. *Quarterly Journal of Experimental Psychology, 37A,* 407-425.

Era, P., & Heikkinen, E. (1985). Postural sway during standing and unexpected disturbance of balance in random samples of men of different ages. *Journal of Gerontology: Medical Sciences, 40,* 287-295.

Ferrandez, A. M., Pailhous, J., & Durup, M. (1990). Slowness in elderly gait. *Experimental Aging Research, 16,* 79-89.

Fleury, M., Bard, C., Audiffren, M., Teasdale, N., & Blouin, J. (1993). The attentional cost of amplitude and directional requirements when pointing to targets. *Quarterly Journal of Experimental Psychology: Human Experimental Psychology, 47A,* 481-495.

Geurts, A. C. H., Mulder, T., Nienhuis, B., & Rijken, R. A. J. (1992a). Dual-task assessment of reorganization of postural control in persons with lower limb amputation. *Archives of Physical Medicine and Rehabilitation, 72,* 1059-1064.

Geurts, A. C. H., Mulder, T. W., Nienhuis, B., & Rijken, R. A. J. (1992b). Postural reorganization following lower limb amputation. *Scandinavian Journal of Rehabilitation, 24,* 83-90.

Girouard, Y. (1980). L'attention et l'acquisition de l'habileté motrice. In C. H. Nadeau, W. R. Halliwell, K. M. Newell, & G. C. Roberts (Eds.), *Psychology of motor behavior and sport* (pp. 535-552). Champaign (Il): Human Kinetics Publishers.

Herman, R., Cook, T., Cozzens, B., & Freedman, W. (1973). Control of postural reactions in man: the initiation of gait. In R. S. Stein, K. G. Pearson, R. S. Smith, & J. B. Redford (Eds.), *Control of posture and locomotion* (pp. 363-388). New York: Plenum Press.

Horak, F. B., Shupert, C. L., & Mirka, A. (1989). Components of postural dyscontrol in the elderly: A review. *Neurobiology of Aging, 10,* 727-738.

Johnson, B. M., Miao, M., & Sadum, A. A. (1987). Age-related decline of human optic nerve. *Age, 10,* 5-9.

Kahneman, D. (1973). *Attention and effort.* Englewood Cliffs, NJ: Prentice-Hall.

Kerr, B., Condon, S. M., & McDonald, L. A. (1985). Cognitive spatial processing and the regulation of posture. *Journal of Experimental Psychology: Human Perception and Performance, 11,* 617-622.

Kolleger, H., Wöber, C., Baumgartner, C., & Deecke, L. (1989). Stabilizing and destabilizing effects of vision and foot position on body sway of healthy young subjects: A posturographic study. *European Neurology, 29,* 241-245.

Lajoie, Y., Teasdale, N., Bard, C., & Fleury, M. (1993). Attentional demands for static and dynamic equilibrium. *Experimental Brain Research, 97,* 139-144.

Larish, D. D., Martin, P. E., & Mungiole, M. (1988). Characteristic patterns of gait in the healthy old. In J. A. Joseph (Ed.), *Central determinants of age-related declines in motor function* (pp. 18-32). New York: The New York Academy of Sciences.

Larsson, L., Grimsby, G., & Karlsson, J. (1979). Muscle strength and speed of movement in relation to age and muscle morphology. *Journal of Applied Physiology, 46,* 451-456.

Lee, D. N., Lishman, J. R., & Thomson, J. A. (1982). Regulation of gait in long jumping. *Journal of Experimental Psychology: Human Perception and Performance, 8,* 448-459.

Lee, W. A. (1989). A control systems framework for understanding normal and abnormal posture. *American Journal of Occupational Therapy, 43,* 291-301.

Lipsitz, L. A., Jonsson, P. V., Kelley, M. M., & Koestner, J. S. (1991). Causes and correlates of recurrent falls in ambulatory frail elderly. *Journal of Gerontology, 46,* M114-122.

Lord, S. R., Clark, R. D., & Webster, W. W. (1991). Physiological factors associated with falls in an elderly population. *Journal of the American Geriatrics Society, 39,* 1194-1200.

Manchester, D., Woollacott, M., Zederbauer-Hylton, N., & Marin, O. (1989). Visual, vestibular and somatosensory contributions to balance control in the older adult. *Journal of Gerontology: Medical Sciences, 44,* M118-127.

McDowd, J. M., & Craik, F. I. M. (1988). Effects of aging and task difficulty on divided attention performance. *Journal of Experimental Psychology: Human Perception and Performance, 14,* 267-280.

McFadyen, B. J., & Winter, D. A. (1991). Anticipatory locomotor adjustments during obstructed human walking. *Neuroscience Research Communications, 9,* 37-44.

McIlroy, W. E., & Maki, B. E. (1993). The influence of atention on postural control. *Society for Neurosciences Abstracts, 19,* 990.

Meeuwsen, H. J., Sawicki, T. M., & Stelmach, G. E. (1993). Improved foot position sense as a result of repetitions in older adults. *Journal of Gerontology: Psychological Sciences, 48,* 137-141.

Mulder, T., Berndt, H., Pauwels, J., & Nienhuis, B. (1993). Sensorimotor adaptability in the elderly and disabled. In G. E. Stelmach & V. Hömberg (Eds.), *Sensorimotor impairments in the elderly* (pp. 413-426). Dordrecht: Kluwer Academic Publishers.

Pailhous, J., & Clarac, F. (1984). Approche comportementale de la locomotion: Eléments d'analyse chez l'homme et chez l'animal. In J. Paillard (Ed.), *Comportements, Vol I. La lecture sensorimotrice et cognitive de l'expérience spatiale : directions et distances* (pp. 135-157. Paris: Editions du C.N.R.S.

Parasuraman, R. (1981). Sustained attention: A multifactorial approach. In J. Long & A. Baddeley (Eds.), *Attention and performance IX* (pp. 493-511). Hillsdale, NJ: Lawrence Erlbaum.

Patla, A. E., Frank, J. S., Winter, D. A., Rietdyk, S., Prentice, S., & Prasad, S. (1993). Age-related changes in balance control system: Initiation of stepping. *Clinical Biomechanics, 8,* 179-184.

Patla, A. E., Prentice, S. D., Robinson, C., & Neufeld, J. (1991). Visual control of locomotion: Strategies for changing direction and for going over obstacles. *Journal of Experimental Psychology: Human Perception and Performance, 17,* 603-634.

Posner, M. I., & Keele, S. W. (1969). Attention demands of movements. In *Proceedings of the 17th International Congress of Applied Psychology*. Amsterdam: Swets & Zeitlinger.

Rosenhall, U., & Rubin, W. (1975). Degenerative changes in the human vestibular sensory epithalia. *Acta Otolaryngologia, 79,* 67-81.

Salthouse, T. A. (1985). *A theory of cognitive aging*. Amsterdam: Elsevier.

Sekuler, R., Hutman, L. P., & Owsley, C. J. (1980). Human aging and spatial vision. *Science, 209,* 1255-1256.

Shumway-Cook, A., Baldwin, M., Kerns, K., & Woollacott, M. (1993). The effect of cognitive demands on postural control in elderly fallers and non-fallers. *Society for Neurosciences Abstracts, 19,* 990.

Skinner, H. B., Barrack, R. L., & Cook, S. D. (1984). Age-related decline in proprioception. *Clinical Orthopaedics and Related Research, 184,* 208-211.

Sokal, R. R., & Rohlf, F. J. (1981). *Biometry* (2nd ed.). New York: W. H. Freeman and Company.

Stelmach, G. E., Teasdale, N., DiFabio, R. P., & Phillips, J. (1989). Age related decline in postural control mecanisms. *International Journal on Aging and Human Development, 29,* 205-223.

Stelmach, G. E., Zelaznik, H. N., & Lowe, D. (1990). The influence of aging and attentional demands on recovery from postural instability. *Clinical and Experimental Aging, 2,* 155-161.

Teasdale, N., Bard, C., Dadouchi, F., Fleury, M., LaRue, J., & Stelmach, G. E. (1992). Posture and elderly persons: evidence for deficits in the central integrative mechanisms. In G. E. Stelmach & J. Requin (Eds.), *Tutorials in motor behavior II* (pp. 917-931). Amsterdam: North Holland.

Teasdale, N., Bard, C., Larue, J., & Fleury, M. (1993). On the cognitive penetrability of posture control. *Experimental Aging Research, 19,* 1-13.

Welford, A. T. (1988). Reaction time, speed of performance, and age. In J. A. Joseph (Ed.), *Central determinants of age-related declines in motor function* (pp. 1-17). New York: The New York Academy of Sciences.

Whanger, A. D., & Wang, A. S. (1974). Clinical correlates of the vibratory sense in elderly psychiatric patients. *Journal of Gerontology, 29,* 39-45.

Wickens, C. D. (1984). Processing resources in attention. In R. Parasuraman & D. R. Davies (Eds.), *Varieties of attention* (pp. 63-102). New York: Academic Press.

Winter, D. A. (1991). *The biomechanics and motor control of human gait: Normal, elderly and pathological* (2nd ed.). Waterloo: University of Waterloo.

Winter, D. A., Patla, A. E., Frank, J. S., & Walt, S. E. (1990). Biomechanical walking pattern changes in the fit and healthy elderly. *Physical Therapy, 70,* 340-347.

Changes in sensory motor behavior in aging
A.-M. Ferrandez and N. Teasdale (Editors)

VISUAL CONTROL OF OBSTACLE AVOIDANCE DURING LOCOMOTION: STRATEGIES IN YOUNG CHILDREN, YOUNG AND OLDER ADULTS

Aftab E. Patla,[1] Stephen D. Prentice,[1] and Lilian T. Gobbi[1,2]

1. University of Waterloo, Canada
2. University of Sao Paulo State, Brazil

Abstract

The focus of this chapter is on understanding how obstacle avoidance during locomotion is affected by normal aging process and how this adaptability in locomotor system develops as children acquire independent bipedal locomotion. Obstacle avoidance paradigms offer a rich source of material for understanding the unique sensorimotor integration common to many visually guided movements. Based on studies on young healthy adults, we have proposed a jigsaw puzzle metaphor summarizing the key ingredients for successful obstacle avoidance. The nature of visual and kinesthetic input and the contribution of the effector system properties form the pieces of the puzzle. Studies on healthy older adults reveal relatively well preserved obstacle avoidance strategies, although there are some differences when compared to the healthy young adults. Deterioration in sensory input and effector system characteristics shows up as adaptive changes in feedforward control of limb trajectory over obstacles. This suggests that the puzzle is relatively

---

Correspondence should be sent to Aftab E. Patla, Neural Control Laboratory, Department of Kinesiology, University of Waterloo, Waterloo, ON N2L 3G1, Canada (e-mail: patla@healthy.uwaterloo.ca).

robust with cracks appearing in some pieces. Preliminary studies on children provide interesting signposts for the development of stable obstacle avoidance strategies. High failure rate and poorer control of limb trajectory over obstacles characterize the gait patterns of young children in a cluttered environment. This suggests that the pieces of the puzzle have to be sculpted and merged into a coherent picture during the development process.

Key words: Obstacle avoidance, locomotion, intersegmental dynamics, aging, development.

INTRODUCTION

Our ability to go over or under obstacles in the travel path, that we share with other legged animals, affords us far greater flexibility in the terrains that we can navigate when compared to vehicles with wheels (Patla, 1994a,b; Raibert, 1986). Besides, obstacle avoidance is a critical component of the proactive balance control strategy (Patla, 1994a,b): Recovery from an accidental trip is possible through the use of phase dependent modulation of reflexes (cf. Eng, Winter, & Patla, 1994) but has to be a last resort rather than a default choice for travel over uneven terrain. Successful obstacle avoidance requires integration of primarily visual and kinesthetic sensory input with the dynamics of the effector system (Patla, Rietdyk et al., 1993; 1994; Patla, Prentice et al., 1994a,b; Armand, Patla, & Huissoon, 1994). Researchers have argued that the precise limb positioning involved during travel over uneven terrains forms the evolutionary building block for the precise reaching movements in upper limbs (Georgopoulos & Grillner, 1989; Kalaska & Drew, 1993). Therefore obstacle avoidance paradigms offer a rich source of material for understanding the unique sensorimotor integration common to many visually guided movements.

In this chapter we focus on how obstacle avoidance is affected by normal aging process, and how this adaptability in the locomotor system develops as children acquire independent bipedal locomotion. The high incidence of falls attributed to tripping (Overstall et al., 1977) as we age clearly suggests that obstacle avoidance strategies are adversely affected. By analysing the obstacle avoidance strategies in healthy elderly adults, we may be able to gain insights into the weak links in this sensorimotor transformation. At the other end of the continuum, is examining this capability in children after they are able to walk independently. Observing children trying to walk in a cluttered environment clearly demonstrates that obstacle avoidance does not automatically accompany independent bipedal locomotion. Studying the emergence of obstacle avoidance capabilities in infants during independent locomotion can provide insights into how the various components of this sensorimotor transformation are brought together. We begin with identifying these components from our studies on young healthy adults, and then explore the effects of aging and development.

INGREDIENTS FOR SUCCESSFUL OBSTACLE AVOIDANCE STRATEGIES IN YOUNG ADULTS

The term strategy in the title implies that we have different options for obstacle avoidance and the selection is based on some cost benefit analysis. Whether the obstacle is on ground or above ground, one choice is to go around provided of course that the obstacle does not obstruct the travel path completely. When subjects were given a choice to go over or around obstacles on ground, they chose to go over obstacles of heights that were less than or equal to the subjects lower leg length (Patla unpublished results). This lawful transition in action mode based on body scaled environmental features is similar to the affordance studies that have examined subjects climbing staircase of different riser heights (Warren, 1984) and going through doorways of different widths (Warren & Whang, 1987). Since it is physically possible to go around obstacles of any height, the relative height specific choice must offer some advantage. In contrast, physical limitations would preclude going over obstacles that are too high. It is evident that subjects decision to go around obstacles when it is a proportion of their height embodies other individual's characteristics such as body mass: Going over obstacles that are too high or around obstacles that are too small, would not be energy efficient.

For going over obstacles, the two major sensory modalities that are critical are vision and kinesthetic system. Vision of course is essential for providing exteroceptive information about the obstacle location, and its characteristics that are directly observable such as height and width, and those properties that are visually inferred such as fragility (Patla & Rietdyk, 1993; Patla, Rietdyk et al., 1993, 1994). Shape of the obstacle does not affect the limb trajectory (Spaulding & Patla, 1991), height and width do (Patla & Rietdyk, 1993). Subjects are able to provide an accurate estimate of the obstacle height (Patla, Rietdyk et al., 1993). Besides the exteroceptive role, vision serves an exproprioceptive role providing for example information about limb orientation and its position and velocity as it goes over obstacles. Our work has shown that lack of expproprioceptive information about the swing limb as it goes over obstacle, introduces larger variability in toe clearance, biases the toe clearance of the leading limb upwards, and biases the foot placement before the obstacle backwards (Patla, Rietdyk et al., 1993, 1994).

Consider next the contributions of the kinesthetic system to the control of limb trajectory over obstacles. Knowledge about the location of the limb endpoint, the toe, clearly would be useful for ensuring obstacle clearance. This information can be determined from relative joint angles of the swing limb and information about the stance limb orientation. Muscle receptors are the most likely candidate for obtaining relative joint angle information (Gandevia & Burke, 1992). We have used muscle vibration to investigate what relative joint angle of the swing limb is monitored by the nervous system (Patla, Rietdyk et al., 1994a). Our study have shown that relative knee joint angular displacement is actively monitored and toe elevation adjusted accordingly, while the relative ankle angle is not sampled by the nervous system. This agrees with the reduced stretch reflex gain in the ankle plantar flexors (Stein, 1991), and the dominant role played by the muscles around the knee joint in the control of swing limb flexion (McFadyen & Winter, 1991; Patla & Rietdyk, 1993; Patla, Prentice et al., 1994). Therefore the nervous system selectively samples specific receptors for feedforward control of limb elevation over obstacles. Correlational analysis between the toe clearance of the leading and trailing limb suggest that kinesthetic and visual expproprioceptive inputs from the leading limb are not used to fine tune the trailing limb trajectory: Leading and trailing limb trajectories are controlled relatively independently (Patla, Rietdyk et al., 1994).

Next we focus on the modulations of the normal locomotor movements necessary for obstacle avoidance. Before we discuss the specific movement features, it is important to focus on the success rate. We have shown that obstacles up to 8 cm can be successfully cleared within the time constraints of one step cycle (Patla, Prentice et al., 1991). When obstacles are clearly visible from the start, young subjects are able to successfully clear the obstacle without tripping with the leading and trailing limb: Although in less than 0.5% of the total trials subjects did accidentally trip over the obstacle with the trailing limb (Patla, unpublished results). This suggests that modulation of the locomotor movements are coarse and prone to error.

We have identified key kinematic markers that capture the salient features of the limb trajectory over obstacles (Patla & Rietdyk, 1993). The most critical parameter is the toe clearance over the obstacle. We have found that subjects provide a relatively large safety margin of about 10 cm average clearance (standard deviation of 2 cm) over the obstacle (Patla & Rietdyk, 1993; Patla, Rietdyk et al., 1993, 1994). This supports the earlier observation that motor changes for obstacle avoidance are not precise; the large safety margin is adequate for obstacle avoidance. A second important kinematic measure is the hip hiking and its contribution to toe clearance. The average contribution of hip hiking (representing the stance limb contribution) to toe clearance for the leading limb is about 20% for the leading limb and about 8% for the trailing limb, and is relatively independent of obstacle height (Patla, Rietdyk et al., 1994). Larger contribution by hip hiking would be destabilising; the small contribution as discussed later serves also to assist in swing limb flexion. The third critical kinematic measure is the foot location at toe-off before the obstacle. This represents adjustments made to the strides before the obstacle. We have found that this distance expressed as a percentage of the stride length is about 60% and is relatively independent of obstacle height (Patla, Rietdyk et al., 1993): This suggests that subjects adjust their strides before the obstacle such that location of the obstacle within the stride subject goes over is fixed.

We now turn our attention to the kinetics of obstacle avoidance. Swing limb flexion is achieved by flexion at the hip, knee and ankle joint (Patla & Rietdyk, 1993). We have calculated area under the translational power profile at the hip joint, and rotational power profile at the hip, knee and ankle joint from toe-off to when the limb is over the obstacle. Analysis of these various energy components revealed that translational energy applied at the hip joint and rotational energy applied at the knee joint vary as a function of obstacle height, while rotational energy applied at the hip and ankle joint are invariant and independent of obstacle height (Patla, Prentice et al., 1994; Patla & Prentice, 1994). This clearly suggests that hip and ankle joint flexion is not achieved by active contribution of the muscle about these joints, but is as a result of passive intersegmental dynamics. By exploiting the passive interactions between segments, the nervous system is able to simplify the control of limb elevation and at the same time make it energy efficient. Energy efficiency is achieved by supplying active energy at few joints and has been confirmed by modelling studies (Patla, Prentice et al., 1994; Armand, Patla & Huissoon, 1994). The large toe clearance over obstacles suggests that the safety rather than energy economy guides the limb trajectory specification. Nevertheless, for a desired toe elevation properties of the effector system are exploited to full advantage. Bernstein (1967) in his seminal work had clearly argued that the effector system dynamics has to be incorporated in any movement control scheme.

We are going to use the jigsaw puzzle metaphor for summarizing the salient features of obstacle avoidance strategies. For successful obstacle clearance, sensory information from the visual and kinesthetic system are used to modulate the normal locomotor pattern. The sensorimotor transformation involves modification in the stance and swing limb muscles and incorporates the intersegmental dynamics to achieve limb elevation with minimal muscle involvement. This is summarized in Figure 1.

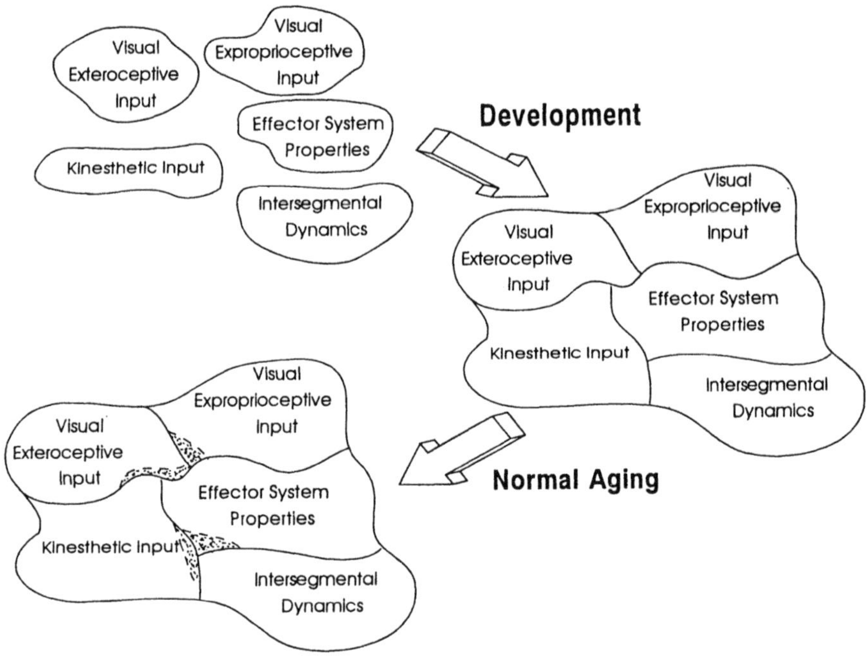

FIGURE 1. Key ingredients for successful, efficient obstacle avoidance strategy are summarized as pieces of a jigsaw puzzle. During development of this strategy, the pieces of the puzzle are sculpted and brought together. Under normal aging process cracks (showed as shaded areas in the pieces) appear in the puzzle. Visual exteroceptive input includes information about obstacle and terrain properties; visual exproprioceptive input provides information about body and limb orientation and velocity; kinesthetic input includes stance and swing limb position and velocity and body orientation referenced to ground; effector system properties include muscle strength and joint range of motion; and intersegmental dynamics refer to among other things the passive forces and moments acting on the multi-linked skeletal system.

EFFECTS OF NORMAL AGING PROCESS ON OBSTACLE AVOIDANCE STRATEGIES

One of the first experiments we have done was placing obstacles of various heights in the travel path and asked the subjects to go from one end of the room to another by either going over or around the obstacle. Whereas the young subjects showed a lawful transition from going over to around when the obstacle height was a certain proportion of their leg length, the results in the elderly were more variable. Only six out of the eighteen subjects tested showed a fixed transition point in the action mode selected; although the transition did not always occur when the obstacle height was equal to subjects lower leg length, but was more variable and more often occurred at lower obstacle height. As stated before, the metric dimension of the body is a natural candidate for arriving at a dimensionless number that captures this transition in action mode since it is correlated with body mass and strength. The body metric, height, is relatively stable as we age, but the body mass, its distribution, strength and range of joint motion (flexibility) can and do change. Therefore the relationship between body height and these other parameters will be modified. Since these other factors can influence the control of limb over obstacles, it is not surprising that the transition point based simply on body metric is not stable. In a recent study on affordances in staircase climbing in young and older adults, Konczak et al. (1992) made similar observations. The results from the other 12 subjects were clearly different from those discussed above. Instead of showing a change in action mode at some obstacle height, this group of subjects chose one strategy (going over or around) for all obstacle conditions: Seven out of the twelve subjects chose to go around for all the obstacles. The choice of going around for all obstacles is understandable since it is a safer strategy. Going over obstacles that are high can be potentially dangerous, not only increasing the chances of tripping but also slipping on surface after the obstacle that is occluded from view by the obstacle. Therefore selection of one strategy for all conditions must offer other advantages. Probably, it eliminates decision making during travel and thereby reduces the demands on the cognitive system.

Next let us examine incidence of tripping over obstacles. Epidemiological studies have shown that tripping over obstacles during walking clearly accounts for the majority of falls in the elderly (Prudham & Evans, 1981), representing the failure of the reactive balance control system to recover from the perturbation. In a recent study by Chen et al. (1991) in which they examined the kinematics of obstacle avoidance in young and older adults, they found that in over 2000 trials four of the older adults accidentally stepped on the obstacle. In our study we also had no incidence of tripping with the leading limb and in less than 0.5% of the trials occasional grazing of the obstacle by the trailing limb. These results clearly suggest some deterioration in the feedforward control of limb elevation over obstacles. The exact locus of the problem is difficult to identify, although we will discuss it later.

Now we examine the three key kinematic markers of limb trajectory over obstacles. In Figure 2, hip and toe spatial trajectories over obstacles of three different heights are shown for one elderly subject. These trajectories look remarkably similar in shape to those of the younger adults. The toe clearance values for the young and older adults are shown in Figure 3. It is clear that toe clearance over level ground is not different between the two groups, and confirms the earlier observation by us (Winter, Patla et al., 1990). In contrast the toe clearance over obstacles is higher in the older adults; it also is a function of obstacle height (Figure 3). The higher toe clearance over obstacles in older adults have been also observed by Watanabe and Miyakawa (1994), although Chen et al. (1991) did not find this to be the case. The discrepancy may be attributed to the use of harness by Chen et al. (1991) which may have resulted in a less conservative strategy by the partici-

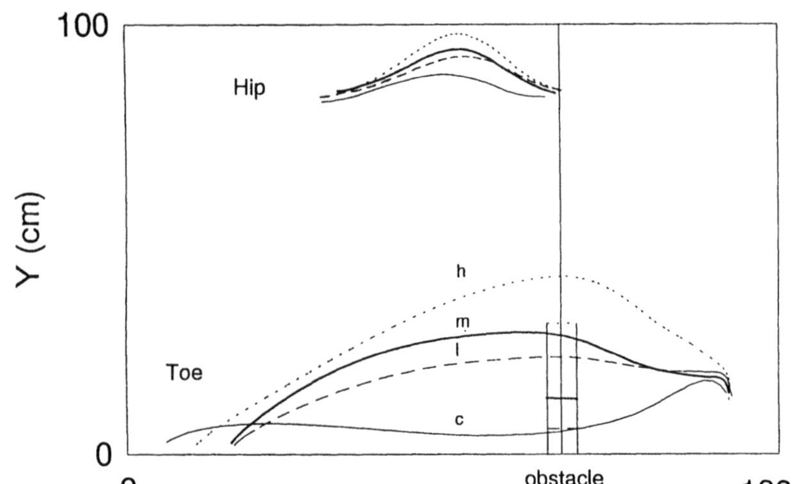

FIGURE 2. Hip and toe spatial trajectories (Y versus X) over obstacles of three different heights (l-7 cm, m-14 cm and h-27 cm) and no obstacle condition (c) are shown for one healthy elderly subject.

pants. The variability in toe clearance over obstacles was not significantly different between the young and older adults. Therefore the higher toe clearance seen for the older adults cannot be attributed to noise in the motor output. Providing a larger toe clearance probably represents a very conservative and safe strategy on the part of the older adults. The increase in toe clearance as a function of obstacle height indicates problems in visual perception. This is confirmed by psychophysical tests on healthy elderly subjects where we asked them to verbally estimate the heights of obstacles presented. In contrast to the young subjects, the healthy elderly subjects showed greater variability in the slopes of the linear regression analyses between the estimated and actual obstacle height. Additional supporting evidence for noise in visual perception affecting the motor output comes from our studies on subjects with visual deficits. Deterioration in visual function such as acuity and contrast sensitivity as we age has been documented. Over 25% of the older adults

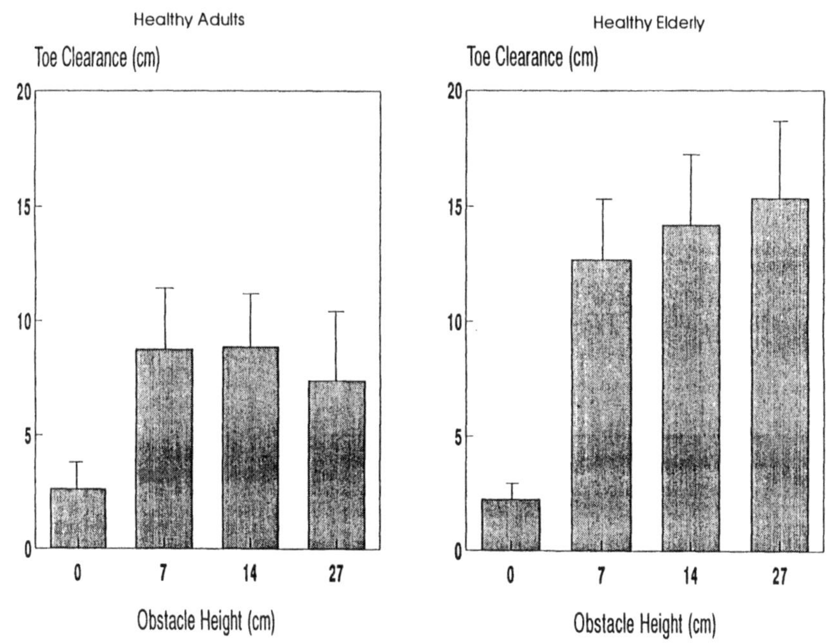

FIGURE 3. Toe clearance (cm) values over obstacles are shown for healthy young and older adults. The bar heights represent mean values across subjects, while the line height on top is equal to one standard deviation. The toe clearance for the no obstacle condition was measured where the obstacle would have been placed.

suffer from macular degeneration (loss of the fovea). In a separate study when we compared subjects with age-related maculopathy (ARM) with age matched controls while going over obstacles of different heights and contrast, we found that ARM subjects had higher toe clearance on small, low contrast obstacles (Patla et al., 1995). This finding supports our earlier contention that increase in toe clearance as a function of obstacle height reflects deterioration in visual judgement of the obstacle height.

The relative contribution of hip hiking to toe clearance shows similar trend as the toe clearance values (Figure 4). The average contribution of hip hiking to toe clearance is considerably higher in the older adults and it increases as a function of obstacle height. Numerical analysis of the data reveal that the higher toe clearance over obstacle is primarily achieved by increasing the hip hiking and not by flexing the swing limb joints more. We have argued that larger contribution by the hip hiking can be potentially destabilising by disturbing the large mass of the upper

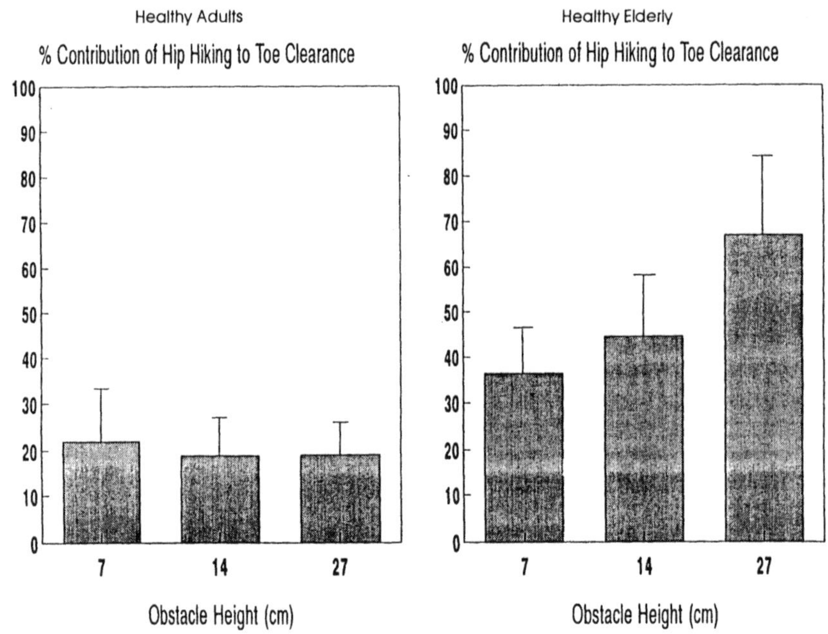

FIGURE 4. The percent contribution of hip hiking to toe clearance are shown for the three obstacle heights for healthy young and older adults. Mean and standard deviation values across subjects are indicated by bar and line heights.

body (Patla & Rietdyk, 1993). It is therefore surprising that the older adults would choose this strategy to achieve higher toe clearance. These results would suggest that the option of increasing swing limb flexion to achieve higher toe elevation was not available. One possible reason for this can be reduced range of motion in the joints, a common observation in the elderly. If greater swing limb flexion was used to achieve higher toe elevation, subsequent landing could be compromised if appropriate action is not taken. In younger adults we know that subjects absorb energy at the hip joint to break the fall of the swing limb under the action of gravity, and supply energy at the ankle joint to orient the foot for proper landing (Patla, Prentice et al., 1994; Patla & Prentice, 1994). It is possible that the older adults used additional hip hiking instead of limb flexion to simplify and ensure stable landing. That the older adults are more concerned about stable landing after going over obstacles is borne out when we examine the kinematic parameters at foot contact. Of particular note is the fact that the older adults landed with negative horizontal velocity indicating that the foot was moving backwards at landing. This strategy only seen in younger adults for very high obstacles would minimize chances of slipping. To achieve such a landing the limb has to be extended out further prior to foot contact; increased swing limb flexion would require additional action to extend the knee joint. This supports our argument that stable landing and not reduced range of motion at the joints is the reason for using additional hip hiking to achieve higher toe clearance.

The last kinematic parameter of interest is the location of the foot position prior to take-off over the obstacles. As discussed before, this positioning reflects adjustments made to the strides before the obstacle. Older adults had their foot further away from the obstacle at take-off compared to the younger adults; although this parameter was not modulated as a function of obstacle height as seen in younger adults (Figure 5). Similar observations have been made by Chen et al. (1991). By adopting this strategy, the older adults would have longer time to monitor and make on-line modification to the leading limb trajectory over the obstacle. It also provides better positioning (not too close to the obstacle) of the trailing limb as it goes over the obstacle. Correlational analysis between the toe clearance of the leading and trailing limb showed no statistically significant relationship similar to those found for the younger adults (Patla, Rietdyk et al., 1994).

The discussion on the obstacle avoidance kinematics shows some interesting feedforward adaptive strategies used by the older subjects to minimize chances of tripping and slipping. Some of the changes, particularly in the toe clearance values, reflect deterioration in the visual input available for locomotor modifications. Consider next the kinetic changes to the normal locomotor patterns in the older adults. The translational energy applied at the hip joint and the rotational energy applied at the hip, knee and ankle joint to elevate the limb from toe-off to over the obstacle are shown in Figure 6 as a function of obstacle height. As these graphs show, only the translational energy applied at the hip joint and rotational energy applied at the knee joint were modulated as a function of obstacle height. These results are similar to those found for the younger adults (Patla, Prentice et al., 1994; Patla & Prentice, 1994). Since swing limb flexion over obstacles involves flexion at all the three joints, these results clearly show that hip and ankle joint flexion is achieved not by active muscle involvement but through passive interaction between segments (Patla, Prentice et al., 1994; Patla & Prentice, 1994; Armand, Patla & Huissoon, 1994). Therefore the exploitation of the intersegmental dynamics to provide simple and efficient control is preserved in the healthy older adults.

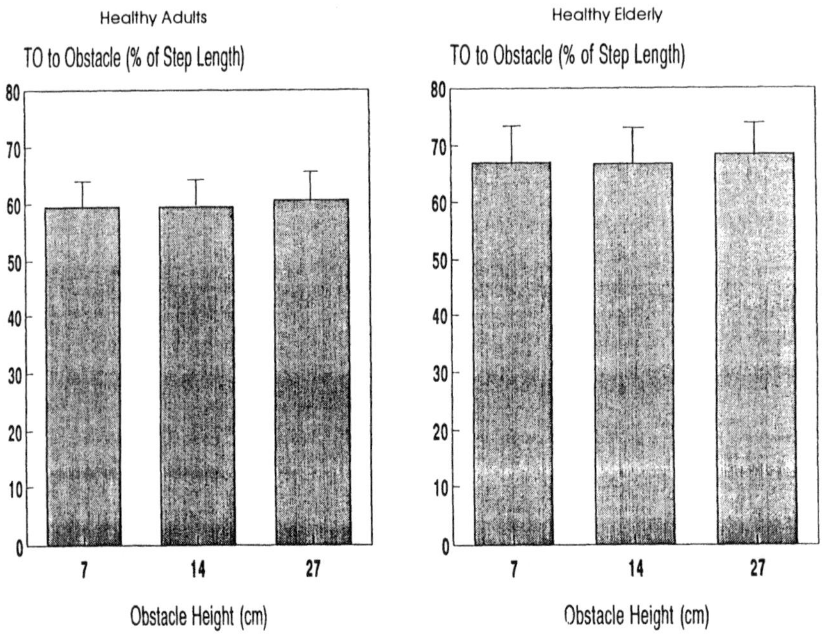

FIGURE 5. The distance from toe-off (TO) to obstacle expressed as a percentage of step length are shown for the three obstacle heights for healthy young and older adults. Mean and standard deviation values across subjects are indicated by bar and line heights.

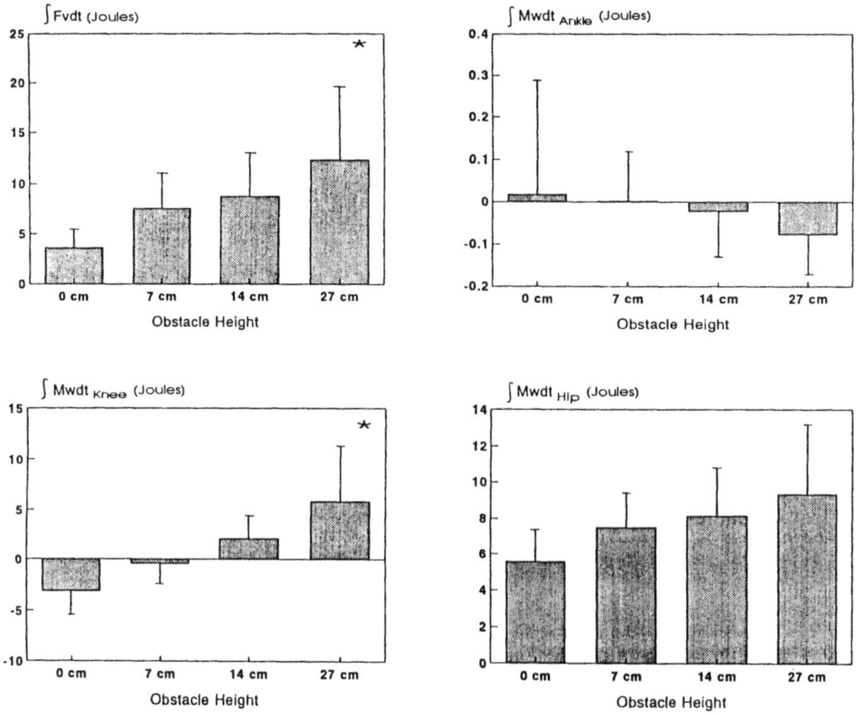

*FIGURE 6. Translational (∫F.dt) and rotational (∫M.ϖdt) energy (Joules) about the hip, knee and ankle joint (calculated by integrating the power profiles (F.v & M.ω) from the toe-off to when the limb is over the obstacle) for the no obstacle and the three obstacle heights are summarized for the healthy elderly subjects. Mean and standard deviation values across subjects are indicated by bar and line heights. A * indicates the energy values were significantly modulated as a function of obstacle height.*

If we return to the jigsaw puzzle metaphor of Figure 1, based on the data presented here we can argue that the obstacle avoidance picture is relatively well preserved as a function of the normal aging process; although there are some cracks appearing in the building blocks shown schematically in Figure 1. Deterioration in sensory input and effector system characteristics show up as adaptive changes in feedforward control of limb trajectory over obstacles. The use of same sensory systems during normal level path locomotion is relatively unaffected by age (see Konczak, 1994). Therefore obstacle avoidance paradigm can offer an early window on changes with age in sensorimotor coupling during locomotion. Any sensory or motor pathology on top of the normal aging process can lead to a breakdown in this critical locomotor adaptive strategy.

DEVELOPMENT OF STABLE AND EFFICIENT OBSTACLE AVOIDANCE STRATEGIES IN CHILDREN

Researchers have studied development of independent bipedal locomotion in young infants from a variety of perspectives, and have documented the kinematics, kinetics and muscle activity patterns primarily over level ground during straight path locomotion (see for example review by Forssberg, 1985). The work by Thelen and her colleagues (see, for example, review by Thelen, 1985) have clearly highlighted among other things the harnessing of the effector system dynamics to the expression of bipedal locomotion. The work by E. J. Gibson and her colleagues have shown how visual and kinesthetic input about the traversability of terrains influences the walking and exploration patterns in infants (see review by Gibson & Schmuckler, 1989). Both of these research programs have a bearing on development of obstacle avoidance strategies. As can be seen in the literature, the terms maturation and development have been used interchangeably. We understand that development is a resultant process of the integrated contribution of maturation and experience. We are interested in determining the characteristics of obstacle avoidance strategies in children following the emergence of voluntary stable bipedal locomotion and charting its progress as children develop. This offers unique opportunity to examine how this particular sensorimotor transformations develop.

We have recently used nested rings to bring together factors that influence the expression of skilled locomotor behaviour (Patla, 1994b). The innermost circle is the effector system dynamics and morphology and is surrounded by a ring representing the core locomotor pattern. The core locomotor pattern reflects the output from the rudimentary neural circuitry present in animals (Bradley & Bekoff, 1989). The emergence of stable alternating pattern of the limbs though requires interaction between the neural circuitry and passive dynamics of the effector system. Researchers studying infant kicking movements (precursor to and primitive building blocks for stepping movements) and supported treadmill stepping movements have very nicely shown that stable and efficient interaction between neural circuitry and effector system dynamics is an integral part of the maturation process (Jensen et al., 1994; Ulrich et al., 1994). This is just as important as the maturation of higher centers specially related to the dynamic equilibrium system (Forssberg, 1985; Leonard et al., 1991) which represent the next two rings in our framework for understanding skilled locomotor behaviour.

The ability to travel in a cluttered environment over different surfaces await the emergence of a stable walking pattern over level ground. The unique intersensory coupling (between vision and kinesthetic system) and sensorimotor coupling necessary for safe and efficient obstacle avoidance is therefore built on the stable normal locomotor pattern (Bril & Brenière, 1993). The pattern of failure in obstacle avoidance suggests deficiency in visual judgement of obstacle height and location vis a vis their stride. Problems with small obstacles is similar to the findings for ARM subjects discussed before. Also, motor nerves and anterior roots mature before dorsal nerve roots and sensory nerves following a cephalo-caudal principle for each segment in the spinal cord. This process is completed between 2 and 5 years of age (Rafalowska, 1979, cited by Sutherland et al., 1988). This has clear implications for the use of kinesthetic input in feedforward and feedback control of limb trajectory over obstacles and also the integration of kinesthetic input with the visual input. Thus safe obstacle avoidance during locomotion, a key marker of skilled locomotor behaviour, has to await maturation of the sensory systems and the motor apparatus and coupling between them during development.

We have begun to examine whether or not children systematically choose to go over or around obstacles of different heights placed in their travel path. Our preliminary work on four children (14 - 30 month old) have shown no consistent pattern even across trials in their choice of strategies when faced with an obstacle during walking. It is probably unreasonable to expect that children will show the lawful transition from going over versus around based simply on the body scaled metric dimension of the obstacle. Besides factors such as motivation and attention which are difficult to control, the developmental changes in anthropometric parameters (such as inertias, masses, strength, etc.) preclude a simple relationship between body height and other factors that can impact whether or not subjects can go over obstacles. If we through enticements force the children to go over obstacles, some interesting results emerge. These are discussed next.

First let us focus on the success rates for going over obstacles. Out of a total of 90 trials (30 trials each in three children), in 17 trials children either hit, stepped on or touched the obstacles with the leading or trailing limb. The probability of hitting the obstacle was lower for the leading limb when compared to the trailing limb (2 versus 4), with the children having the greatest difficulty with the smallest obstacle (0.5 cm high). These failure rates are high when we compare them to the results from the young and older adults. These observations highlight a key point. Successful negotiation of obstacles do not occur simultaneously with stable bipedal walking pattern.

Consider next the key kinematic parameters that we have been using to describe obstacle avoidance. The toe clearance values for the leading and trailing limb are shown in Figure 7. Toe clearance values show similar trends as the older subjects (for 0.5 and 6 cm high obstacles). For the highest obstacle since not all subjects could go over, the numbers are based on smaller n's (trials and subjects). Leading and trailing limb show similar toe clearance values (except for the high obstacle); this agrees with the data on younger adults (Patla, Rietdyk et al., 1994). Hip and toe spatial trajectories over obstacles of different heights are shown in Figure 8 for one subject. If we examine these trajectories some other differences emerge. The relative contribution of hip hiking to toe clearance is higher than for the young adults (over 30% compared to around 20% for young adults) and is similar to the results from the older adults. Based on relatively small changes in the angular displacement of the

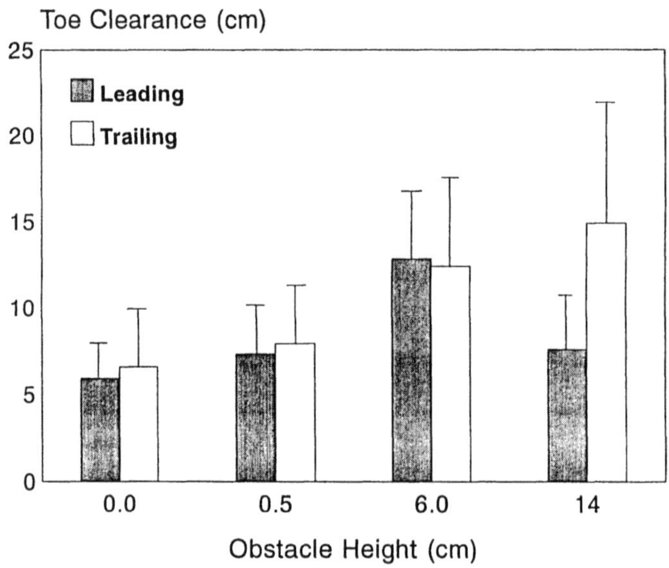

FIGURE 7. Toe clearance (cm) values for the leading and trailing limb are shown for the no obstacle condition and the three obstacle heights. Mean and standard deviation values are indicated by bar and line heights.

hip, knee and ankle joint, we speculate that the larger contribution of hip hiking to toe clearance reflects inability to exploit the passive inter-segmental dynamics to achieve swing limb flexion over obstacles. Definitive proof for this will have to wait for the kinetic analysis on these children's swing limb trajectory. Studies on infant supported stepping (Ulrich et al., 1994) clearly indicate that stable efficient pattern wether it is for level walking or for going over obstacles involve exploitation and control of the effector system dynamics. It is reasonable that this neural-effector system dynamics coupling is not a single entity but is rather context specific: Emergence of a stable interaction in one context may facilitate but does not guarantee stable interaction in another context. The second aspect of the limb trajectories that merit attention is the high variability seen in the location of the foot prior to take-off over the obstacle. For example, the values for one subject ranged all the way from 14% (foot very close to the obstacle) to 65% (closer to the values for young adults). This variability reflects poorer stride adjustments prior to going over the obstacle.

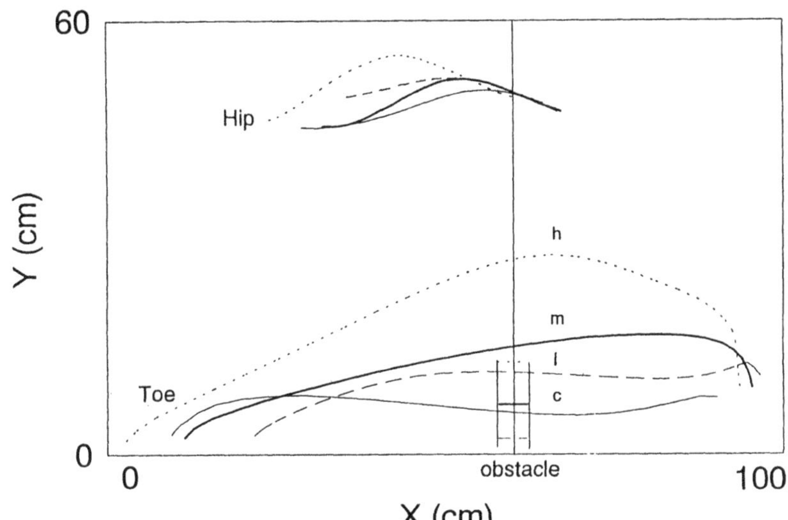

FIGURE 8. Hip and toe spatial trajectories (Y versus X) are shown for the no obstacle condition and the three obstacle heights for one young child. The legends are: c - no obstacle condition; l - 0.5 cm obstacle; m - 6 cm obstacle; and h - 14 cm obstacle.

Even though the data on children are preliminary, they provide some interesting signposts for the development of a stable obstacle avoidance strategy. The pieces of the jigsaw puzzle have not been sculpted and brought together (Figure 1), resulting in high failure rate and poorer control of limb trajectory over obstacles in children who have just begun to walk independently. The challenge is to expand our data base, and carry out further analysis to map the time course of development of how the visual input (exteroceptive and exproprioceptive), and kinesthetic input are used along with the properties of the effector system (passive interaction between segments) to control limb trajectory over obstacles.

ACKNOWLEDGEMENTS

The financial support from NSERC Canada is gratefully acknowledged. Lilian T. Gobbi is supported by a scholarship from CAPES, Brazil.

REFERENCES

Armand, M., Patla, A. E., & Huissoon, J. P. (1994, submitted). The role of active torques and forces and intersegmental dynamics in the control of swing phase over level ground and obstacles: Biomechanical modelling approach. *IEEE Transactions on Systems, Man and Cybernetics.*

Bernstein, N. (1967). *The co-ordination and regulation of movements.* Oxford, UK: Pergamon Press.

Bradley, N. S., & Bekoff, A. (1989). Development of locomotion: Animal models. In M. H. Woollacott & A. Shumway Cook (Eds.), *Development of posture and gait across the life span* (pp. 48-73). Columbia, SC: University of South Carolina Press.

Bril, B., & Brenière, Y. (1993). Posture and independent locomotion in early childhood: Learning to walk or learning dynamic postural control? In G. J. P. Salvesberg (Ed.), *The development of coordination in infancy* (pp. 337-358). Amsterdam: Elsevier.

Chen, H. C., Ashton-Miller, J. A., Alexander, N. R., & Schultz, A. R. (1991). Stepping over obstacles: Gait patterns of healthy young and old adults. *Journal of Gerontology, 46* (6), M196-203.

Eng, J. J., Winter, D. A., & Patla, A. E. (1994). Neuromuscular strategies for recovery from a trip in early and late swing during human walking. *Experimental Brain Research, 102,* 339-349.

Forssberg, H. (1983). Ontogeny of human locomotor control. I. Infant stepping supported locomotion and transition to independent locomotion. *Experimental Brain Research, 57,* 480-493.

Gandevia, S. C., & Burke, D. (1992). Does the nervous system depend on kinesthetic information to control natural limb movements. *Behavioral and Brain Sciences, 15,* 614-632.

Georgopoulos, A. P., & Grillner, S. (1989). Visuomotor coordination in reaching and locomotion. *Science, 245,* 1209-1210.

Gibson, E. J., & Schmuckler, M. A. (1989). Going somewhere: An ecological and experimental approach to development of mobility. *Ecological Psychology, 1* (1), 3-25.

Jensen, J. L., Ulrich, B. D., Thelen, E., Schneider, K., & Zernicke, R. F. (1994). Adaptive dynamics of the leg movement patterns of human infants. I. The effects of posture on spontaneous kicking. *Journal of Motor Behavior, 26* (4), 303-312.

Kalaska, J. F., & Drew, T. (1993). Motor cortex and visuomotor behavior. *Exercise and Sport Sciences Reviews,* 397-436.

Konczak, J. (1994). Effects of optic flow on the kinematics of human gait: A comparison of young and older adults. *Journal of Motor Behavior, 26* (3), 225-236.

Konczak, J., Meeuwsen, H. J., & Cress, M. E. (1992). Changing affordances in stair climbing: The perception of maximum climbability in young and older adults. *Journal of Experimental Psychology: Human Perception and Performance, 18* (3), 691-697.

Leonard, C. T., Hirschfeld, H., & Forssberg, H. (1991). The development of independent walking in children with cerebral palsy. *Development Medicine and Child Neurology, 33,* 567-577.

Maurer, D., & Maurer, C. (1988). *The world of the newborn.* New York: Basic Books.

McFadyen, B. J., & Winter, D. A. (1991). Anticipatory locomotor adjustments during obstructed human walking. *Neuroscience Research Communications, 9* (1), 37-44.

Overstall, P. W., Exton-Smith, A. N., Imms, F. L., & Johnson, A. L. (1977). Falls in the elderly related to postural imbalance. *British Medical Journal, 1,* 261-264.

Patla, A. E. (1993). Age-related changes in visually guided locomotion over different terrains. In G.E. Stelmach & V. Homberg (Eds.), *Sensory-motor impairments in the elderly* (NATO ASI Series volume) (pp. 231-252). Dordrecht: Kluwer.

Patla, A. E. (1995a). Mobility problems in the elderly: Diagnosis and rehabilitation strategies. In R. L. Craik & C. A. Oatis (Eds.), *Gait analysis: Theory and application* (pp. 436-449). St. Louis, MO: Mosby.

Patla, A. E. (1995b). The neural control of locomotion. In B. S. Spivack (Ed.), *Evaluation and management of gait disorders* (pp. 53-78). New York: Marcel Dekker Inc.

Patla, A. E. (1995, in press). Neurobiomechanical Bases for the Control of Human Locomotion. In A. Bronstein, T. H. Brandt, & M. Woollacott (Eds.), *Clinical aspects of balance and gait disorders.* City, UK: Edward Arnold.

Patla, A. E., Elliott, D. B., Flanagan, J., Rietdyk, S., & Spaulding, S. (1995, in press). Effects of age-related maculopathy on strategies for going over obstacles of different heights and contrast. *Proceedings of the 2nd North American Clinical Gait Conference, Gait and Posture.*

Patla, A. E., & Prentice, S. D. (1995 submitted). The role of active forces and intersegmental dynamics in the control of limb trajectory over obstacles during locomotion in humans. *Experimental Brain Research.*

Patla, A. E., Prentice, S. D., Armand, M., & Huissoon, J. P. (1994). The role of effector system dynamics on the control of limb trajectory over obstacles during locomotion: Empirical and modelling approaches. In K. Taguchi, M. Igarashi, & S. Mori (Eds.), *Vestibular and neural front* (XIIth International Symposium on Posture and Gait) (pp. 33-336). Amsterdam: North-Holland/ Elsevier.

Patla, A. E., Prentice, S. D., Robinson, C., & Neufeld, J. (1991). Visual control of locomotion: Strategies for changing direction and for going over obstacles. *Journal of Experimental Psychology: Human Perception and Performance, 17* (3), 603-634.

Patla, A. E., & Rietdyk, S. (1993). Visual control of limb trajectory over obstacles during locomotion: Effect of obstacle height and width. *Gait and Posture, 1* (1), 45-60.

Patla, A. E., Rietdyk, S., Martin, C., & Prentice, S. (1995, in press). Locomotor patterns of the leading and trailing limb while going over solid and fragile obstacles: Some insights into the role of vision during locomotion. *Journal of Motor Behavior.*

Patla, A. E., Rietdyk, S., Prentice, S., Unger-Peters, G., & Gobbi, L. (1993). Understanding the roles of sensory inputs in the control of limb trajectory over obstacles during locomotion. *Society for Neuroscience Abstracts, 19.*

Prudham, D., & Evans, J. G. (1981). Factors associated with falls in the elderly: A community study. *Age and Ageing, 10,* 141-6.

Raibert, M. H. (1986). *Legged robots that balance.* Cambridge, MA: MIT Press.

Spaulding, S., & Patla, A. E. (1991). Obstacle avoidance during locomotion: Effect of obstacle shape on gait modifications. *International Brain Research Organisation Conference.*

Stein, R. B. (1991). Reflex modulation during locomotion. In A. E. Patla (Ed.), *Adaptability of human gait: Implications for the control of locomotion* (pp. 21-36). Amsterdam: North-Holland/Elsevier.

Sutherland, D., Olslen, R., Biden, E., & Wyat, M. (1988). *The development of mature walking.* London, UK: Mac Keith.

Thelen, E. (1985). Development origins of motor coordination: Leg movements in human infants. *Developmental Psychobiology, 18* (1), 1-22.

Ulrich, B. D., Jensen, J. L., Thelen, E., Schneider, K., & Zernicke, R. F. (1994). Adaptive dynamics of the leg movement patterns of human infants: II. Treadmill stepping in infants and adults. *Journal of Motor Behavior, 26* (4), 313-324.

Warren, W. H., Jr. (1984). Perceiving affordances: Visual guidance of stair climbing. *Journal Experimental Psychology: Human Perception and Performance, 10,* 683-703.

Warren, W. H., Jr., & Whang, S. (1987). Visual guidance of walking through apertures: Body scaled information for affordances. *Journal of Experimental Psychology: Human Perception and Performance, 13,* 371-383.

Watanabe, K., & Miyakawa, T. (1994). Gait analysis during stepping over the different height of obstacles in aged persons. In K. Taguchi, M. Igarasha, & S. Mori (Eds.), *Vestibular and neural front* (pp. 195-198). Amsterdam: North-Holland/Elsevier Science.

Winter, D. A., Patla, A. E., Frank, J. S., & Walt, S. E. (1990). Biomechanical walking pattern changes in the fit and healthy elderly. *Physical Therapy, 70* (6), 340-347.

Changes in sensory motor behavior in aging
A.-M. Ferrandez and N. Teasdale (Editors)

CONSTRAINTS ON PREHENSION: A FRAMEWORK FOR STUDYING THE EFFECTS OF AGING

Eric A. ROY,[1] Patricia L. WEIR[2] and Jack L. LEAVITT[2]

1. University of Waterloo, Canada
2. University of Windsor, Canada

Abstract

One of the characteristic changes in performance seen with aging is a slowing in cognitive and motor processes. Work using a variety of motor tasks reveals longer processing times for the elderly on measures reflecting response selection and programming (reaction time) and movement execution (movement time), suggesting that aging affects each stage in processing a motor response. Much of this work on aging has been limited to simple flexion/extension and/or pointing movements which do not involve the more intricate, complex hand movements used in activities of daily living. Since both daily living and clinical assessment require more complex prehension movements, we are focusing our studies on these more complex functional movements. We begin by examining the various theories of aging, contrasting in particular hardware with software explanations. Models of prehension are then discussed, with a specific emphasis on the movement constraints framework proposed by MacKenzie and Iberall (1994). We then review our work on aging and prehension and conclude with a discussion of how these findings might be interpreted using the constraints framework.

Key words: Aging, attention, motor control, movement time, prehension, reaction time, spatial variability of movement.

Correspondence should be sent to Dr. Eric A. Roy, Department of Kinesiology, University of Waterloo, Waterloo, Ontario N2L 3G1, Canada (e-mail: eroy@healthy.uwaterloo.ca).

INTRODUCTION

Changes over the adult lifespan in movements such as prehension have been analyzed using an information processing framework as this paradigm is particularly well-suited to study any shifts in cognitive processing in later life (Klatzky, 1988). While there is no agreement on the specifics of a particular model for information processing, a common conceptual framework does exist (Lovelace, 1990). The general tenets of this approach include the following notions: information from the environment that serves as input to the perceptual-motor system is processed through a number of stages resulting in an observable motor response; identification of the stages is derived from observing performance in several experimental conditions, each of which is thought to require a particular stage. This approach relies heavily on such temporal measures as reaction time, and assumes that this interval is controlled by sequential and/or parallel information processes. If an experiment is designed so the time of all processing stages but the one of interest is held constant, it is possible to infer that any change in reaction time is attributable to the particular processing stage being studied. In psychomotor performance three stages have been identified (Schmidt, 1982). The relative contribution of each stage to performance depends on the task. The first, stimulus identification, involves stimulus detection and pattern recognition. The second and third stages, response selection and response programming, encompass selecting the appropriate response and organizing and initiating movement, respectively. Reaction time, the interval of time between stimulus presentation and response initiation, reflects the summation of the stages of information processing.

Although the major variables that affect information processing occur prior to movement, once initiated the processes of movement execution are also reflected in derivations of a temporal measure, movement time. Again, as with the reaction time paradigm, experimental manipulations serve to provide insights into the processes involved in controlling the movement. In general, movement of the arm to a target involves two principle stages, a ballistic or pre-programmed stage and a feedback-based stage (Woodworth, 1899). Traditionally, movement time, the interval of time between movement initiation and response completion, has been used to describe these two stages of movement. The recent advent of advanced optoelectric movement analysis systems, however, permits an opportunity to partition movement time into portions that reflect these stages more directly (e.g., acceleration and deceleration portions) through movement kinematics. These measures which include linear and angular displacements, velocities and accelerations describe the movement pattern independent of the forces that cause the movement (Winter, 1979). In motor control, kinematic analyses have been used to infer motor planning and control processes (Atkeson & Hollerbach, 1985; Hollerbach & Atkeson, 1987), thereby providing a window into response programming and the processes underlying movement execution (Abend, Bizzi, & Morasso, 1982; Annett, 1988; Goggin & Stelmach, 1990; Hay, Bard, Fleury, & Teasdale, 1991; Hollerbach, 1982; Hollerbach & Atkeson, 1987). For example, invariance in the shape of a movement trajectory (e.g., velocity profile) across different movement conditions (e.g., movement amplitude) is thought to reflect the operation of the same motor program across the different conditions. Thus, a combination of traditional temporal measures with kinematic measures will provide a complete description of the movement pattern produced.

This chapter focuses on the last stage in the information processing paradigm, response programming, and on the control processes involved during movement execution. We begin by examining the various theories of aging, contrasting in particular hardware with software explanations of cognitive slowing. Work on the effects of aging on response programming and movement execution is then reviewed. Models of prehension, the focus of our work on aging, are then examined, with a specific emphasis on the movement constraints framework proposed by MacKenzie and Iberall (1994). We then review our work on aging and prehension and conclude with a discussion of how these findings might be interpreted using the constraints framework.

THEORIES OF AGING

A number of authors have pointed out (Bates & Goulet, 1971; Bates, Reese, & Nesselroade, 1978; Birren, 1959, 1974; Birren, Bengston, & Deutchman, 1988; Botwinick 1978; Kausler, 1982; Kuhlen, 1963; Salthouse, 1982; Wohlwill, 1970) that age is not a causal variable. Consequently, the passage of time in and of itself is responsible for nothing. Explanations of changes associated with age therefore must rely on variables that exert their effects over time, not on time itself. Obviously, the mechanics of the body will limit strength, endurance, agility, speed and range of movement, but often the major limitation to the performance of activities of daily living arises from a reduction in central processing capabilities (e.g., attention, response selection and programming, e.g., Welford, 1985). There are several theoretical explanations for age-related differences in motor function, ranging from the very broad to the very specific, and they differ in whether they consider age-related changes in motor function as real or as ancillary to a general deterioration of the central nervous system.

Different cognitive processes, and by inference motor processes, decline with age at different rates within and between individuals. Thus, the resulting slowing in movement for the elderly is not necessarily the same from person to person (Walsh, 1982; Welford, 1985). Further this pattern of change may be dependent on the nature of the task. For example, tasks involving motor skills that afford automatic processing (Posner & Snyder, 1975; Schneider & Shiffrin, 1977; Shiffrin & Schneider, 1977), often show age-equivalent performance (Hasher & Zacks, 1979). However, this may be true only when these tasks are learned prior to old age (Fisk, McGee, & Giambra, 1988). Thus, in considering changes in motor performance with age, movement control must be understood as a complex set of interactions between the performer and the constraints faced by the performer.

There are two general hypotheses of motor slowing, and a meaningful metaphor to explain this slowing is the computer (Charness, 1985, 1991; Salthouse, 1985a). The major distinction to be made is between a computer's hardware and software. Hardware explanations focus upon neuroanatomical changes occurring with aging that may underlie the observed concomitant cognitive changes (Petit, 1982). If the speed of cognitive operations is determined by the integrity of the nervous system and if the neural network supporting cognition is impaired by aging as Cerella (1990) suggests, then slowing is an inevitable result. Such neuroanatomical mechanisms can be compared to the hardware or circuitry of a computer system. Software explanations focus upon computational efficiency, as Charness (1991) notes.

> "But, as many of those with experience with microcomputer software recognize, different programs can have vastly different efficiency. A tightly coded program running at 8 Mhz can outperform a sloppy one running at 12 Mhz. That is, older adults operating with efficient cogni-tive routines, software, can easily outperform younger adults who don't have access to the same efficient programs" (Charness, 1991, p. 205).

Each hypothesis (hardware vs software) can be used to argue that what appear to be age differences in movements are really manifestations of more fundamental age differences. The first position argues that age-related effects are simply expressions of a general slowing of cognitive operations in old age (Cerella, 1985; Salthouse, 1985a,b). There are three different versions of this 'hardware' position. These can be labelled the 'Input/Output' (Salthouse, 1985a), the 'Birren' (as presented in Salthouse, 1982), and the 'Neural noise' hypotheses (Welford, 1982) of cognitive slowing. The second position, software, argues that age-related effects are the result of different and potentially less efficient programming and/or control strategies.

Hardware explanations

We might envisage that changes in the hardware (e.g., the nervous system) which arise with age can be central (within the central nervous system) or peripheral, affecting processes within the peripheral nervous system, such as muscles, joints or sensory receptors. The "strong versions" (as noted by Hartley, 1992) attempt to ascribe changes in function with age to particular changes in hardware which appear with age. The Neural Noise Hypothesis, one of the strong versions of the hardware hypothesis, argues that slowing with age results from increased neural noise. There are a number of possible reasons for the increase in neural noise, for example, dendritic atrophy, loss of neural tissue, decreased cerebral blood flow, increased lipofuscin, but all result in signals being less recognizable in the central nervous system of older adults. (Salthouse, 1982). This explanation for age-related differences in cognitive function is presented by Welford (1982, p. 163). He states

> "if the signal-to-noise ratio is low, performance is inaccurate, either because low signal levels cause errors of omission (forget to perform an operation) or because noise causes errors of commission (perform an operation out of order or at the wrong time)".

For older adults, in order to compensate for their low signal-to-noise ratio, more time is taken to examine the signal and average out the noise. Thus, with additional time older adults can have similar signal-to-noise ratios as young adults, and be as accurate in the performance of a task.

The Input\Output Hypothesis (Salthouse, 1985a) is another of the strong versions of the hardware hypothesis. It specifies a number of peripheral mechanisms which may be responsible for the slowing observed with age. Uniform slowing of synaptic transmission (Birren, 1974; Birren, Woods, & Williams, 1980) or information loss at each transmission (Myerson, Hale, Wagstaff, Poon, & Smith, 1990) are two mechanisms thought to be important.

The "weak versions" (as noted by Hartley, 1992) of the hardware theory of aging propose that changes in function with age arise from changes in hardware without specifying what these changes might be.

The Birren hypothesis is an example of a weak version that proposes that the time of all neural processes becomes slower with advanced age: older adults can use the same behaviour processes, but are simply slower (Amrhein, Stelmach, & Goggin, 1991; Gottsdanker, 1980a,b, 1982a,b; Stelmach, Goggin, & Amrhein, 1988; Stelmach, Goggin, & Garcia-Colera, 1987). A cousin of the Birren hypothesis, the Cycle-Time hypothesis (Salthouse & Somberg, 1982; Simon & Pouraghabagher, 1978), attributes the slowing in older adults to all stages of the information processing system, indicating that slowing is a generalized phenomenon.

These hardware theories of aging of cognitive, and by inference motor, function present a picture of cognitive aging where older adults experience a global slowing due to a reduction in central resources (Salthouse, 1985a, 1987, 1988a,b,c). Salthouse (1985b) has pointed out that there are several ways to conceptualize these resource limitations, two of which are of particular relevance here. First is the idea that there is a limited short-term memory capacity, and second, the total amount of mental energy or attention available for the execution of operations may be reduced in old age.

Software explanations

The hardware explanations reflect changes in central processing which are not under the direct control of the individual. Software explanations on the other hand reflect processes over which the aging individual does have control. These would include: a) the efficiency of programming and/or control processes, and b) the use of different cognitive strategies (paths to solve the movement problem). Within the context of movement control, inefficient control would exist when, as some suggest, older adults operate as a closed-loop system. Welford (1981) says older adults spend more time monitoring their responses than do young adults, which as Rabbitt (1982) explains may be due to the method of control used. Feedforward control involves initiating changes in the movement in anticipation of changes that may occur in the future. Feedback control, on the other hand, involves utilizing current information to initiate corrective patterns of an ongoing movement. Rabbitt suggests that young adults are able to use either type of control when necessary, while older adults lose the option of utilizing feedforward control mechanisms and must rely on feedback control. As a consequence older adults are disadvantaged in two ways: they lose control options and, thus, are left using the less efficient control process. Thus, older adults are slower because of a reliance upon sensory feedback to correct ongoing movements. Some support for this proposal is found in work by Haaland, Harrington, and Grice (1993) who found that their older subjects were more dependent on visual feedback in a visual pointing task, particularly for the longer movements.

Strategic causes of slowing in older adults in the context of reaction time tasks have also been proposed by Welford (1981, 1982, 1984a). He attributes the slowing to cautionary behaviour and an increased emphasis upon accuracy of response. He has suggested that older adults set a more cautious criterion for accepting or rejecting the presence of a signal. This strategy allows the older adults to be more accurate in their responses because they are more certain of the signal. Second, a more cautious criteria may be set because of an inability (attentional resource limitations) to adjust or shift the criterion from moment to moment in order to create a balance between speed and accuracy. In fact, Rabbitt (1979) suggests that older adults will initially react faster and faster until an error is made, and then will slow down significantly. Thus, they are likely to keep their speed within a small range just below where an error may occur.

Several observations can be made from our examination of these theories of human slowing with aging. The first is that behavioural slowing may occur because of both unintentional (Hardware) and intentional (Software) reasons. The neuroanatomical and neurophysiological changes that occur over the adult age span are well documented (Petit, 1982) and it follows that concomitant behavioural changes should accompany these physical changes. Such evidence provides a great deal of credibility to the 'strong' versions (Neural-Noise Hypothesis and the Input/Output Hypothesis) of the hardware explanation, and invite studies which would attempt to correlate changes in neural structures with those observed in behaviour. The 'weak' versions (Birren Hypothesis and the Cycle-Time Hypothesis) of the hardware explanation are somewhat more difficult to examine since no clear neurological mechanisms are identified for the slowing seen in older adults. Rather inferences such as those proposed by Salthouse (1985a,b, 1988a) are made as to the central mechanisms which might be involved. In this work, reaction time has often been used to reflect hardware changes. The software (inefficient movement programming and control, and the adoption of different strategies for solving movement problems) explanation is intuitively appealing, but is potentially very difficult to refute without considerable insight into the various strategies available to the performer in any given task.

In sum inferences as to whether changes with age are attributable to hardware or software mechanisms are made, within the context of the information processing model, by examining reaction time, movement time and its associated kinematic derivatives. Given that reaction time reflects central processes (e.g., stimulus identification, response selection and response programming) and movement time more peripheral processes (e.g., feedback processing and movement control) it is tempting to argue that changes in reaction time measures with age are indicative of changes in the central nervous system (central hardware changes), while those in movement time reflect peripheral hardware changes. Such simplicity is tenuous. First, we know from studies of neuropathology that damage to the central nervous system such as that seen in stroke or Parkinson's disease can affect both reaction time and movement time (e.g., Jeannerod, 1986; Marsden, 1989; Stelmach, Worringham, & Strand, 1986). Secondly, both of these performance measures, but particularly movement time, are sensitive to the strategies used by a person in performing a task. With this in mind considerable evidence reveals that some behavioural changes observed with brain damage (e.g., slowness in gait) may not be a direct consequence of the damage. Rather, they represent compensations for other changes (e.g., poor balance control) which are more directly attributable to the damage (e.g., Marsden, 1982). In a sense these observed changes in performance represent software changes designed to compensate for changes in the hardware. Such "software solutions" attest to the interactive complexity of the human movement system and to the need for much better descriptions of cause and effect relationships in aging.

These examples from pathology are instructive in that they reveal that attributing observed changes in behaviour with aging to hardware and software changes cannot be made using an either/or solution. One approach to examining the contribution of hardware changes in aging would involve correlating changes in central or peripheral hardware (e.g., decreased proprioceptive sensitivity in the hand) with specific cognitive or motor processes (e.g., control of grasp in reaching) such as has been done in the neurosciences (e.g., Jeannerod, Michel, & Prablanc, 1984). While this approach may provide the information necessary to forge the links between hardware changes and behaviour in aging, specific hardware changes which occur outside the context of neuropathology in the normal aging process may be difficult to identify and measure. Recognizing this limitation an alternative approach involves using available behavioural measures to infer hardware changes. In this regard reaction time has often been used to reflect hardware changes (e.g., Salthouse, 1985a,b) in that it represents the time taken to complete central processing and, relative to other measures such as movement time, is less sensitive to the effects of strategies (software) available to the subject. We will discuss this notion at greater length later on in the chapter. We now turn to an examination of the research examining movement slowing in elderly adults.

AGING EFFECTS ON MOVEMENT EXECUTION

Two measures are used to reflect the control processes involved in movement execution, movement time and movement kinematics. Movement time reflects the overall timing of movement execution. Movement kinematics provide insight into control processes which are not apparent in overall temporal measures of motor performance such as movement time.

Studies using movement time have typically employed the Fitts' (1954) paradigm to examine age differences in pointing movements. To the extent that aging differentially affects the capability of the motor system to adapt to increased processing demands, one might expect to see a larger movement time in older people in response to the increased index of difficulty (target width and amplitude). Welford (1984b) and Cooke, Brown, and Cunningham (1989) reported that movement time does increase as processing demands increase. However, this increase was constant across age groups. These findings suggest that with aging there is a generalized slowing of movement, but the system remains sensitive to the processing demands.

A number of studies using kinematic measures to examine performance of these pointing movements have provided additional insights into the effects of aging (Darling, Cooke, & Brown, 1989; Haaland et al., 1993; Goggin & Stelmach, 1990; Murrell & Entwistle, 1960; Roy, Winchester, Weir, & Black, 1993; Warabi, Noda, & Kato, 1986). These studies suggest that the increased movement time for elderly subjects arises from more time being spent in deceleration (possibly reflecting more time needed to process feedback information) and smaller peak velocities (possibly reflecting reduced force generation at movement initiation). In addition, the elderly were found less able to scale velocity to the amplitude of the movement (potentially reflecting a reduced capability for modulating force generation, Goggin & Stelmach, 1990; Haaland et al., 1993).

This work suggests that kinematic measures do provide more insight into the motor processes occurring in movement execution than do chronometric measures such as movement time. Movement kinematics allow one to determine the movement pattern underlying the observed movement time. Much of the this work on upper limb function in the elderly, however, has been limited to flexion/extension movements (Darling et al., 1989), and pointing movements (Goggin & Stelmach, 1990; Murrell & Entwisle, 1960; Roy et al., 1993; Warabi et al., 1986; Welford, Norris, & Shock, 1969). These movements do not require the intricate, complex hand movements used in activities of daily living such as grooming, cooking, eating, and creative endeavours such as painting and sculpting. Thus, given the necessity of prehension in activities of daily living, and the degree to which these activities are used in clinical assessments (Guralnik, Branch, Cummings, & Curb, 1989; Jacobsen-Sollerman & Sperling, 1977), we have begun to focus on these movements in our studies of changes in motor behaviour that occur with age.

EFFECTS OF AGING ON PREHENSION

Theories of prehension

Prehension refers to capturing an object in a stable grasp using the hand and fingers. Jeannerod (1981, 1984), Arbib (1981, 1985, 1987, 1990), and Paillard (1982) suggest that prehension involves two components: a transport or reaching component (involving proximal musculature) which moves the limb to an appropriate spatial location, and a grasp component (involving distal musculature) which orients and postures the hand. Neurologically these ideas are supported by Kuypers (1962, 1964) who has identified different pathways in the central nervous system controlling these two types of musculature. While the transport and grasp components ensure that the arm and hand are brought to the correct location in the correct orientation their role ends at the point of object contact. In order to effect a stable grasp, forces must be applied to the object by the hand (Johansson & Westling, 1984). Thus, one can consider a kinematic phase up to the point of contact, and a kinetic-kinematic phase occurring subsequent to initial contact. Incorporating these ideas, MacKenzie and Iberall (1994) have defined prehension as "the application of functionally effective forces by the hand to an object for a task, given numerous constraints" (p. 15).

Coordination of the components of prehension. Jeannerod's (1981, 1984) seminal work was aimed at examining the presence of two distinct visuomotor channels operating simultaneously to control the transport and grasp components. He posited that extrinsic object properties (e.g., location, amplitude) influence only the transport component, while intrinsic object properties (e.g., size, shape, weight) influence only the grasp component. His initial work supported that there were separate parallel pathways that independently controlled the two components. In addition he reported a temporal invariance such that the time of peak aperture (grasp) correlated with the time of peak deceleration which he took as support for a temporal linkage between the two components. Over the years, some controversy appeared as to the independence of visuomotor channels controlling the two components. Several studies support this notion (Wallace & Weeks, 1988; Wing & Fraser, 1983), while others report that intrinsic object properties can affect both the grasp and transport components (Gentilucci, Castiello, Corradini, Scarpa, Umilta, & Rizzolatti, 1991; Jakobson & Goodale, 1991; Marteniuk, Leavitt, Mackenzie, & Athenes, 1990; Soechting & Flanders, 1993). While the correlational evidence is weak (Gentilucci et al., 1991; Marteniuk et al., 1990), theoretically the two components must be linked in order for the movement to unfold in the correct sequence. The hand must open while the arm is being transported to the object. If the hand opens too early or too late, the timing is off and the object will not be successfully grasped.

Wing and colleagues provided further evidence as to how the two components operate together. Using data from an artificial limb and hand (Wing & Fraser, 1983), and data resulting from a manipulation of movement speed (Wing, Turton, & Fraser, 1986) they proposed that the linkage between the two components went beyond timing. They reported that spatial variability in arm transport is compensated for by an increase in the size of the grasp aperture, thereby suggesting a spatial linkage between the two components. This approach adds to the theoretical knowledge of how the coupling between the arm and hand may change, based on the context of the setting or task. This idea was further developed by Marteniuk et al. (1990) who examined the influence of object size. They reported that as the size of the object increased, the maximum aperture increased, and the percentage of movement time decreased. They proposed that the changing relationship between the arm and hand reflected a functional linkage between the two components. Moreover, this relationship was free to vary in order that the goal of the task be met, thereby acknowledging the flexibility of the motor system.

Conceptual models. Some of these theoretical ideas have been captured in Arbib's (1981, 1985, 1987, 1990) coordinated control program (CCP). The CCP was initially developed to formalize Jeannerod's (1981) early findings. The program's basic premise is that the control system is composed of both perceptual and motor schemas. The perceptual schemas are activated to gather information about environmental parameters, while the motor schemas control different aspects of the movement. The information from the perceptual schemas is used to assign response parameters to the motor schemas. The motor schemas operate in parallel and are coordinated in time (Wallace & Weeks, 1988) and space (Jeannerod, 1981). Because the CCP involves the continuous interplay between the perceptual and motor schemas as we interact with our environment, the temporal interaction (time dependency) between the transport and the grasp components as reported by Jeannerod (1981, 1984) would be an example of a CCP at work. One limitation of these conceptual models is that they do not address the underlying question of the sequencing of the entire grasping movement.

Analyses have traditionally been limited to movement prior to contact with the object. MacKenzie and Iberall (1994) have taken these conceptual ideas past the point of object contact. They have developed an opposition-space-model derived from Iberall, Bingham, and Arbib's (1986) notion of oppositions. Iberall et al. (1986) described three basic directions along which the human hand can apply forces: pad, palm and side. Most experimental work has focused on pad opposition which "occurs between hand surfaces along a direction generally parallel to the palm. This usually occurs between volar surfaces of the fingers and thumb, near or on the pads" (MacKenzie & Iberall, 1994, p. 31). Thus, the opposition space model relies on the interface between the hand and the object. Using the opposition-space-model, the prehension task can be divided into multiple phases from planning the opposition through to releasing the opposition at the completion of the task. This model ties together the serialization of multiple sub-tasks, such as transporting the hand, preshaping the hand, acquiring the object in a stable grasp, manipulating the object, and releasing the object. From Jeannerod's (1981, 1984) initial work through to the conceptualization of the opposition space model, two themes are common. First, the control system is distributed involving parallel activation and coordinated control of several components, and second, there are different phases as the act of grasping unfolds.

Constraints framework. In their consideration of the mechanisms involved in prehension MacKenzie and Iberall (1994) place considerable emphasis on the notion of constraints. Constraints are those variables that limit the use of feedback, as well as the structural variables that affect the preparation and the execution of movement goals (Marteniuk, MacKenzie, Jeannerod, Athenes, & Dugas, 1987). Different levels of constraints must be manipulated in order to study the complex interactions among movement goals, the environment that surrounds the performer, object properties, and the knowledge and experience the performer brings to the task. These constraints fit into three categories: sensorimotor, physical, and high level (MacKenzie & Iberall, 1994).

Sensorimotor constraints refer to temporal and spatial limitations of the central nervous system because of insufficient neural, perceptual or physiological information to sustain the reach and/or the grasp components of prehension. The availability of visual information during a reaching movement, for example, serves to reduce the peak velocity and increase the time spent in deceleration (Roy, Elliott & Kalbfleisch, Note 1), while impairments to the kinesthetic/proprioceptive system serve to increase the performer's reliance on visual information to successfully complete the grasp (Jeannerod, 1986).

Physical constraints are determined by the properties (extrinsic and intrinsic) of the object-to-be-grasped as well as the biomechanical limitations of the performer. For example, the transport and grasp formation over the approach to the object is influenced by object properties (e.g., size, location; Jeannerod, 1981; Marteniuk et al., 1990), as is the force generation once contact is made (e.g., texture, weight; Johansson & Westling, 1984; Westling & Johansson, 1984).

At the top of this hierarchy are the high level constraints that are reflected in the informational and/or functional knowledge base of the performer as well as the performer's intentions (movement goals). Prior knowledge of object characteristics affect the control of reaching movements. For example, movement to grasp a light bulb is slower than to grasp a tennis ball of comparable size. Kinematic analyses also revealed that movements toward the more fragile light bulb involved lower peak velocities and more time in deceleration (Marteniuk et al., 1987; Wing et al., 1986). Task goals have also been shown to influence prehension. Tasks that require a greater precision (placing versus throwing; lifting versus transporting) result in a longer deceleration portion prior to contact (Marteniuk et al., 1987; Weir & MacKenzie, Note 2).

This framework, when combined with knowledge about the effects of aging as reflected in hardware and software changes, provides a workable paradigm in which to interpret our current findings on prehension and to develop predictions for future work. In this combined framework (as depicted in Figure 1) we envisage that hardware and software differences between the young and elderly contribute to age differences on a given measure of performance and that the degree to which these differences contribute to age effects depends on the task constraint.

In the case of sensorimotor constraints hardware differences make the largest contribution. For example, the relative effects of the availability of visual information on reaching in the young and elderly depends, to a

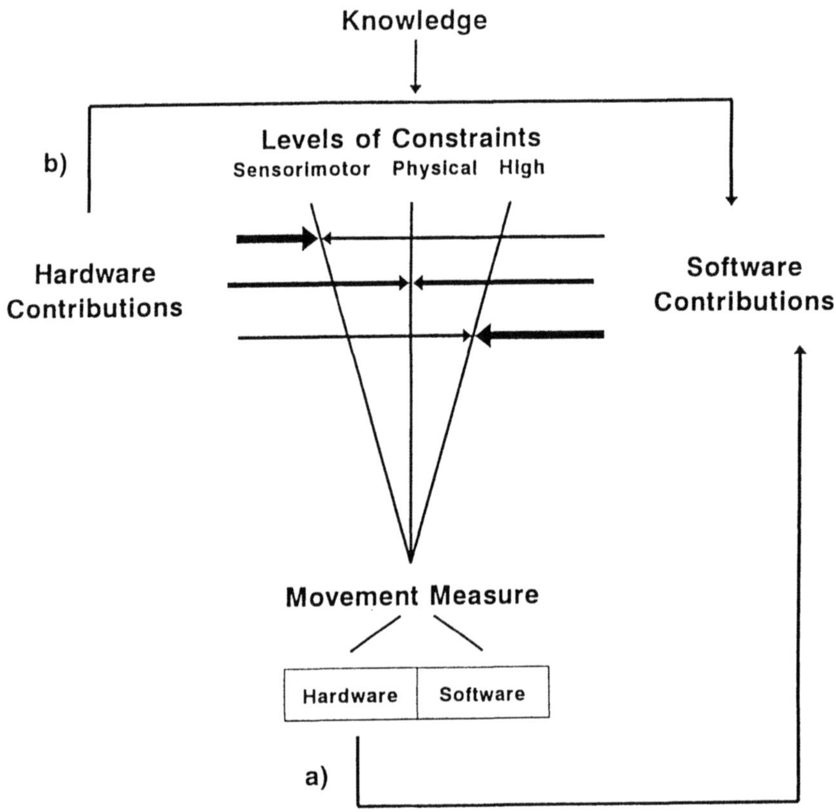

FIGURE 1. Age differences on a performance measure reflect age differences in both hardware (e.g., neuromuscular capabilities) and software (e.g., strategies). The magnitude of the contribution that these age differences have on performance is mediated by the constraints of the task. The thickness of the arrows leading from hardware and software contributions reflect the influence each is thought to have on each constraint. Sensorimotor constraints are most strongly influenced by hardware differences between the young and old; high level constraints are most affected by software differences, while physical constraints are thought to receive an equal contribution from hardware and software differences. The interactive complexity of task constraints with hardware and software contributions to age differences on a given measure of performance is depicted in the two pathways. Pathway "a" represents software solutions (i.e., strategies) an older performer may adopt to compensate for the influence that hardware changes have on performance. The integrity of the hardware, however, may constrain the potential compensatory solutions available to the performer (pathway "b", see text for details).

large extent, on the relative integrity of the visual processing system. Changes in the visual system with age (hardware) which reduce the visual processing capacity and speed contribute very substantially to age differences in the effect of the availability of visual information in reaching movements (e.g., Haaland et al., 1993).

Software differences, on the other hand, would appear to make the largest contribution to high level constraints. The ability to formulate goals and to use various strategies in performance (software) make a very important contribution to the relative effects of high level constraints on performance in the young and elderly. In serial reaction time tasks, for example, Rabbitt (1979) has shown that the slowness observed in the elderly arises not from basic differences in the speed of processing (hardware) but rather from differences in the decision rule used to respond to the presence of a target (software). The elderly adopted a more conservative strategy.

Hardware and software differences between the young and elderly would appear to contribute equally to the effects that physical constraints have on performance differences between the young and elderly. For example, in grasping objects of varying weight subjects spend a longer time enclosing (e.g., moving the index finger and thumb to grasp the object) heavier objects to effect a stable grasp. This increased time appears to reflect the time needed to generate the increased force required to lift the heavier object. This pattern of grasping could be affected by hardware characteristics, that is, the ability to generate force. Alternately software factors could be important. As one learns about the physical characteristics of the object as they pertain to its weight (e.g., size) a strategy may be adopted whereby one spends more time approaching and holding on to the object before picking it up. Changes with age in either of these hardware or software factors could affect performance of the elderly relative to the younger subjects.

While age differences in hardware and software may contribute to age differences in performance, knowledge of a particular hardware problem may prompt the subject to adopt a particular strategy to compensate for the problem, what we have termed a software solution to a hardware problem (see [a] in Figure 1). For example, changes in tactile sensitivity in the finger tips may prompt the older person to generate greater force while picking up an object so as to avoid dropping it. The integrity of the hardware, however, will impact on the compensatory strategies available to the subject (see [b] in Figure 1). For example, a concomitant weakness in the intrinsic hand muscles (e.g., Cole, 1991) may prevent the subject from exerting greater force to compensate for the reduced tactile sensitivity.

Such potential hardware-software interactions attest to the complexity of the relationship between aging and motor performance and suggest that it may be difficult to ascribe age differences in performance to either hardware or software differences alone. Nevertheless, this model does provide a useful framework in which to examine the contributions made by hardware and software differences to age differences in performance. For example, the contribution of software differences will be most apparent through manipulating high level constraints, while the contribution of hardware differences will be most obvious when sensorimotor constraints are manipulated.

Studies of prehension and aging

Our initial work on prehension has adopted the framework of physical and high-level constraints. In these studies only healthy young and elderly women and men have participated as subjects. The young subjects were undergraduate students between the ages of 20 and 25. The elderly subjects were between 65 and 75 years of age, representing a group of young-elderly. On average there was a 50 year age spread between the young and elderly subjects. All subjects were right-handed, had normal or corrected to normal vision and were free of any neurological or physiological impairments that might influence their motor behaviour. The elderly subjects were screened using the Digit Symbol Substitution Test (Subtest of the Wechsler Adult Intelligence Scale-Revised, Wechsler, 1981), and they all performed within the norm for their age group.

Physical constraints: object size. Initially, Desjardins-Denault and Roy (Note 3) examined the effects of object size, using three metal disks of different diameters (2.5, 5.5, and 7.5 cm). Contrary to what was expected, elderly subjects had shorter movement times, higher peak velocities, and shorter times in deceleration. Weir, Adkin, and Leavitt (Note 4) continued this line of work, but made the task more ecologically valid by presenting four light bulbs of different diameters (3.4, 5.1, 6.9, and 9.7 cm) that were placed in a standard light socket. Contrary to Desjardins-Denault and Roy (Note 3), there were no age differences in movement kinematics over the approach phase to grasp the light bulb. However, over the transport phase to the light bulb socket young subjects moved more quickly than the elderly. Two procedural differences may explain why these two studies did not concur. First, the subjects in Desjardins-Denault and Roy (Note 3) were aware that they were being compared to a younger sample and second, all their subjects were instructed to move as quickly as possible. Common to both studies was the finding that the movement trajectories, as determined by relative timing measures, were similar between the two age groups. This finding suggests that the young and elderly performed the tasks in a similar manner, but scaled them differently in the time dimension.

Physical constraints: object size and movement amplitude. More recently Desjardins-Denault, Winchester, Roy, and Weir (Note 5) examined the influence of index of difficulty (amplitude and object width) on pointing and grasping. Both young and elderly subjects pointed to or grasped two objects (2.5, 7.5 cm) over two movement amplitudes (15, 30 cm). For pointing, the influence of movement amplitude and object width was the same for young and elderly subjects, but the elderly were slower. Movements to smaller objects had lower peak velocities and longer movement and deceleration times, whereas larger amplitude movements resulted only in longer movement times.

However, for grasping, the influence of index of difficulty was different across the age groups. Movement amplitude exerted the major effect on the reaching component for the elderly subjects at 30 cm. Subjects exhibited significantly lower velocities, longer movement times, and more time in deceleration (see Figure 2). Similar to the findings of our earlier work (Desjardins-Denault & Roy, Note 3; Weir et al., Note 4), the movement patterns used by the young and elderly did not differ.

For the grasp component, on average, age did not influence the ability to appropriately scale the hand to match the size of the object to be grasped. The young subjects reached peak aperture sooner, but there were no differences in the relative time following peak aperture. Thus, setting up the opposition space for making contact with an object is accomplished in the same way by young and elderly subjects.

Physical constraints: object motion. Leavitt and Mallat (Note 6) manipulated a physical constraint in a somewhat different way, by requiring the subject to capture a moving object. Subjects were required to reach forward and grasp a dowel located 30 cm in front of them. The dowel (2.2 cm in diameter and 2 meters long) was either stationary (self-paced) or dropping vertically (externally-paced) from a height of 47.5 cm above the work space at a velocity of 38.72 cm/sec. Interestingly, the movement times between the young and elderly subjects did not differ, although the elderly subjects exhibited lower peak velocities, and when externally paced, less time in deceleration (see Figure 3). This shorter deceleration time suggests that the elderly subjects delayed the onset of movement until the dowel was closer to the table top. This delayed onset is also reflected in a longer time to reach peak aperture. However, there were no differences in peak apertures between the age groups. Again, the movement trajectories used by young and elderly subjects were similar. In terms of the pacing, several kinematic variables were influenced. Peak apertures were significantly larger in the externally paced condition, which were accompanied with a smaller percentage of time spent in deceleration, and less time spent in closing the hand onto the dowel. Thus the demands of the pacing had the same influence on the movement patterns executed by both the young and elderly subjects.

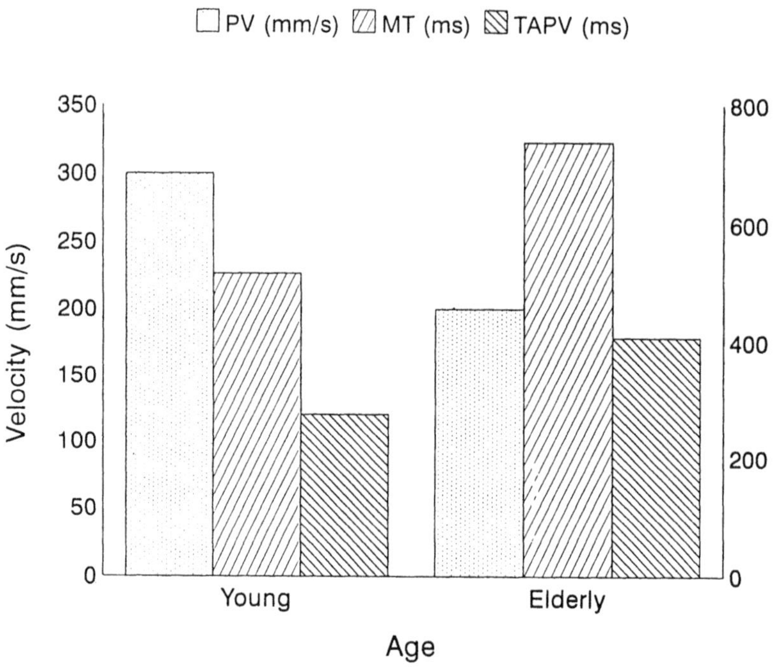

FIGURE 2. The influence of age on grasping performance over 30 cm. Note that the peak velocity is lower for the elderly, while their movement time and time in deceleration are longer.

These studies suggest that physical constraints have little differential influence on the movement trajectories executed by young and elderly subjects, although the elderly appear to be more affected by movement amplitude.

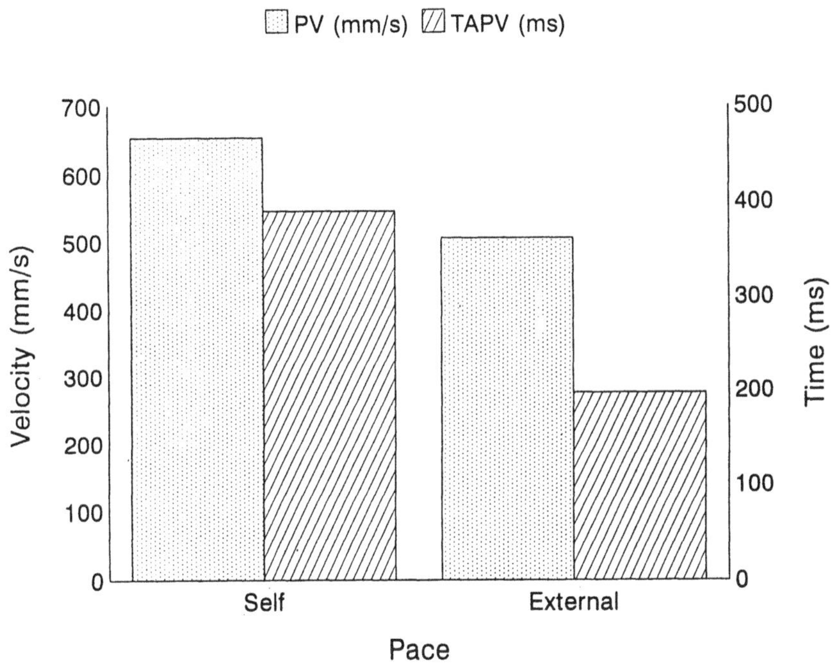

FIGURE 3. The influence of pace on grasping performance in the elderly. Note that in the externally paced condition, the elderly exhibit lower peak velocity and a shorter time in deceleration.

High level constraints. Our other work has sought to examine the effects of high level constraints by manipulating the goal of the movement or the intention of the subject. Desjardins-Denault et al. (Note 5) contrasted pointing and grasping movements. They found that for both age groups grasping movements resulted in longer movement times, lower peak velocities and a larger percentage of movement time spent in deceleration. In addition, for peak velocity, there was an age by task interaction. The young subjects showed a significantly greater peak velocity when pointing as compared to grasping, whereas the elderly subjects produced the same velocity for both tasks, suggesting the use of a conservative movement control strategy by the elderly. Weir,

MacDonald, Mallat, Leavitt, and Roy (Note 7) extended this finding showing that the same grasping movement is dramatically affected by what one does with the object after it is grasped. They examined the influence of the subject's intention by having young and elderly subjects reach to grasp a 4.5 cm diameter, 1 cm thick disk, and then transport the disk 30 cm to: a) place the disk into a tight fitting well (5.0 cm diameter, 0.5 cm deep, PLACE-WELL), b) place the disk into a large square box (20 × 20 × 1 cm, PLACE-BOX), or c) throw the disk into the box (THROW-BOX). The task was analyzed over two phases, first the approach to contact the disk, and second, transporting of the disk to the appropriate target location (e.g, the well or the box).

In the approach phase, reaching to grasp the disk prior to placing resulted in longer movement times than prior to throwing, and greater percentages of movement time were spent in deceleration, for both age groups. In general these findings show that when the current task (e.g., grasp vs point) or the upcoming task (e.g., place vs throw) requires more precision, the movement pattern executed reflects a lengthened deceleration portion. While the young and elderly did not differ on the basis of movement time over the approach phase, the elderly subjects reached peak velocity sooner than the young subjects, and spent a greater relative time in deceleration following peak velocity (see Figure 4a). This is the first prehension study to differentiate between the age groups on the basis of the shape of the movement trajectories. The grasp component, as reflected in measures of peak aperture and time to peak aperture, was not influenced by this high level constraint. However, paralleling the lengthened deceleration portion, elderly subjects spent a greater relative time enclosing the hand to grasp the disk.

In examining the transport phase that required subjects to place or throw the object, the elderly subjects were able to compensate for the increased task demands in a manner similar to that of the young subjects. Movements that required more precision (e.g., placing the object in the tight fitting well as opposed to the large box) exhibited longer movement times, more time after peak velocities, and greater relative times following peak velocities. However, in this phase, the elderly subjects produced longer movement times, but with the same relative timing as the younger subjects; the opposite of findings in the approach phase, suggesting the use of similar movement patterns (see Figure 4b).

The lack of age by task interactions suggest that the elderly respond to the precision demands of both phases of the task in a manner similar to the young subjects. It would appear, however, that elderly subjects are more cautious in the approach to contact the disk.

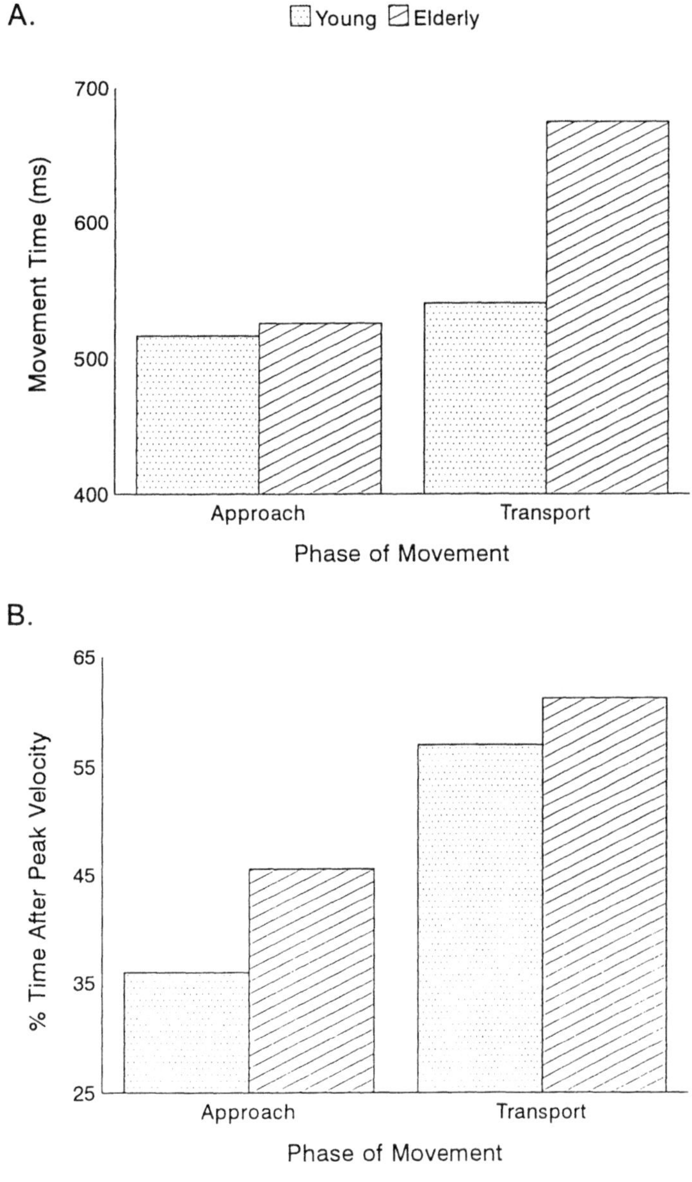

FIGURE 4. Time spent in the approach and transport phases as a function of age (Panel A). Note that age differences exist only in the transport phase with elderly subjects moving more slowly. However, in terms of the relative time spent in deceleration (Panel B), the young and the elderly differ in the approach phase with elderly subjects spending more time in deceleration.

Summary of findings relative to aging

The manipulation of physical and high-level constraints support previous work that has shown that task-related factors influence the grasp and reaching components in prehension (Marteniuk et al., 1987; Wing et al., 1986). These findings also concur with work that has focused on the kinematics of pointing movements. Our research substantially extends this previous work on pointing movements, since our studies have examined more complex reaching and grasping movements and have investigated the effect of movement precision in the context of a more functional serial reaching task. The question of particular interest here is to what extent are these effects influenced by aging?

Physical Constraints. Both age groups spent more time in the acceleration phase when reaching for the moving object than reaching for the stationary object. They may have used this additional time to acquire information about the movement of the dowel. Recently, these findings have been further examined by Desjardins-Denault (Note 8) in which the application of grasping forces was also examined. She found that the young subjects use a higher rate of grip force application when acquiring a dowel in a stable grasp, but when manipulating and releasing the dowel there were no differences between the age groups, either kinematically or in terms of the forces applied to the dowel. In addition, none of the examined physical constraints differentially influenced the grasp component or the relative timing of the kinematic profiles.

Despite the similarities between the kinematics of prehension in the young and elderly some differential effects were apparent. Two effects pertain to physical constraints, movement amplitude and object movement. First, with respect to movement amplitude, the elderly exhibited larger increases in movement and deceleration times with increased movement amplitude. Roy et al. (1993) have suggested that this effect may relate to the relationship between movement amplitude and spatial variability; the greater the amplitude the greater the spatial variability. Larger forces (reflected in higher peak velocities) associated with longer movements have been shown to result in greater variability in both the movement trajectory and the movement end point (e.g., Zelaznik, Schmidt, & Geilen, 1986). Since the older subjects have some difficulty scaling force to meet the amplitude demands of movement (as reflected in the smaller increase in peak velocity with movement amplitude) (cf., Desjardins-Denault et al., Note 5; Goggin & Stelmach, 1990), perhaps these force demands have a greater effect on variability in the older subjects. The older subjects then, may have spent more time in deceleration in order to reduce both the spatial variability in the reach and the variability in the final position of the hand. While the movement trajectory for older people has been shown to be more variable (e.g., Darling et al., 1989), the relationship between movement amplitude, trajectory variability and time in deceleration has not been examined. Clearly, further research is necessary to clarify these deceleration time effects in the older subjects and to interpret the differences between the age groups in deceleration time.

The second effect, object movement, was also more dramatic for the elderly despite the lack of difference in movement times. Elderly subjects spent more time in acceleration than did the young subjects when reaching for the moving as opposed to the stationary object, perhaps representing the time required to sample the characteristics of the moving object. The physical constraint that seems important here pertains to target movement, where the rate of movement of the target plays a role in constraining movement time. Regardless of the performer's age his/her movement must be made in a certain overall time in order to accurately intercept the moving object. This context of object movement may have served to decrease the movement time of the elderly such that even when the targets were not moving they moved in a time comparable to that for the younger subjects.

High level constraints. The influence of a high level constraint depended on the phase of the movement (approach versus transport). In the single phase movement to contact the object (Desjardins-Denault et al., Note 5), the kinematic profiles of the young and elderly did not differ based on the goal of the task (point versus grasp). Further, in the Weir et al. (Note 7) study similarities in the kinematic profiles were seen in the second phase when transporting the disk to the box or well. Thus, it would appear that in completing the task (contact or place on target), the young and elderly subjects produce movement patterns of the same relative shape.

The goal of the task is a source of an age difference. For the approach phase, Desjardins-Denault et al. (Note 5) reported greater movement times for elderly subjects. In contrast, in the Weir et al. (Note 7) study there were no differences in movement time while approaching the disk; however, the relative time spent in deceleration was greater for the elderly subjects. When making simple, single movements the elderly subjects generally move more slowly. For more complex and serial movements the elderly subjects' movement slowing is centred in the deceleration portion, suggesting a fundamentally different means of controlling the movement. During the transport phase of the Weir et al.

study the elderly subjects exhibited lower peak velocities and longer movement times both overall and within the acceleration and deceleration phases of the movement.

INFERRING HARDWARE AND SOFTWARE CONTRIBUTIONS TO OBSERVED AGE DIFFERENCES IN PERFORMANCE

Given the framework outlined earlier in the chapter (see Figure 1) how might we infer potential contributions of hardware and software differences between the young and elderly to the observed age differences in performance. Such inferences should involve focusing on the effects of task constraints on performance and, in particular, identify any age by task constraint interactions. Two types of interactions seem plausible, one in which the effect of the constraint is seen in both age groups, although differing in magnitude, and the other where the effect is seen in one group but not the other. Of the first type of interaction two effects are apparent in our findings, one for each type of constraint.

Looking first at physical constraints the interaction involved the effect of movement amplitude. In this case the elderly exhibited a smaller effect of amplitude on peak velocity, but larger effects on movement time and time in deceleration. The fact that these effects are in the same direction as those for the young subjects suggests that this difference in magnitude likely arises from a difference in the way the elderly subjects controlled their movements (e.g., a software difference). As we discussed in the previous section the older subjects may have spent more time in deceleration so as to reduce the effect of spatial variability in the movement trajectory on variability in the final position of the hand.

From the standpoint of high level constraints the interaction involving the task goal of pointing versus grasping with age, revealed that for the pointing movements peak velocities were significantly greater for the young than for the elderly subjects (Desjardins-Denault et al., Note 5). Thus, this effect for the task goal likely arises from a difference in the movement strategies used by the two age groups (e.g., a software difference).

How might this framework be useful in providing insight into hardware/software differences when no age-task constraint interactions exist, but there are overall main effects of age on performance? Such a pattern, evident in our findings, suggests that aging tends to affect performance (e.g., generally slower movements) regardless of the nature of the high level constraint. With this global effect it would seem important to attempt to manipulate overall movement strategies in order to make inferences about the relative contributions of hardware and software differences. An example is the overall slowing of movement with age. Insight into the hardware/software basis for this slowing might be provided by equating young and elderly subjects on movement time. This can be accomplished by requiring the older subjects to move faster and the young subjects to move slower. If the slowness observed in the older subjects arises from a learned movement strategy (software), requiring them to speed up their movements may have relatively little effect on their performance. If, on the other hand, the slowness is due to a more fundamental problem associated with how they control their movements (e.g., time to process feedback, rate of force generation), one might expect some degree of deterioration in performance with increased speed of movement (e.g., reduced accuracy, increased spatial variability of movement).

This approach was recently adopted in a study by Morgan, Phillips, Bradshaw, Mattingley, Iansek, and Bradshaw (1994). Subjects were required to point to targets in a zig-zag pattern. They performed at their own speed or were required to speed up (elderly subjects) or slow down (young subjects). Thus, each group was forced to move like the other group. This paradigm allowed the researchers to determine if the slow movements exhibited in elderly subjects was simply a function of strategy, or actual slowing of information processing. When strategic differences were controlled, the kinematics of the elderly subjects' movements demonstrated hesitancy and a larger number of submovements, suggesting the decline in motor behaviour was not simply due to movement time, since these had been equated. They concluded that the elderly subjects suffered a decline in motor coordination.

In examining potential hardware and software contributions to aging it is important to consider the distinction between process and product which derives out of work in cognitive neuropsychology (e.g., Rapp & Caramazza, 1991; Roy, 1990). Product refers to the goal of the performer, while the process refers to the means of achieving that goal. The information processing approach which forms the basis of our work in motor behaviour tends to focus on the component processes that are involved in achieving a particular behaviour or movement (e.g., product). This direction is particularly evident in the work on movement kinematics reviewed in this chapter: picking up an object (the product) is examined in exquisite kinematic detail (the motor control processes). This focus on process, however, tends to blind us to the fact that aging often does not adversely affect the behavioural product, in this example, picking up the object. That is, elderly people are able to pick up and manipulate objects, although the motor control process may be different from that employed by younger people. In a sense this product-process distinction is similar to the ability-competence distinction alluded to by Rabbitt (1979) and Salthouse (1990). The question for Rabbitt (1979, p. 623) arising from this distinction is not "why are old people so bad at [motor] tasks?" but rather, "how, in spite of growing disabilities, do old people preserve such relatively good performance?". In the context of our discussion this distinction invites us to consider how the processes we are measuring through kinematics and kinetics relate to the elderly person's capability to achieve particular movement goals such as placing a tea cup on a saucer. The former measures might be seen to reflect certain more basic movement abilities, while the latter measures are representative of more general movement competencies such as are examined on tests of independent activities of daily living (IADL, e.g., Myers, 1992; Myers, Holliday, Harvey, & Hutchinson, 1993). In our work this relationship is being examined in the following way. We have simulated an IADL skill, placing a cup on a saucer, using the task developed by Weir et al. (Note 7). The intent is 1) to examine, in this closer to real life reaching task, the influence of high level constraints pertaining to the movement goal (e.g., the precision requirements of the placing task) and 2) to determine how these effects relate to the person's self-rated and actual performance on a series of IADL skills. Using this approach we hope to gain insight into how constraints affect reaching performance in the elderly, and how these effects relate to the elderly person's competence in daily living activities.

CONCLUSION

Slowing in cognitive and motor processes is one of the characteristic changes in performance seen with aging. Using the information processing approach a number of studies involving a variety of motor tasks have revealed longer processing times for the elderly on measures reflecting response selection and programming (reaction time) and movement execution (movement time), suggesting that aging affects each stage in processing a motor response. The recent advent of advanced optoelectric movement analysis systems has permitted the partitioning of movement time using kinematic analyses. A number of studies (Darling et al., 1989; Haaland et al., 1993; Goggin & Stelmach, 1990; Murrell & Entwistle, 1960; Roy et al., 1993; Warabi et al., 1986) suggest that these kinematic measures provide greater insight into the motor processes occurring in response programming and movement execution than do chronometric measures such as movement time. The increased movement time for elderly subjects arises from more time being spent in deceleration (possibly reflecting more time needed to process feedback information) and smaller peak velocities (possibly reflecting reduced force generation at movement initiation).

Much of the work on aging has been limited to simple flexion/extension and/or pointing movements which do not demand the more intricate, complex hand movements used in activities of daily living. Such prehension movements, however, are routinely involved in daily living activities and are often used in clinical assessments. Thus, our studies of aging have focused on these more complex functional movements.

Our search for the effects of aging on prehension began by examining the various theories of aging, contrasting in particular hardware with software explanations. The two general hypotheses of motor slowing, hardware and software, derive from the computer metaphor (Charness, 1985, 1991; Salthouse, 1985a). Hardware explanations focus upon neuroanatomical changes occurring with aging that may underlie the observed concomitant cognitive changes (Petit, 1982). Software explanations focus upon computational efficiency and are thought to reflect the strategies adopted in performing a task. Both hardware and software changes occur with aging and both have been shown to explain performance differences between the young and the elderly. One of the principal questions addressed in this chapter was how do we gain insight into the contribution made by these two types of change to age differences observed in task performance.

Within the context of prehensile movements we argued that these hardware and software contributions might be dependent on the constraints of the task as defined by MacKenzie and Iberall (1994). Software changes with age might make their greatest contribution through high level constrains which reflect the strategies used in performing a task. Hardware changes may be observed most clearly through sensory motor constraints which reflect the sensory (e.g., the availability and timing of visual information during movement) and motor (e.g., the force required at movement initiation) demands of the task. Physical constraints reflecting the environmental characteristics of the task (e.g., the amplitude of the movement or the size of the object) may receive an equal contribution from hardware and software changes. We argued that inferences as to the contribution of hardware and software changes to age differences in performance require an examination of task constraints on performance with a particular focus on age by task constraint interactions. A review of our own work examining the effects of all three types of constraint revealed evidence of both hardware and software contributions to age differences in prehension.

In future work examining potential hardware and software contributions to aging we emphasized the importance of considering the distinction between process and product as derived from work in cognitive neuropsychology (e.g., Rapp & Caramazza, 1991; Roy, 1990), where product refers to the goal of the performer, and process refers to the means of achieving that goal. This distinction invites us to consider how the processes we are measuring through kinematics and kinetics relate to the elderly person's capability to achieve particular movement goals such as placing a tea cup on a saucer. Using an approach where we focus on the relationship between process and product we hope to gain insight into how constraints affect reaching performance as one ages and how these effects relate to the person's competence in functional daily living activities.

ACKNOWLEDGEMENTS

Funding for the research reported in this manuscript was provided by the Natural Sciences and Engineering Research Council of Canada (E.A.R and P.L.W.), the Ontario Mental Health Foundation (E.A.R.), the Parkinson Foundation of Canada (E.A.R.) and the University of Windsor Research Board (J.L.L.)

REFERENCE NOTES

1. Roy, E.A., Elliott, D. & Kalbfleisch, L. (1991). The role of vision in pointing. Unpublished manuscript, Department of Kinesiology, University of Waterloo. Available from Dr. E. Roy, Department of Kinesiology, University of Waterloo, Waterloo, Ontario, Canada, N2L 3G1.
2. Weir, P.L., & MacKenzie, C.L. (1994 - submitted). Phases of prehension: The influence of dowel weight and task intent. Available from Dr. P. Weir, Department of Kinesiology, University of Windsor, Windsor, Ontario, Canada, N9B 3P4.
3. Desjardins-Denault, S. & Roy, E.A. (1991). Prehension in elderly individuals. Unpublished manuscript, University of Waterloo. Available from Dr. E. Roy, Department of Kinesiology, University of Waterloo, Waterloo, Ontario, Canada, N2L 3G1.
4. Weir, P.L., Adkin, A., & Leavitt, J.L. (1991). The effects of object size and age on kinematics of prehension. Paper presented at the An-

nual Conference of the Canadian Society for Psychomotor Learning and Sport Psychology. London, Ontario. Available from Dr. P. Weir, Department of Kinesiology, University of Windsor, Windsor, Ontario, Canada, N9B 3P4.
5. Desjardins-Denault, S., Winchester, T., Roy, E.A., & Weir, P.L. (1994 - submitted). Kinematic variation in pointing in young and elderly subjects. Available from Ms. S. Desjardins-Denault, Department of Kinesiology, University of Waterloo, Waterloo, Ontario, Canada, N2L 3G1.
6. Leavitt, J.L., & Mallat, B. (1993). A kinematic analysis of age related differences in grasping stationary and moving objects. Paper presented at the Annual Conference of the Canadian Society for Psychomotor Learning and Sport Psychology. Montreal, Quebec. Available from Dr. J. Leavitt, Department of Kinesiology, University of Windsor, Windsor, Ontario, Canada, N9B 3P4.
7. Weir, P.L., MacDonald, J.R., & Mallat, B, Leavitt, J.L., & Roy, E.A. (1994 - submitted). Age related differences in prehension: The influence of task goals. Available from Dr. P. Weir, Department of Kinesiology, University of Windsor, Windsor, Ontario, Canada, N9B 3P4.
8. Desjardins-Denault, S. (1994). How changing the frequency of visual information influences reaching and grasping performance in young and elderly subjects. Unpublished Master's Thesis, University of Windsor. Available from Ms. S. Desjardins-Denault, Department of Kinesiology, University of Waterloo, Waterloo, Ontario, Canada, N2L 3G1.

REFERENCES

Abend, W., Bizzi, E., & Morasso, P. (1982) Human arm trajectory formation. *Brain, 105,* 331-348.

Amrhein, P. C., Stelmach, G. E., & Goggin, N. L. (1991). Age differences in the maintenance and restructuring of movement preparation. *Psychology and Aging, 6* (3), 451-466.

Annett, J. (1988). Discussion: programming and coordination. In A. M. Colley & J. R. Beech (Eds.), *Cognition and action in skilled behavior* (pp. 145-153). Amsterdam: North-Holland Co.

Arbib, M. A. (1981). Perceptual structures and distributed motor control. In V. B. Brooks (Ed.), *Handbook of physiology, Section 1: The nervous system* (Vol. 2, pp. 1449-1480). Maryland, American Physiology Society.

Arbib, M. A. (1985). Schemas for the temporal organization of behavior. *Human Neurobiology, 4,* 63-72.

Arbib, M. A. (1987). Levels of modelling of mechanisms of visually guided behavior. *The Behavioral and Brain Sciences, 10,* 407-465.

Arbib, M. A. (1990). Programs, schemas and neural networks for the control of hand movements: Beyond the RS framework. In M. Jeannerod (Ed), *Attention and performance XIII* (pp. 111-138). New York: Erlbaum.

Atkeson, C. G., & Hollerbach, J. M. (1985). Kinematic features of unrestrained vertical arm movements. *The Journal of Neuroscience, 5,* 2318-2330.

Bates, P. B., & Goulet, L. R. (1971). Exploration of developmental variables by manipulation and simulation of age differences in behavior. *Human Development, 14,* 149-170.

Bates, P. B., Reese, H. W., & Nesselroade, J. R. (1978). *Life-span developmental psychology: Introduction to research methods.* Monterey, CA: Brooks-Cole.

Birren, J. E. (1959). Principles of research in aging. In J. E. Birren (Ed.), *Handbook of aging and the individual.* Chicago: University of Chicago Press.

Birren, J. E. (1974). Translations in gerontology: From lab to life: Psychophysiology and speed of response. *American Psychologist, November,* 808-815.

Birren, J. E., Bengtson, V. L., & Deutchman, D. E. (1988). *Emergent theories of aging*. New York: Springer Publishing Company.

Birren, J. E., Woods, A. M., & Williams, M. V. (1980). Behavior slowing with age: Causes, organization, and consequences. In L.W. Poon (Ed.), *Aging in the 1980s* (pp. 293-308). Washington, DC: American Psychological Association.

Botwinick, J. (1978). *Aging and behavior.* New York: Springer.

Cerella, J. (1985). Information processing rates in the elderly. *Psychological Bulletin, 98,* 67-83

Cerella, J. (1990). Aging and information processing rate. In J. E. Birren & K. W. Schaie (Eds.), *Handbook on the psychology of aging* (3rd ed., pp. 201-221). San Diego: Academic Press.

Charness, N. (Ed.) (1985). *Aging and human performance.* Chichester, England: Wiley.

Charness, N. (1991). Cognition and aging. In C. Blais (Ed.), *Aging into the Twenty-First Century* (pp. 204-222). North York: Captus University Publications.

Cole, K. J. (1991). Grasp force control in older adults. *Journal of Motor Behavior, 23* (4), 251-258.

Cooke, J. D., Brown, S. H., & Cunningham, D. A. (1989). Kinematics of arm movements in elderly humans. *Neurobiology of Aging, 10,* 159-165.

Darling, W., Cooke, J., & Brown, S. (1989). Control of simple arm movements in elderly humans. *Neurobiology of Aging, 10,* 149-157.

Fisk, A. D., McGee, N. D., & Giambra, L. M. (1988). The influence of age on consistent and varied semantic-category search performance. *Psychology and Aging, 3,* 323-333.

Fitts, P. M. (1954). The information capacity of the human motor system in controlling the amplitude of movement. *Journal of Experimental Psychology, 47,* 381-391.

Gentilucci, M., Castiello, U., Corradini, M. L., Scarpa, M., Umilta, C., & Rizzolatti, G. (1991). Influence of different types of grasping on the transport component of prehension movements. *Neuropsychologia, 29* (5), 361-378.

Goggin, N., & Stelmach, G. E. (1990). Age-related differences in a kinematic analysis of precued movements. *Canadian Journal on Aging, 9,* 371-385.

Gottsdanker, R. (1980a). Aging and the use of advance probability information. *Journal of Motor Behavior, 12,* 133-143.

Gottsdanker, R. (1980b). Aging and the maintenance of preparation. *Experimental Aging Research, 6,* 13-27.

Gottsdanker, R. (1982a). Effort of preparation and age. *Perceptual and Motor Skills, 59,* 527-538.

Gottsdanker, R. (1982b). Age and simple reaction time. *Journal of Gerontology, 37,* 342-348.

Guralnik, J., Branch, L., Cummings, S., & Curb, J. (1989). Physical performance measures in aging research. *Journal of Gerontology, 44* (5), M141-M146.

Haaland, K. Y., Harrington, D. L., & Grice, J. W. (1993) Effects of aging on planning and implementing arm movements. *Psychology and Aging, 8,* 617-632.

Hartley, A. A. (1992). Attention. In I. M. Craik & T. A. Salthouse (Eds.), *The handbook of aging and cognition* (pp. 3-49). Hillsdale, NJ: Lawrence Erlbaum Associates Publishers.

Hasher, L., & Zacks, R. T. (1979). Automatic and effortful processes in memory. *Journal of Experimental Psychology: General, 108,* 356-388.

Hay, L., Bard, C., Fleury, M., & Teasdale, N. (1991). Kinematics of aiming in direction and amplitude: A developmental study. *Acta Psychologica, 77,* 203-215.

Hollerbach, J. M. (1982). Computers, brains and the control of movement. *Trends in Neuroscience, 5,* 189-192.

Hollerbach, J. M., & Atkeson, C. G. (1987). Deducing planning variables from experimental arm trajectories: Pitfalls and possibilities. *Biological Cybernetics, 44,* 64-77.

Iberall, T., Bingham, G., & Arbib, M. A. (1986). Opposition space as a structuring concept for the analysis of skilled hand movements. Coins Technical Report 85-19.

Jacobson-Sollerman, & Sperling, L. (1977). Grip function of the healthy hand in a standardized hand function test. *Scandinavian Journal of Rehabilitation Medicine, 9,* 123-129.

Jakobson, L. S., & Goodale, M. A. (1991). Factors affecting higher-order movement planning: A kinematic analysis of human prehension. *Experimental Brain Research, 86,* 199-208.

Jeannerod, M. (1981). Intersegmental coordination during reaching at natural visual objects. In J. Long & A. Baddeley (Eds.), *Attention and performance IX* (pp. 153-168). Hillsdale, NJ: Erlbaum.

Jeannerod, M. (1984). The time of natural prehension movements. *Journal of Motor Behavior, 16,* 235-254.

Jeannerod, M. (1986). The formation of finger grip during prehension. A cortically mediated visuomotor pattern. *Behavioral Brain Research, 19,* 99-116.

Jeannerod, M., Michel, F., & Prablanc, C. (1984) The control of hand movements in a case of hemianesthesia following a parietal lesion. *Brain, 107,* 899-920.

Johansson, R. S., & Westling, G. (1984). Roles of glabrous skin receptors and sensorimotor memory in automatic control of precision grip when lifting rougher or more slippery objects. *Experimental Brain Research, 56,* 550-564.

Kausler, D. H. (1982). *Experimental psychology and human aging.* New York: John Wiley and Sons.

Klatzky, R. (1988). Theories of information processing and theories of aging. In L. L. Light & D. M. Burke (Eds.), *Language, memory, and aging* (pp. 1-16). New York: Cambridge University Press.

Kuhlen, R. G. (1963). Age and intelligence: The significance of cultural change in longitudinal vs cross-sectional findings. *Vita Humana, 6,* 113-124.

Kuypers, H. G. J. M. (1962). Corticospinal connections: Postnatal development in the Rhesus monkey. *Science, 138,* 678-680.

Kuypers, H. G. J. M. (1964). The descending pathways to the spinal cord, their anatomy and function. In J. C. Eccles & J. C. Shade (Eds.), *Organization of the spinal cord* (pp. 182-202). Amsterdam: Elsevier.

Larish, D., & Stelmach, G. E. (1982). Preprogramming, programming, and reprogramming of aimed hand movements as a function of age. *Journal of Motor Behavior, 14,* 322

Lovelace, E. A. (1990). Basic concepts in cognition and aging. In E.A. Lovelace (Ed.), *Aging and cognition: Mental processes, self-awareness and interventions. Advances in Psychology, Volume 72* (pp. 1-28). Amsterdam: North-Holland Co.

MacKenzie, C. L., & Iberall, T. (1994). *The grasping hand.* Amsterdam: Elsevier.

Marsden, C. D. (1982) The mysterious function of the basal ganglia. *Neurology, 32,* 514-539.

Marsden, C. D. (1989). Slowness of movement in Parkinson's disease. *Movement Disorders, 4* (Suppl. 1), S26-S37.

Marteniuk, R. G., Leavitt, J. L., MacKenzie, C. L., & Athenes, S. (1990). Functional relationships between grasp and transport components in a prehension task. *Human Movement Science, 9* (2), 149-176.

Marteniuk, R. G., MacKenzie, C. L., Jeannerod, M., Athenes, S., & Dugas, C. (1987). Constraints on human arm trajectories. *Canadian Journal of Psychology, 41,* 365-378.

Morgan, M., Phillips, J. G., Bradshaw, J. L., Mattingley, J. B., Iansek, R., & Bradshaw, J. A. (1994). Age-related motor slowness: Simply strategic? *Journal of Gerontology: Medical Sciences, 49* (3), 133-M139.

Murrell, K. F., & Entwisle, D. G. (1960). Age differences in movement pattern. *Nature, 185,* 948-949.

Myers, A. M. (1992). The clinical swiss army knife: Empirical evidence on the validity of IADL functional status measures. *Medical Care, 30* (No. 5, Supplement), MS96-MS111.

Myers, A. M., Holliday, P. J., Harvey, K. A., & Hutchinson, K. S. (1993). Functional performance measures: Are they superior to self-assessments? *Journal of Gerontology: Medical Sciences, 28,* 196-206.

Myerson, J., Hale, S., Wagstaff, D., Poon, L. W., & Smith, G. A. (1990). The information-loss model: A mathematical theory of age-related cognitive slowing. *Psychological Review, 97,* 475-487.

Paillard, J. (1982). The contribution of peripheral and central vision to visually guided reaching. In D. J. Ingle, M. A. Goodale, & R. S. W. Mansfield (Eds.), *Analysis of visual behavior* (pp. 367-385). Cambridge, MA: MIT Press.

Petit, T. L. (1982). Neuroanatomical and clinical neuropsychological changes in aging and senile dementia. In F. I. M. Craik & S. Trehub

(Eds.), *Aging and cognitive processes* (pp. 1-21). New York: Plenum Press.

Posner, M. I., & Snyder, C. R. R. (1975). Attention and cognitive control. In R. L. Solso (Ed.), *Information processing and cognition: The Loyola Symposium* (pp. 55-85). Hillsdale, NJ: Erlbaum.

Rabbitt, P. (1979). How old and young subjects monitor and control responses for accuracy and speed. *British Journal of Psychology, 70,* 305-311.

Rabbitt, P. (1982). Breakdown of control processes in old age. In T. M. Field, A. Huston, H. C. Quay, L. Troll, & G. Finley (Eds.), *Review of human development* (pp. 540-550). New York: John Wiley & Sons.

Rapp, B. C., & Caramazza, A. (1991) Cognitive neuropsychology: From impaired performance to normal cognitive structure. In R. G. Lister & H.J. Weingartner (Eds.), *Perspectives on cognitive neuroscience* (pp. 384-404). New York: Oxford University Press.

Roy, E. A. (1990). The interface between normality and pathology in understanding motor function. In G. Reid (Ed.), *Problems in movement control* (pp. 3-30). Amsterdam: North-Holland Co.

Roy, E. A., Winchester, T., Weir, P., & Black, S. (1993). Age differences in the control of visually aimed movements. *Journal of Human Movement Studies, 24,* 71-81.

Salthouse, T. A. (1982). *Adult cognition: An experimental psychology of human aging.* New York: Springer-Verlag.

Salthouse, T. A. (1985a). *A theory of cognitive aging.* Amsterdam: North-Holland Co.

Salthouse, T. A. (1985b). Speed of behavior and its implication for cognition. In J. E. Birren & K. W. Schaie (Eds.), *Handbook of psychology of the aging* (2nd Ed., pp. 400-426). New York: Van Nostrand Reinhold.

Salthouse, T. A. (1987). Adult age differences in integrative spatial ability. *Psychology and Aging, 2,* 254-260.

Salthouse, T. A. (1988a). Resource-reduction interpretations of cognitive aging. *Developmental Review, 8,* 238-272.

Salthouse, T. A. (1988b). The role of processing resources in cognitive aging. In M. L. Howe & C. J. Brainerd (Eds.), *Cognitive development in adulthood* (pp. 185-239). New York: Springer-Verlag.

Salthouse, T. A. (1988c). Initiating the formalization of theories of cognitive aging. *Psychology and Aging, 3,* 3-16.

Salthouse, T. A. (1990). Cognitive competence and expertise in aging. In J.E. Birren & K.W. Schaie (Eds.), *Handbook of psychology of the aging* (3rd ed., pp. 310-329). New York: Van Nostrand Reinhold.

Salthouse, T. A., & Somberg, B. L. (1982). Isolating the age deficit in speeded performance. *Journal of Gerontology, 37,* 59-63.

Schmidt, R. A. (1982). *Motor control: A behavioral emphasis.* Champaign, IL: Human Kinetics Publishers.

Schmidt, R. A. (1985). The search for invariance in skilled movement behavior. *Research Quarterly for Exercise and Sport, 56,* 188-200.

Schneider, W., & Shiffrin, R. M. (1977). Controlled and automatic human information processing: I. Detection, search, and attention. *Psychological Review, 84,* 1-66.

Shiffrin, R. M., & Schneider, W. (1977). Controlled and automatic human information processing: II. Perceptual learning, automatic attending, and a general theory. *Psychological Review, 84,* 127-190.

Simon, J. R., & Pouraghabagher, A. R. (1978). The effect of aging on the stages of processing in a choice reaction time task. *Journal of Gerontology, 33* (4), 553-561.

Stelmach, G. E., Goggin, N. L., & Amrhein, P. C. (1988). Aging and reprogramming: The restructuring of planned movements. *Psychology and Aging, 3,* 151-157.

Stelmach, G. E., Goggin, N. L., & Garcia-Colera, A. (1987). Movement specification time with age. *Experimental Aging Research, 13,* 39-46.

Stelmach, G. E., Worringham, C. J., & Strand, E. A. (1987). The programming and execution of movements sequences in Parkinson's disease. *International Journal of Neuroscience, 36,* 55-65.

Soechting, J. F., & Flanders, M. (1993). Parallel, interdependent channels for location and orientation in sensorimotor transformations for reaching and grasping. *Journal of Neurophysiology, 70,* 1139-1150.

Wallace, S. A., & Weeks, D. L. (1988). Temporal constraints in the control of prehensile movement. *Journal of Motor Behavior, 20,* 81-105.

Walsh, D. A. (1982). The development of visual information processes in adulthood and old age. In F. I. M. Craik & S. Trehub (Eds.), *Aging and cognitive processes: Advances in the study of communication and effect* (Vol. 8, pp. 99-124). New York: Plenum Press.

Warabi, T., Noda, H., & Kato, T. (1986). Effect on aging on sensorimotor functions of eye and hand movements. *Experimental Neurology, 92,* 686-697.

Wechsler, D. (1981). *Wechsler Adult Intelligence Scale-Revised.* New York: Psychological Corporation.

Welford, A. T. (1981). Signal, noise, performance, and age. *Human Factors, 23,* 97-109.

Welford, A. T. (1982). Motor skills and aging. In J. Mortimer, F. Pirozzolo, & G. Maletta (Eds.), *Aging motor system* (pp. 152-187). New York: Praeger Publishers.

Welford, A. T. (1984a). Between bodily changes and performance: Some possible reasons for slowing with age. *Experimental Aging Research, 10,* 73-88.

Welford, A. T. (1984b). Psychomotor performance. In C. Eisdorfer (Ed.), *Annual review of gerontology and geriatrics* (pp. 237-273) New York: Springer Publishing Co.

Welford, A. T. (1985). Changes in performance with age: An overview. In N. Charness (Ed.), *Aging and human performance* (pp. 333-369). New York: John Wiley and Sons.

Welford, A. T., Norris, A. H., & Shock, N. W. (1969). Speed and accuracy of movement and their changes with age. *Acta Psychologica, 30,* 3-15.

Westling, G., & Johansson, R. S. (1984). Factors influencing the force control during precision grip. *Experimental Brain Research, 53,* 277-284.

Wing, A. M., & Fraser, C. (1983). The contribution of the thumb to reaching movements. *Quarterly Journal of Experimental Psychology, 35A,* 297-309.

Wing, A. M., Turton, A., & Fraser, C. (1986). Grasp size and accuracy of approach in reaching. *Journal of Motor Behavior, 18,* 245-260.

Winter, D. A. (1979). *Biomechanics of human movement.* Toronto: Wiley & Sons.

Wohlwill, J. F. (1970). Methodology and research strategy in the study of developmental change. In L. R. Goulet & P. B. Bates (Eds.), *Life-span developmental psychology: Research and theory.* New York: Academic Press.

Woodworth, R. S. (1899). The accuracy of voluntary movement. *Psychological Review, Monograph Supplement 3,* Whole 13/3.

Zelaznik, H., Schmidt, R. A., & Geilen, S. (1986). Kinematic properties of rapid aimed hand movements. *Journal of Motor Behavior, 18,* 353-372.

Changes in sensory motor behavior in aging
A.-M. Ferrandez and N. Teasdale (Editors)

AGE, PERCEIVED HEALTH, AND SPECIFIC AND NONSPECIFIC MEASURES OF PROCESSING SPEED

Timothy A. SALTHOUSE and Julie L. EARLES

Georgia Institute of Technology, Atlanta

Abstract

Many measures presumed to reflect the duration of specific information processes are currently being used by researchers examining aging and cognition. In this chapter we examine empirical relationships between these measures of specific information processing speed and measures of nonspecific processing efficiency, as well as the influence of adult age and health status on both types of speed measures. The influences of health on both specific and non-specific speed measures were small in the data sets examined, and health status had little or no moderating effects on the relations between age and measures of processing speed. The age-related influences on speed were substantial, but the effects on the specific speed measures were not independent of those on the nonspecific speed measures. Recommendations concerning analyses of measures hypothesized to reflect specific information processes are discussed.

Key words: Aging, cognition, health, information processes, speed.

Correspondence should be sent to Timothy A. Salthouse, School of Psychology, Georgia Institute of Technology, Atlanta, GA 30332-0170, U.S.A. (e-mail: tim.salthouse@psych.gatech.edu).

INTRODUCTION

As indicated by the title, this chapter is concerned with the effects of age-related and health-related influences on specific and nonspecific measures of processing speed. Relations between adult age and measures of speed of performance have been investigated at least since the time of Galton (e.g., Ruger & Stoessiger, 1927), and the topic has been the focus of nearly continuous research over the last 40 years. Much of the early research was summarized in an edited book published in 1965 (Welford & Birren, 1965), and subsequent reviews have appeared in 1977 (Welford, 1977), 1979 (Birren, Woods, & Williams, 1979), 1985 (Salthouse, 1985), and 1990 (Cerella, 1990).

Research in this area has gone through several different stages corresponding to different perspectives on the interest in, or value of, speed measures in the context of aging. For example, an early view was that age-related slowing was relatively uninteresting, at least from the perspective of higher-order cognitive functioning, because it was assumed to be a peripheral limitation somewhat analogous to declines in visual and auditory sensitivity. In the 1960s and 1970s, however, many researchers became convinced that central nervous system factors were involved in the age-related slowing phenomenon, and hence that slowing of internal processes could be expected to have consequences for a wide variety of cognitive operations. Because this was also the period when the information processing framework became popular within cognitive psychology, a great number of studies were conducted in which time measures were used to compare adults of different ages in the durations of specific processes such as memory scanning and mental rotation.

Within the last 10 years there has been considerable interest in possible relations among the age differences in different measures of speed, primarily by extending a method introduced by Brinley (1965). This procedure consists of plotting the mean times of one age group (e.g., older adults) against those of a different age group (e.g., young adults), and then examining the parameters of the regression function relating the two sets of means. Many analyses of this type have revealed that the relations are often highly systematic, with high correlations and slopes frequently in the range of 1.5 to 2.0 for contrasts of adults in their 20s with adults in their 60s. Although interpretation of these relations remains controversial (e.g., Cerella, 1994; Fisk & Fisher, 1994; Myerson, Wagstaff, & Hale, 1994; Perfect, 1994), the existence of the systematic relations has raised questions about the independence of the age-related influences on different speed measures, and has focused interest on the issue of general or common factors contributing to age-related slowing.

In this chapter we describe a different approach to the investigation of the interrelations of the age-related effects across different types of speed measures. Because the distinction between specific and non-specific speed measures is a central theme in the chapter, we will begin with a discussion of these two terms.

Types of speed measures

Some speed measures are hypothesized to reflect quite specific processes or components because they are derived from the manipulation of a theoretically relevant factor, and are therefore postulated to represent the duration of a particular process sensitive to that manipulation. For example, a researcher might vary the number of items presented in the memory set in the Sternberg memory-scanning paradigm, and then compute the slope of the function relating reaction time (RT) to the number of memory set items. Because the slope of that function has been interpreted as representing the time to search or access information from memory, it can be categorized as a specific speed measure. Another type of specific speed measure is the difference between the RTs in two conditions presumed to differ in some critical process because the difference could be interpreted as reflecting the duration or speed of the critical process. For example, the difference in RT in a Sternberg paradigm with setsizes of 4 and 2 can be interpreted as the time required to search two additional items in memory. More generally, whenever a difference score or a slope measure is computed to provide what is hypothesized to be a purer or more precise measure of some process or component, it can be classified as a specific measure of processing speed.

In contrast, other measures are often postulated to be nonspecific, or relatively general, because they are presumed to reflect the duration of many processes, and not merely the duration of the critical process. For example, when a difference score is computed, the time in the simplest (or fastest) task condition might be considered to reflect nonspecific processes because that condition presumably involves a mixture of sensory, motor, and other processes with only a minimal amount of the critical process. Another possible nonspecific measure is the intercept from a linear regression equation relating RT to the quantitative value of the manipulated variable because the intercept is often postulated to represent the duration of all processes except for the critical one reflected in the slope parameter. Finally, the average of speed measures across all conditions in the task could be interpreted as representing a mixture of many processes, including the critical one. For this reason, the mean time across all task conditions could be classified as a nonspecific speed measure because it presumably reflects the aggregate duration of a variety of different processes.

The primary question we address in this chapter is the nature of the influence of adult age, and of self-perceived health status, on these two categories of speed measures. At the outset we should acknowledge that the relation between specific and nonspecific speed measures is likely to vary according to the type of nonspecific speed measure. For example, the measure of speed in the simplest condition and the intercept from a regression equation are usually assumed to reflect theoretically distinct processes from those represented by a difference score or by the slope parameter. These particular nonspecific and specific measures might therefore be expected to be largely independent of one another. However, because the mean or average includes the duration of the critical process in addition to other processes, it might be expected to have a moderate to strong positive relation with specific measures based on a difference score or slope. (See Chapman and Chapman, 1988, for a discussion of the mathematical relations between difference scores and measures of overall performance.) Although the mathematical relations may vary according to the particular combination of specific and nonspecific speed measures, the focus here is on empirical rather than theoretical relations among the measures and thus several combinations of specific and nonspecific measures are examined.

Health status

Health was assessed in the data to be described by a self-rating on a 5-point scale ranging from 1 for Excellent to 5 for Poor. That is, the research participants were simply asked to classify their own health status with a number between 1 for the highest level and 5 for the lowest level. This is obviously a very crude method of assessment, and the results involving this variable will have to be interpreted cautiously. Nevertheless, the available evidence suggests that self-ratings of health have at least moderate validity as an index of health status. For example, self-ratings of health have been found to be significantly related to: (a) physician assessments of overall health (Heyman & Jeffries, 1963; LaRue, Bank, Jarvik, & Hetland, 1979; Maddox, 1962, 1964; Maddox & Douglass, 1973; Suchman, Phillips, & Streib, 1958); (b) reported medical problems or number of prescription medications (Fillenbaum, 1979; Kaplan & Camacho, 1983; Liang, 1986; Linn & Linn, 1980;

Mossey & Shapiro, 1982; Pilpel, Carmel, & Galinsky, 1988; Salthouse, Kausler, & Saults, 1990; Tissue, 1972); and (c) longevity or survival (Botwinick, West, & Storandt, 1978; Heyman & Jeffers 1963; Kaplan & Camacho, 1983; LaRue, et al., 1979; Mossey & Shapiro, 1982; Pfeiffer, 1970; Singer, Garfinkel, Cohen, & Srole, 1976; Suchman et al., 1958).

RESEARCH QUESTIONS

In the analyses to be reported we examine the relation of age and self-reported health status, both alone and in combination, to specific speed measures before and after consideration of the nonspecific speed measure. The conceptual framework of our investigation is illustrated in Figure 1. Notice that health is postulated to function as a potential mediator of the age-related influences on one or both measures of speed, and that at least some of the age-related influences on the specific speed measure are hypothesized to be mediated through the nonspecific speed measure. The goal of the analyses to be described is to determine the relative strength of each of the different paths in this figure for various combinations of specific and nonspecific measures of processing speed.

Among the possible outcomes of the analyses are: (a) that almost all of the age-related effects on the speed measures are mediated through the health variable; (b) that health has little or no effects on either speed measure; (c) that age and health status have independent influences on the nonspecific and specific measures, with substantial unique relations of age and health on both the nonspecific and specific speed measures; (d) that the influences on the two speed measures are completely overlapping, in that all of the age-related and health-related effects on the specific speed measures are mediated through the influences on the nonspecific speed measures; and (e) that the nonspecific and specific speed measures have a suppression relation with one another, such that the influences related to age or health on one measure are obscured or suppressed by the influence on the other speed measure and the relation between the two measures. The latter three outcomes are not necessarily qualitatively distinct because they could be viewed as different points along a continuum. That is, age-related and health-related effects on the specific speed measure could be independent of, mediated through, or suppressed by, the effects on the nonspecific speed measure. However, it is important to note that only if the influences on the nonspecific and specific speed measures were largely independent would it be meaningful to consider the two types of speed measures separately, or in isolation, as is often the case in much contemporary research in the field of aging and cognition.

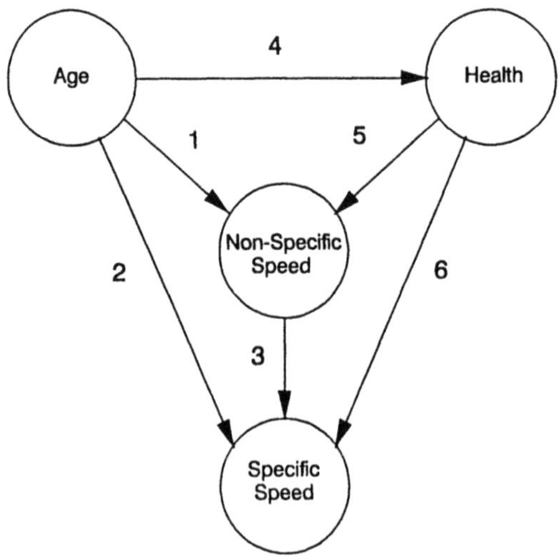

FIGURE 1. Diagram illustrating possible relations among age, health, and nonspecific and specific measures of processing speed.

Analyses

Two analytical methods will be used in the current investigation. The primary analytical method is hierarchical multiple regression, which yields squared semi-partial correlations representing independent portions of variance. Of particular interest is the amount of age-related (or health-related) variance in the specific speed measure before and after control of the variance in the nonspecific measure. If there is no difference in the magnitude of the variance in the before and after comparisons, then one could infer that the influences on the two speed measures are independent. However, if there is substantial reduction in the age-related (or health-related) variance after control of the nonspecific measure, then one could infer that a large proportion of the influences are common or shared. Finally, if the magnitude of the age-related variance in the specific speed measure increases when the nonspecific measure is controlled, then suppression can be inferred to exist.

The second analytical method to be employed in this project is path analysis. The goal of the path analyses is to indicate the relative strength of each of the paths portrayed in Figure 1. Because the outcome of the path analyses will be standardized regression coefficients, which represent the amount of change in standard deviation units in one variable corresponding to a change of one standard deviation in another variable, the magnitude of different paths can be compared with one another. If many of the age-related and health-related influences on the specific speed measure are mediated through the nonspecific measure, then paths 1, 5, and 3 in Figure 1 should be relatively strong, and paths 2 and 6 should be weak.

Data sets

The data for the analyses were derived from several recent studies conducted by Salthouse and colleagues, with the data sets briefly described in Table 1. All of the studies involved adults from a wide range of ages, although the samples in Data Sets B and E consisted of only young (age 18 to 25) and old (age 55 to 80) adults rather than a continuous distribution of ages as in the other studies. In all cases research participants were asked to evaluate their health status by a self-rating on a 5-point scale ranging from 1 for excellent to 5 for poor. This rating served as the health index in the present analyses.

Data Sets A and B involved two computer-administered versions of the Digit Symbol Substitution test (Salthouse, 1992a). This is a choice reaction time (RT) task involving a pair of visually presented stimuli and keypress responses. In the Digit Symbol version of the task, one member of the pair is a digit and the other is a symbol, and the decisions are based on whether the digit-symbol pair matches according to a code table presented at the top of the display. In the Digit Digit version of the task, both members of the pair are digits, and hence the decision is based on physical identity. A code table is still presented in this version of the task, but because it merely contains pairs of identical digits, it is redundant and unnecessary for performance of the task.

Three measures of performance were obtained from these tasks. Because the difference between the Digit Symbol and Digit Digit times presumably reflects the duration of processes specifically associated with the substitution of symbols and digits (e.g., search of the code table, or retrieval of learned associations), it can serve as the specific speed measure.[1] The nonspecific measures are the time in the simplest

---

1. It should be noted that although difference scores often have low reliability, that is not necessarily the case for the current measures. For example, estimated reliabilities for the Digit Symbol - Digit Digit difference score were .79 and .86 in two studies reported in Salthouse (in press).

condition, which in this case is the Digit Digit condition, and the mean of the Digit Digit and Digit Symbol measures.

TABLE 1.
Description of data sets.

Set	Source	n	Sample	Task/Measures
A	Earles & Salthouse (in press)	744	Continuous 18-87	Digit Digit and Digit Symbol. Time for Digit Digit, mean, and difference.
B	Assorted studies	694	Young/old	Digit Digit and Digit Symbol. Time for Digit Digit, mean, and difference.
C	Salthouse (1994) Study 1	246	Continuous 18-84	Digit Symbol with 0-9 symbols. Time for 0 and for 3 symbols, mean, intercept, and slope with 3-9 symbols.
D-1	Salthouse (1994) Study 2	258	Continuous 20-87	Memory Search with 1-4 digits. Time for 1 item, mean, intercept, and slope.
D-2	Salthouse (1994) Study 2	258	Continuous 20-87	Memory Search with 1-4 letters. Time for 1 item, mean, intercept, and slope.
E	Salthouse & Coon (1994)	80	Young/old	Arithmetic with 0-7 operations. Time for 0 and 1 operation, mean, intercept, and slope for 1-7 operations.
F-1	Salthouse et al. (in press) Study 2	131	Continuous 17-79	Arithmetic with 0-4 operations under single task conditions. Time for 0 and 1 operation, mean, intercept, and slope for 1-4 operations.
F-2	Salthouse et al. (in press) Study 2	131	Continuous 17-79	Arithmetic with 0-4 operations under dual task conditions. Time for 0 and 1 operation, mean, intercept, and slope for 1-4 operations.

Data Set C involved tasks designed to represent an extension of the Digit Digit and Digit Symbol tasks. In this data set those tasks are referred to as involving 0 and 9 symbols, respectively, and new conditions with 3 and 6 symbols were also administered. The decisions in these new conditions involved a mixture of physical identity and associational equivalence judgments because both types of trials were intermixed in these conditions. The slope of the function relating RT to number of symbols (with values of 3, 6, and 9) served as the specific, substitution, measure. The condition with 0 symbols was not included in the regression equation because of the possibility that it might differ from the other conditions in qualitative (i.e., presence/absence) rather than only quantitative (i.e., how many) dimensions. Measures of nonspecific speed were the time with 0 symbols and with 3 symbols as alternative measures of the simplest condition, the mean time across conditions with 3, 6, and 9 symbols, and the intercept of the linear regression function relating RT to number of symbols (with values of 3, 6, and 9).

Data Sets D-1 and D-2 involved a Sternberg memory search task with either digits (1) or letters (2) as stimuli. The same research participants performed both tasks, and in each case the number of memory set stimuli ranged from 1 to 4, and a single item served as the probe stimulus. The task for the subject was to decide as rapidly as possible, by pressing one key for YES and another key for NO, whether the probe stimulus had been presented in the memory set. The specific speed measure (presumably representing the speed of memory scanning) was the slope of the regression equation relating RT to number of memory set items. The time with 1 item, the mean time across conditions with 1, 2, 3, or 4 items, and the intercept of the regression equation, served as the nonspecific measures.

Data Sets E and F involved a verification arithmetic task in which problems varied in the number of arithmetic operations. The task for the subject was to decide as rapidly as possible, by pressing one key for YES and another key for NO, whether the arithmetic equation was correct or incorrect. The number of addition or subtraction operations ranged from 0 (i.e., physical identity decision) to 7 in Data Set E, and from 0 to 4 in Data Sets F-1 and F-2. The same research participants contributed data to Data Sets F-1 and F-2. Data Set F-1 involved performance of the arithmetic task in isolation, and Data Set F-2 involved performance of the arithmetic task while simultaneously attempting to remember four letters. In each case, the specific speed measure was the slope of the regression equation either with 1 to 7 operations (Data Set E), or with 1 to 4 operations (Data Sets F-1 and F-2). The nonspecific measures were the time with 0 operations and the time with 1 operation as alternative measures of the simplest condition, the mean time across conditions with 1 or more operations, and the intercept of the regression equation.

Results

Table 2 contains the means, standard deviations, and correlations with age and self-reported health for the relevant variables in each data set. Several points should be noted about the values in this table. First, there were moderate to large (i.e., .38 to .76) correlations with age for all nonspecific speed measures except the intercept in Data Set F-2. Second, the age correlations with the specific speed measures ranged from moderately positive (i.e., A, B, C, E, F-2) to near zero (i.e., D-1, D-2, F-1). Third, correlations indicating the fit of the regression equations to the data were generally high, with only those for the memory search task in Data Sets D-1 and D-2 averaging less than .9. Although these correlations were not used in the subsequent analyses, the moderately high values provide some assurance that the slopes are meaningful reflections of the duration of a process related to the manipulated variable because the square of the correlation represents the proportion of variance accounted for by the linear regression equation. Furthermore, the small to nonexistent relations between age and the correlations suggest that the fit of the regression equations did not vary systematically as a function of age. The fourth and final point to note from the values in Table 2 is that correlations involving the perceived health variable were much smaller than those with the age variable, and many of them were not significantly different from zero.

TABLE 2.
Means, standard deviations, and correlations with age and health for all speed measures.

				Correlations		
Data set	Type	Mean	SD	Age	Health	A-H
A						.14*
Digit Digit	Nonspecific	807	259	.44*	.17*	
Mean	"	1215	327	.56*	.20*	
Difference	Specific	818	317	.43*	.14*	
B						.09
Digit Digit	Nonspecific	689	224	.59*	.11*	
Mean	"	1078	312	.76*	.10	
Difference	Specific	778	345	.60*	.03	

* $p < .01$

TABLE 2 (continued).

				Correlations		
Data set	Type	Mean	SD	Age	Health	A-H
C						.13
DigSym-0	Nonspecific	759	234	.51*	.23*	
DigSym-3	"	1194	324	.59*	.29*	
Mean	"	1357	350	.61*	.28*	
Intercept	"	1049	304	.53*	.31*	
Slope	Specific	51	25	.34*	.02	
(Correlation)	-	.95	.08	.08	-.01	
D-1						.06
1-item	Nonspecific	973	457	.41*	.04	
Mean	"	1071	432	.43*	.05	
Intercept	"	919	478	.41*	.03	
Slope	Specific	61	62	-.05	.03	
(Correlation)	-	.73	.39	-.17*	-.05	
D-2						.06
1-item	Nonspecific	909	399	.39*	.08	
Mean	"	1004	380	.42*	.09	
Intercept	"	849	404	.38*	.10	
Slope	Specific	62	54	.03	-.04	
(Correlation)	-	.73	.41	-.12	-.05	
E						.00
0-operations	Nonspecific	1259	465	.76*	.00	
1-operation	"	1739	652	.74*	-.03	
Mean	"	5775	1360	.64*	.09	
Intercept	"	274	707	.44*	-.04	
Slope	Specific	1375	337	.41*	.12	
(Correlation)	-	.98	.06	.11	.11	
F-1						.08
0-operations	Nonspecific	1219	531	.46*	.05	
1-operation	"	1673	569	.56*	.15	
Mean	"	3415	904	.42*	.09	
Intercept	"	389	637	.41*	.04	
Slope	Specific	1210	365	.13	-.01	
(Correlation)	-	.98	.03	-.07	-.15	
F-2						.08
0-operations	Nonspecific	1403	479	.72*	.13	
1-operation	"	1934	628	.65*	.16	
Mean	"	3988	1404	.49*	.10	
Intercept	"	505	999	.20	.15	
Slope	Specific	1393	745	.26*	-.00	
(Correlation)	-	.92	.22	-.04	-.00	

Table 3 contains the results of the hierarchical regression analyses on the nonspecific speed measures. The mean for the entries in column 3, corresponding to health-related variance alone, was .022, and that for the entries in column 4, corresponding to health-related variance after control of age, was .012. These values indicate that self-reported health

TABLE 3.
Results of hierarchical regression analyses on nonspecific speed measures.

1	2	3	4	5	6
	Non specific	Health-related variance		Age-related variance	
Data set	speed measure	Alone	After age	Alone	After health
A	Digit Digit	.028*	.011*	.195*	.178*
	Mean	.039*	.015*	.309*	.285*
B	Digit Digit	.012*	.004	.353*	.345*
	Mean	.009	.001	.574*	.566*
C	0 symbols	.055*	.029*	.262*	.236*
	3 symbols	.087*	.049*	.351*	.313*
	Mean	.078*	.041*	.372*	.335*
	Intercept	.097*	.060*	.284*	.247*
D-1	1 item	.001	.000	.172*	.171*
	Mean	.002	.000	.187*	.185*
	Intercept	.001	.000	.164*	.163*
D-2	1 item	.007	.003	.149*	.145*
	Mean	.009	.004	.173*	.168*
	Intercept	.010	.006	.145*	.141*
E	0 operations	.000	.000	.581*	.581*
	1 operation	.001	.001	.549*	.549*
	Mean	.008	.008	.549*	.549*
	Intercept	.002	.002	.193*	.193*
F-1	0 operations	.003	.000	.213*	.210*
	1 operation	.023	.012	.310*	.299*
	Mean	.008	.003	.179*	.174*
	Intercept	.020	.013	.167*	.160*
F-2	0 operations	.018	.006	.521*	.509*
	1 operation	.026	.012	.425*	.411*
	Mean	.011	.005	.239*	.233*
	Intercept	.021	.017	.039	.035

* $p < .01$

status was associated with a little over 2% of the total variance in the nonspecific speed measure, and that only about 1% of the variance was independent of age. Self-reported health status therefore does not appear to be an important factor in the nonspecific speed measures in these data sets.

The mean for the entries in column 5, representing age-related variance alone, was .289, and that for the entries in column 6, corresponding to age-related variance after control of health, was .278. Age was therefore associated with an average of nearly 29% of the variance in the nonspecific speed measures, and it was still associated with an average of 28% after statistical control of the measure of health status. The results in Table 3 can thus be succinctly summarized as follows: Health status had little or no effect on the nonspecific speed measures, but the age-related influences on these measures were substantial, and largely independent of health status.

Results of the hierarchical regression analyses on the specific speed measures are presented in Table 4. Column 4 in this table contains the correlation between the nonspecific and specific speed measures described in columns 2 and 3. Because the results in Table 3 indicated that there were little or no independent influences of health on the nonspecific speed measure, only the proportions of age-related variance in the specific speed measures are represented in the entries in columns 5, 6, 7, and 8.

Contrasts of column 5 with 7 and column 6 with 8 are informative about the influence of health status on the age-related effects in the specific speed measure. That is, the first value in each pair indicates the age-related variance before control of health status, and the second value in the pair indicates the age-related variance after health status was controlled. Although there is some variation in the individual values, the means of the relevant comparisons were identical (i.e., the means of columns 5 and 7 were both .099, and the means of columns 6 and 8 were both .059). On the average, therefore, there is no evidence of health-related mediation of the adult age differences in the specific speed measures in these data sets.

Evidence relevant to the mediational influence of nonspecific speed on the age differences in the specific speed measures is available in a contrast of columns 5 and 6 (and also in a contrast of columns 7 and 8 after the influence of health has been removed). Entries in column 5 represent the amount of age-related variance in the specific speed measure before consideration of the nonspecific speed measure, and entries in column 6 represent the amount of age-related variance in the specific measure after control of the variance in the nonspecific measure. Means of these values across all entries in Table 4 were .099 and .059, respectively, indicating that an average of about 40.4% (i.e., [.099-.059]/.099) of the age-related variance in the specific speed measure was shared with the nonspecific speed measure.

TABLE 4.
Results of hierarchical regression analyses on specific speed measures.

1 Data set	2 Specific	3 Non-specific	4 *r*	5 Age-related variance in specific measure Alone	6 After NS	7 After Health	8 After Health, NS
A	Difference	Digit Digit	.18*	.181*	.150*	.169*	.144*
	"	Mean	.63*	"	.009*	"	.009*
B	Difference	Digit Digit	.23*	.360*	.333*	.359*	.333*
	"	Mean	.72*	"	.008*	"	.008*
C	Slope	0 Sym.	.30*	.115*	.046*	.116*	.047*
	"	3 Sym.	.31*	"	.037*	"	.035*
	"	Mean	.50*	"	.002	"	.002
	"	Intercept	.08	"	.124*	"	.124*
D-1	Slope	1 item	-.37*	.002	.014	.003	.013
	"	Mean	-.14	"	.000	"	.000
	"	Intercept	-.45*	"	.022*	"	.021*
D-2	Slope	1 item	-.24*	.001	.018	.002	.019
	"	Mean	.00	"	.001	"	.002
	"	Intercept	-.34*	"	.030*	"	.030*
E	Slope	0 operat.	.39*	.169*	.032*	.169*	.032*
	"	1 operat.	.44*	"	.016*	"	.015
	"	Mean	.86*	"	.032*	"	.032*
	"	Intercept	-.24	"	.334*	"	.331*
F-1	Slope	0 operat.	.17	.018	.003	.018	.004
	"	1 operat.	.33*	"	.003	"	.003
	"	Mean	.75*	"	.042*	"	.040*
	"	Intercept	-.36*	"	.095*	"	.094*
F-2	Slope	0 operat.	.40*	.069*	.001	.070*	.002
	"	1 operat.	.44*	"	.001	"	.001
	"	Mean	.85*	"	.030*	"	.029*
	"	Intercept	-.67*	"	.162*	"	.159*

* $p < .01$; NS: Non-specific.

Inspection of Table 4 reveals that there was a wide range of discrepancies between the values in columns 5 and 6, and hence the results were re-analyzed according to the type of nonspecific speed measure. One striking difference across the simple (i.e., Digit Digit, 0 symbols, 1 item, 0 operations, 1 operation), mean, and intercept nonspecific measures was the average magnitude of the correlation between the nonspecific and specific speed measures. That is, the average correlation was .22 for the simple measures, .52 for the mean measures, and -.33 for the intercept measures.

Perhaps not surprising in light of the differences in the direction and magnitude of the nonspecific-specific relation, the pattern of change in the age-related variance in the specific speed measure after statistical control of the variance in the nonspecific speed measure also varied according to the type of nonspecific measure. The age-related variance decreased by an average of 48.6% (from .107 to .055) after control of the simple nonspecific speed measure, decreased by an average of 86.0% (from .114 to .016) after control of the mean nonspecific speed measure, but *increased* by an average of 106.5% (from .062 to .128) after control of the intercept nonspecific speed measure. At least within these data sets, then, several of the intercept and slope measures from the regression equations appeared to have a suppression relation with one another in that the age-related effects on the slope measure were obscured by the combination of negative age-related effects on the intercept, and a negative relation between the intercept and the slope.

Values of the standardized path coefficients for the paths portrayed in Figure 1 are presented in Table 5. It can be seen that the paths from the health variable to the speed measures (i.e., 5 and 6) were generally quite small, and they differed from 0 by more than two standard errors only in Data Sets A and C. These results are thus consistent with those from the hierarchical regression analyses in suggesting that health status as measured in these studies was not a very important factor contributing to the age-related differences in either the specific or nonspecific measures of processing speed.

Of greatest interest in Table 5 are the entries in the column labeled Path 2, representing the age-related influences on the specific speed measure after taking into consideration both health and the nonspecific speed measure. Because these values had substantial variability, with a range from -.23 to +.72, they were re-analyzed according to the type of nonspecific speed measure. The average age-specific coefficient was .19 when the simple measure was used as the nonspecific speed variable, it was -.04 when the mean was used as the nonspecific speed measure, and it was .36 when the intercept was the nonspecific speed measure.

TABLE 5.
Standardized path coefficients corresponding to the path model portrayed in Figure 1.

Data set	Nonspecific speed measure	Path 1 Age NS	2 Age Spec	3 NS Spec	4 Age Health	5 Health NS	6 Health Spec
A	Digit Digit	.43*	.43*	-.02	.14*	.11*	.08*
	Mean	.54*	.11*	.56*	.14*	.12*	.01
B	Digit Digit	.59*	.72*	-.20	.09*	.06	-.01
	Mean	.76*	.14*	.62*	.09*	.03	-.04
C	0 symbols	.49*	.25*	.19*	.13*	.17*	-.06
	3 symbols	.56*	.24*	.19*	.13*	.22*	-.07
	Mean	.58*	.05	.51*	.13*	.20*	-.13*
	Intercept	.50*	.42*	-.15*	.13*	.25*	.01
D-1	1 item	.41*	.13	-.42*	.06	.01	.01
	Mean	.43*	.01	-.15*	.06	.02	.02
	Intercept	.41*	.16*	-.51*	.06	.01	.03
D-2	1 item	.38*	.15*	-.30*	.06	.06	-.03
	Mean	.41*	.04	-.01	.06	-.07	-.04
	Intercept	.38*	.19*	-.41*	.06	.08	-.01
E	0 operations	.76*	.28*	.17	.01	.00	.11
	1 operation	.74*	.18	.31*	.01	-.03	.12
	Mean	.64*	-.23*	1.00*	.01	.09	.03
	Intercept	.44*	.64*	-.52*	.01	-.05	.09
F-1	0 operations	.46*	.07	.14	.08	.02	-.02
	1 operation	.55*	-.07	.37*	.08	.11	-.06
	Mean	.42*	-.22*	.86*	.08	.06	-.07
	Intercept	.40*	.34*	-.51*	.08	.11	.03
F-2	0 operations	.72*	-.05	.44*	.08	.08	-.06
	1 operation	.64*	-.05	.49*	.08	.11	-.08
	Mean	.48*	-.20*	.96*	.08	.07	-.08
	Intercept	.19*	.41*	-.76*	.08	.13	.08

* Coefficients more than 2 standard errors different from zero; NS: Nonspecific.

In order to interpret these values, it is informative to contrast them with the average correlation between age and the specific speed measures. That is, the correlations represent the total age-related effects on the specific speed measure, including those mediated through health and the nonspecific speed measure. Because the standardized path coefficients represent the direct effects of age, the difference between the correlation (total effects) and the standardized path coefficient (direct effects) can be interpreted as an estimate of the magnitude of the contribution of indirect age-related influences on the specific speed measure. In other words, the correlation corresponds to the sum of the direct and indirect effects of age on the specific speed measure, but the path coefficient represents only the direct effects that are independent of the other variables under consideration. The average correlations, with the average path coefficients in parentheses, were .27 (.19) with the simple nonspecific measure, .31 (-.04) with the mean nonspecific measure, and .17 (.36) with the intercept nonspecific measure. As with the hierarchical regression analyses, then, the age relations on the specific speed measure were reduced when the simple or mean nonspecific measures were controlled, but were increased when the intercept nonspecific measure was controlled.

DISCUSSION

Before discussing implications of the results, it is important to consider limitations of the manner in which health status was evaluated in these data sets. As noted above, each individual was asked to evaluate his or her health status on a 5-point scale. Although the literature cited earlier indicates that measures of this type have been found to be related to other relevant measures, questions can nonetheless be raised about the sensitivity and validity of self-report measures of health status. It is not yet obvious which particular types of objective measures would provide accurate and efficient indices of overall health status, but alternative methods of evaluating health status should be explored to supplement, if not completely replace, the self-report measures.

Perhaps because of the crude manner in which it was assessed, health status was found to have relatively little effect on the age-related differences in either the specific or the nonspecific speed measures. Similar results have been reported by Earles and Salthouse (in press) with a broader assessment of health status, and with different analytical procedures and measures of speed. It is possible that larger influences of health status would be found in samples in which there was a greater variation in health status. In all of the data sets described above the research participants were relatively healthy, and nearly all reported that their health was in the good to excellent range. Therefore, if the samples had included a wider range of health status, the health variable may have had a greater influence on the speed measures, and possibly also on the influence of age on those measures.

Despite reservations about the adequacy of the health assessment in these data sets, it is important to emphasize that substantial relations between age and measures of speed were found in each data set. Whether the lack of influence of health status on these relations was attributable to insensitive measurement or to a restricted range of variation, the fact remains that large age-related influences were observed in samples in which there was relatively little evidence of health-related influences. The apparent implication is that health status is not a major determinant of the age-related variation in at least certain measures of processing speed.

Relations between specific and nonspecific speed measures

Several types of evidence from the analyses reported above indicate that the specific and nonspecific speed measures in these data sets were not independent. For example, many of the correlations in column 4 of Table 4 were significantly different from zero, as were many of the coefficients for the path between the nonspecific and specific speed measures in the column labeled Path 3 in Table 5.

In addition to sharing total variance, the nonspecific and specific speed measures also shared moderate amounts of their age-related variance. This is evident in a contrast of the age-related variance in the specific speed measure before and after control of the nonspecific speed measure (i.e., columns 5 and 6 in Table 4). Although several of the entries in column 5 are quite small, which may be a consequence of low reliability of measures based on slopes or difference scores, many of the values in column 6 are substantially different from those in column 5. For example, the age-related variance in the slope measure in Data Set C was reduced from .115 to .046 (i.e., a reduction of 60%) after control of the 0-symbol nonspecific speed measure, and the age-related variance in the slope measure from Data Set E was reduced from .169 to .032 (i.e., 81%) after control of the 0-operation nonspecific speed measure.

Many of the estimates for the residual amount of age-related variance are significantly greater than zero, and thus it can be inferred that there are age-related influences on some of the specific measures that are independent of those on the non-specific measures. However, the important point from our perspective is that quantitative values of the estimates for the age-related effects are quite different if the specific and non-specific measures are assumed to be independent and considered in isolation.

Perhaps the most interesting patterns in Tables 4 and 5 are those in which the nonspecific and specific measures were negatively related to one another. For example, the 1-item and intercept nonspecific measures were negatively correlated with the slope specific measure in Data Sets D-1 and D-2, and the intercept and slope measures were negatively correlated with one another in Data Sets E, F-1, and F-2. Patterns of this type may be a consequence of an "undifferentiated set" in which, possibly because of low amounts of practice, subjects respond in nearly the same manner across all conditions within the task. That is, rather than altering response time according to the particular demands of each condition in the task, the subject may respond to all conditions in the same, relatively slow, fashion. Subjects who perform in this manner will tend to have high values of the nonspecific speed measures but low values of the specific speed measure. A negative correlation between the nonspecific and specific speed measures may therefore result if a substantial proportion of the participants in a sample exhibit this pattern of responding.

Because of the negative relations between the nonspecific and specific measures, the age relations on one of these measures may suppress the age relations on the other measure. That is, the actual magnitude of the age-specific relation may be an underestimate of the true relation because of the negative nonspecific-specific relation. In fact, when the age-related variation in the mean or the intercept was eliminated, the age-related variance in the slope measure increased in Data Sets D-1,D-2, E, F-1, and F-2.

Theoretical implications

Whenever two measures share a moderate amount of variance with one another, questions arise regarding the independence of the underlying theoretical constructs. That is, even though the measures might be postulated to represent distinct processes, the existence of a significant correlation indicates that the measures are not independent.

Similar reasoning applies to the age-related variance in the measures. In other words, if the measures share an appreciable amount of their age-related variance, then the age-related influences on the measures cannot be considered independent, or wholly distinct. Furthermore, to the extent that two measures share age-related variance, then they can be inferred to have age-related influences in common. The discovery that certain combinations of nonspecific and specific speed measures shared much of their age-related variance therefore suggests that at least part of the age-related effects on some specific speed measures may have been mediated through the age-related effects on the nonspecific speed measures. This finding is consistent with the proposal (e.g., Salthouse, 1992b) that an age-related reduction in a fairly general construct related to processing speed contributes to the age differences in many speeded, and also nonspeeded, cognitive tasks.

Methodological implications

The existence of significant relations between the nonspecific and specific speed measures, and the demonstration that the age relations on the latter are frequently dramatically altered by statistical control of the former, have important implications for how variables such as these should be analyzed. Regardless whether the variables are treated as mediators or as confounds of one another, relations between these variables and other variables such as age and possibly health status will be misleading if the variables are examined separately, or in isolation. It is therefore strongly recommended that multivariate analyses, such as hierarchical regression, path analysis, analysis of covariance, etc., be employed whenever attempting to investigate age-related influences on two or more measures of speed. As is evident in the contrasts between the entries in columns 5 and 6 in Table 4, inferences about the magnitude of age-related influences can be quite inaccurate when the assumption of variable independence is violated.

Although there is frequently considerable theoretical interest in the relations of age or other factors on measures hypothesized to reflect specific processes, we suggest that some modifications of the procedures used to analyze those relations are desirable. In particular, we propose that analyses of specific speed measures should routinely include nonspecific speed measures to allow accurate assessment of the independent and distinct influences of the variables of interest on the specific measures. Unless attempts are made to separate the shared and unique age-related variance in the specific speed measures, only equivocal interpretations of the presence or absence of age-related influences on those measures may be possible.

REFERENCES

Birren, J. E., Woods, A. M., & Williams, M. V. (1979). Speed of behavior as an indicator of age changes and the integrity of the nervous system. In F. Hoffmeister & C. Muller (Eds.), *Brain function and old age* (pp. 10-44). New York: Springer-Verlag.

Botwinick, J., West, R., & Storandt, M. (1978). Predicting death from behavioral test performance. *Journal of Gerontology, 33,* 755-762.

Brinley, J.F. (1965). Cognitive sets, speed and accuracy of performance in the elderly. In A. T. Welford & J. E. Birren (Eds.), *Behavior, aging, and the nervous system* (pp. 114-149). Springfield, IL: Charles C. Thomas.

Cerella, J. (1990). Aging and information processing rates in the elderly. In J. E. Birren & K. W. Schaie (Eds.), *Handbook of the psychology of aging* (3rd Ed., pp. 201-221). New York: Academic Press.

Cerella, J. (1994). Generalized slowing in Brinley plots. *Journal of Gerontology: Psychological Sciences, 49,* P65-P71.

Chapman, L. J., & Chapman, J. P. (1988). Artifactual and genuine relationships of lateral difference scores to overall accuracy in studies of laterality. *Psychological Bulletin, 108,* 127-136.

Earles, J. L., & Salthouse, T. A. (in press). Interrelations of age, health, and speed. *Journal of Gerontology: Psychological Sciences.*

Fillenbaum, G. G. (1979). Social context and self-assessment of health among the elderly. *Journal of Health and Social Behavior, 20,* 45-51.

Fisk, A. D., & Fisher, D. L. (1994). Brinley plots and theories of aging: The explicit, muddled, and implicit debates. *Journal of Gerontology: Psychological Sciences, 49,* P81-P89.

Heyman, D., & Jeffers, F. (1963). Effect of time lapse on consistency of self-health and medical evaluations of elderly persons. *Journal of Gerontology, 18,* 160-164.

Kaplan, G. A., & Camacho, T. (1983). Perceived health and mortality: A nine-year follow-up of the human population laboratory cohort. *American Journal of Epidemiology, 117,* 292-304.

LaRue, A., Bank, L., Jarvik, L., & Hetland, M. (1979). Health in old age: How do physicians' ratings and self-ratings compare? *Journal of Gerontology, 34,* 687-691.

Liang, J. (1986). Self-reported physical health among aged adults. *Journal of Gerontology, 41,* 248-260.

Linn, B. S., & Linn, M. W. (1980). Objective and self-assessed health in the old and very old. *Social Science and Medicine, 14A,* 311-315.

Maddox, G. L. (1962). Some correlates of differences in self- assessments of health status among the elderly. *Journal of Gerontology, 17,* 180-185.

Maddox, G.L. (1964). Self-assessment of health status. *Journal of Chronic Disease, 17,* 449-460.

Maddox, G. L., & Douglass, E. (1973). Self-assessment of health: A longitudinal study of elderly subjects. *Journal of Health and Social Behavior, 14,* 87-93.

Mossey, J. M., & Shapiro, E. (1982). Self-rated health: A predictor of mortality among elderly Americans. *American Journal of Public Health, 72,* 800-808.

Myerson, J., Wagstaff, D., & Hale, S. (1994). Brinley plots, explained variance, and the analysis of age differences in response latencies. *Journal of Gerontology: Psychological Sciences, 49,* P72-P80.

Perfect, T. J. (1994). What can Brinley plots tell us about cognitive aging? *Journal of Gerontology: Psychological Sciences, 49,* P60-P64.

Pfeiffer, E. (1970). Survival in old age: Physical, psychological, and social correlates of longevity. *Journal of the American Geriatric Society, 18,* 273-285.

Pilpel, D., Carmel, S., & Galinsky, D. (1988). Self-rated health among the elderly. *Comprehensive Gerontology: Sect. B, 2,* 110-116.

Ruger, H. A., & Stoessiger, B. (1927). Growth curves of certain characteristics in man. *Annals of Eugenics, 2,* 76-111.

Salthouse, T. A. (1985). Speed of behavior and its implications for cognition. In J. E. Birren & K. W. Schaie (Eds.), *Handbook of the psychology of aging* (2nd Ed., pp. 400-426). New York: Van Nostrand Reinhold.

Salthouse, T. A. (1992a). What do adult age differences in the Digit Symbol Substitution Test reflect? *Journal of Gerontology: Psychological Sciences, 47,* P121-P128.

Salthouse, T. A. (1992b). *Mechanisms of age-cognition relations in adulthood.* Hillsdale, NJ: Lawrence Erlbaum Associates.

Salthouse, T. A. (1994). The nature of the influence of speed on adult age differences in cognition. *Developmental Psychology, 30,* 240-259.

Salthouse, T. A. (in press). Aging associations: Influence of speed on adult age differences in associative learning. *Journal of Experimental Psychology: Learning, Memory and Cognition.*

Salthouse, T. A., & Coon, V. E. (1994). Interpretation of differential deficits: The case of aging and mental arithmetic. *Journal of Experimental Psychology: Learning, Memory, and Cognition, 20,* 1172-1182.

Salthouse, T. A., Kausler, D. H., & Saults, J. S. (1990). Age, self-assessed health status, and cognition. *Journal of Gerontology, 45,* P156-P160.

Singer, E., Garfinkel, R., Cohen, S. M., & Srole, L. (1976). Mortality and mental health: Evidence from the midtown Manhattan restudy. *Social Science and Medicine, 10,* 517- 525.

Suchman, E. A., Phillips, B. S., & Streib, G. F. (1958). An analysis of the validity of health questionnaries. *Social Forces, 36,* 223-232.

Tissue, T. (1972). Another look at self-rated health in the elderly. *Journal of Gerontology, 27,* 91-94.

Welford, A.T. (1977). Motor performance. In J.E. Birren & K.W. Schaie (Eds.), *Handbook of the psychology of aging* (pp. 450-496). New York: Van Nostrand Reinhold.

Welford, A. T., & Birren, J. E. (Eds.) (1965). *Behavior, aging, and the nervous system.* Springfield, IL: Charles C Thomas.

Changes in sensory motor behavior in aging
A.-M. Ferrandez and N. Teasdale (Editors)

BALANCE CONTROL IN OLDER ADULTS: TRAINING EFFECTS ON BALANCE CONTROL AND THE INTEGRATION OF BALANCE CONTROL INTO WALKING

Pei-Fang TANG and Marjorie H. WOOLLACOTT

University of Oregon

Abstract

In this chapter, we review changes in balance control abilities in older adults and the effects of various exercise training programs designed to improve these abilities. To optimize training effects, the systems approach emphasizes the identification of subsystems that contribute to the deterioration of balance control abilities, followed by the implementation of training programs targeted at the remediation of the deterioration of specific subsystems. Training studies using this approach have reported significant improvement in the balance control abilities of older adults. By targeting the body subsystems to be trained, this approach is cost-effective. Examples of sensory balance training during standing are reviewed. While sensory training during standing has shown effects on older adults' standing balance, it is not clear whether these effects transfer to other dynamic balance tasks such as those required during walking. Thus, the incorporation of dynamic balance tasks into a systems approach is suggested. The multifactorial approach to improve balance abilities of older adults often includes exercises targeted at several sub-

---

Authors' address: Department of Exercise and Movement Science, and Institute of Neuroscience, University of Oregon, Eugene, OR 97403-1240, U.S.A. (e-mail: pftang@oregon.uoregon.edu & mwool@oregon.uoregon.edu).

systems at the same time. Results from studies using this approach have not consistently revealed a positive training effect. A training protocol targeted at a population with similar types of balance impairments is suggested.

Key words: Balance, training, standing, walking, older adults.

INTRODUCTION

Balance and gait disorders are two of the primary risk factors contributing to repeated falls in older adults (Campbell, Borri, & Spears, 1989; Tinetti, Speechley, & Ginter, 1988). In a one-year prospective study on 336 community-dwelling older adults, Tinetti et al. (1988) reported that the number of balance and gait abnormalities that an older person had was significantly associated with the likelihood of his/her having multiple falls in the follow-up year. While only 16% of those who presented zero to two of these abnormalities fell in the follow-up year, the percentage increased to 48% for those who showed six or seven of these abnormalities. The balance abnormalities investigated consisted of unsteadiness when attempting to sit, to turn, or after being pushed at the sternum, and an inability to stand on one leg. The gait abnormalities included increased trunk sway and weaving during walking, and an inability to increase walking pace. These gait abnormalities may imply impaired balance control during walking.

Approximately one-third of community-dwelling older adults, aged 65 years and older, fall at least once a year (Campbell, Reinken, Allan, & Martinez, 1981; Prudham & Evans, 1981; Tinetti et al., 1988). A large number of the falls result in serious injuries and require hospital admission (Gryfe, Amies, & Ashley, 1977; Sattin et al., 1990). About 50% of these serious fall injuries eventually lead to institutionalization, indicating loss of self-care abilities (Sattin et al., 1990). The fall rate increases by two (Gryfe et al., 1977) to four times (Baker & Harvey, 1985) in the more frail older adults who live in supervised senior residences or nursing homes.

The high falling rate and the threatening consequences of falls among the elderly indicate an important need for the development of preventive or therapeutic intervention programs aimed at reducing falls in this population. Researchers have suggested that an efficient fall-prevention strategy would be to identify the risk factors that cause falls in the older adult, followed by implementation of specific training programs aimed at decreasing the risks (Hadley, Radebaugh, & Suzman, 1985; Tinetti et al., 1988). Since balance and balance-related gait disorders contribute significantly to fall incidence in the older adult, exercise training that focuses on improving balance and balance control during walking is expected to have effects on decreasing the susceptibility to falls.

Balance impairments in the older adult can be the result of accumulated aging processes and/or dysfunction of multiple body systems (Horak, Shupert, & Mirka, 1989; Woollacott & Shumway-Cook, 1990). According to a recently evolved model of motor control and rehabilitation, the systems model, three factors must be considered when determining the optimal solution to any motor task: the constraints of the individual, the requirements of the task and the characteristics of the environment in which the task will be performed (Figure 1, Shumway-Cook & Woollacott, 1995). The motor task is then accomplished by the interaction of the body subsystems of the individual organized around the functional goals of the task in a given environment (Horak, 1991;

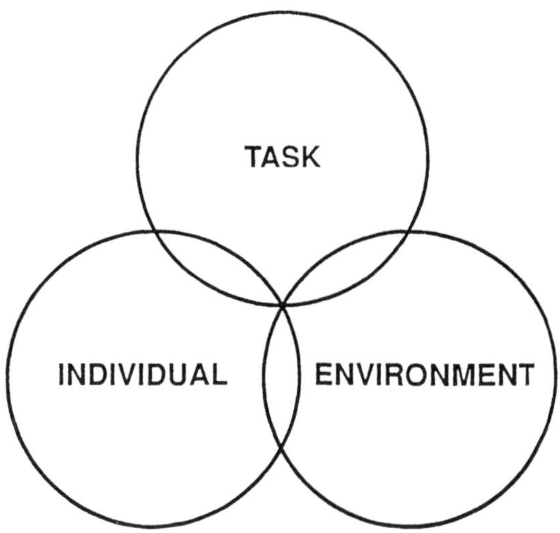

FIGURE 1. The systems model of motor control: three factors must be considered when determining the optimal solution to any motor task: the constraint of the individual, the requirement of the task and the characteristics of the environment in which the task will be performed (adapted from Shumway-Cook & Woollacott, 1995).

Woollacott & Shumway-Cook, 1990). The body subsystems include the nervous system (sensory, motor, and cognitive systems) and the musculoskeletal system.

Two important implications can be drawn from the systems model as it relates to balance training in the older adult. First, because normal movement relies on the complex interactions among multiple subsystems, the greater the number of subsystems that remain intact in an older adult, the better the balance system functions. Accordingly, it is suggested that training programs be targeted toward the subsystems which show age- or disease-related deterioration in function in order to give rise to a better training result. The second implication of this theory is that therapeutic intervention should be task-specific since the way the subsystems are organized is highly related to the functional goal of the motor task. For example, if the purpose of balance training is to evoke improvement in dynamic balance abilities during walking, then the training program may have to include walking activities rather than simple quiet standing.

In addition to the systems model approach, some researchers have used a multifactorial approach to improve balance control abilities in the older adult. The multifactorial approach involves training of multiple body subsystems at the same time, such as muscle strength, muscle endurance, and joint flexibility, despite the fact that some of these subsystems may not necessarily contribute to impaired balance in an older individual.

The purpose of this chapter is to review the literature on the effects of various training studies aimed at improving balance during standing and walking in the older adult. Because of the dearth of literature on effects of balance training on the integration of balance control during walking, the discussion will primarily consider balance training during standing. Further, comparisons of training effects will be made between those using the systems model versus those using a multifactorial approach. Before addressing these training-related issues, we will briefly review literature on balance control abilities in older adults and their relationship with the sensory and motor subsystems. Knowledge of these aspects of balance control difficulty in aging could provide useful guidelines for designing balance training programs. For a more thorough review on age- or disease-related changes in other subsystems of balance control, readers may refer to the work of Alexander (1994), Horak et al. (1989), Woollacott (1989), and Woollacott and Shumway-Cook (1990).

BALANCE CONTROL DURING STANDING AND WALKING IN THE OLDER ADULT: CONTRIBUTION OF SENSORI-MOTOR DEFICITS

In the review of studies of balance control in the older adult, one often encounters two unsolved issues: 1) researchers to date have not reached a consensus regarding the definition of "*normal aging*" in humans, and 2) there is no single measurement for balance control. With regard to the first issue, two different aging models of the human central nervous system have been proposed (Woollacott, 1989). The genetic model states that the function of the nervous system is programmed to gradually decline with advancing age. Once the decline reaches a certain level, functional impairment, such as postural instability, becomes evident. Using this aging model to study balance control in healthy older adults, older subjects with common subclinical degeneration of single or multiple body systems, such as arthritis, hearing loss, cataracts, or decreased memory, may be considered "*healthy*". Accordingly, the selection criteria of "*healthy*" older adults may vary among different research studies.

On the other hand, the catastrophe aging model hypothesizes that the nervous system continues to function at a high level as one ages, unless a person encounters environmental catastrophe(s) or disease(s) which then results in a rapid decline in behavioral function. Subscribing to this model in the study of balance control in aging will lead a researcher to use very stringent criteria to recruit healthy older subjects. In this case, the criterion of being a "*healthy*" older adult is likely to be the same as that for a healthy young adult. Therefore, older adults who show minor degeneration in body systems, such as mild arthritis, may be excluded even though they in fact represent the majority of the aging population (Woollacott, 1989).

To date, there is still a controversy concerning which model best explains human aging. This issue is complicated by many other factors such as diseases that are prevalent in older age, gender, and life style (Wolfson et al., 1992). Apparently, research employing different aging models will lead to very different results. For instance, by incorporating careful clinical examinations of subjects enrolled in a study, Manchester, Woollacott, Zederbauer-Hylton, and Marin (1989) showed that older adults with subclinical neurological signs were more likely to have balance impairments than those who did not.

Since the majority of aging research on balance control employs the genetic aging model and subjects with subclinical signs are often not excluded, our discussions about balance control abilities in older adults are based on studies using older adults with and without subclinical signs.

As to the balance measurement issue, researchers (Patla, Frank, & Winter, 1990) have pinpointed that there is no one single balance measure that represents one's balance abilities in daily life. It has been suggested that balance measurements be done in both static and dynamic contexts (Winter, Patla, & Frank, 1990) in order to obtain a full understanding of one's balance abilities. Due to the limitation of space here, we will only review balance control during standing and walking in older adults in this chapter.

Balance control during standing

Declines in sensory organization abilities

Several previous studies have shown that older adults are highly susceptible to instability during standing when the sensory information related to balance control is reduced or distorted (Woollacott, Shumway-Cook, & Nashner, 1986; Manchester et al., 1989). These findings suggest that deterioration in the visual, vestibular, and somatosensory subsystems may have a great impact on balance control abilities in the older adult.

Using a movable platform paradigm, Woollacott et al. (1986) systematically evaluated the relative contribution of the visual, somatosensory, and vestibular inputs to standing balance control in older and young adults. Subjects were standing on a movable plate surrounded by a visual enclosure. Postural sway, as measured from the hips relative to the ankle joints, was tested in six sensory conditions: eyes-open/ support surface normal (VnSn), eyes-closed/support surface normal (VcSn), visual enclosure swaying with the subject/support surface normal (VsSn), eyes-open/support surface swaying with the subject (VnSs), eyes-closed/ support surface swaying with the subject (VcSs), and visual enclosure and support surface both swaying with the subject (VsSs). In this sensory organization test, when only visual input or somatosensation was reduced or distorted (VcSn, VsSn, VnSs), older adults showed a slightly greater amount of sway than the young adult, but the difference between the young and older adults was nonsignificant. Older adults, however, swayed significantly more than young adults or lost their balance, when both visual inputs and somatosensory inputs from the ankle were reduced and/or unreliable as in the VcSs and VsSs conditions (Figure 2).

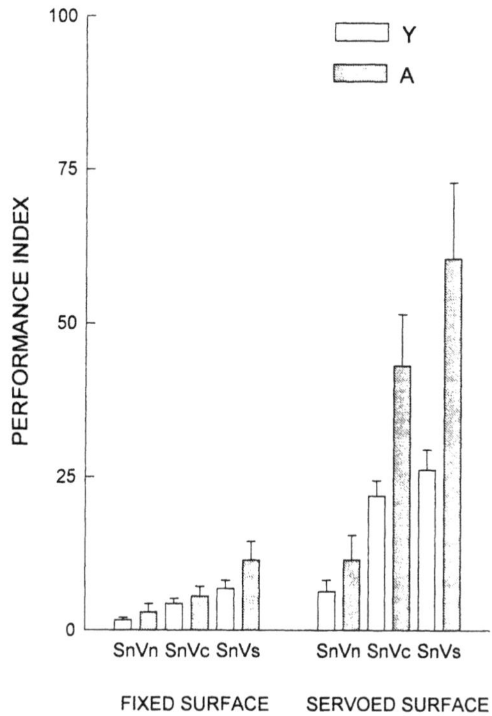

FIGURE 2. Standing stability of young (Y) and older adults (A) in six sensory conditions. The first three conditions were performed while subjects were standing on a firm surface, and thus the somatosensation from the ankle was normal (Sn). In the next three conditions, the support surface was rotated following body sway to maintain the ankle joint at 90 degrees. Thus, somatosensation from the ankle in the next three conditions was inaccurate (Ss). In each set of somatosensory conditions, vision was manipulated by having the subjects open (Vn) or close (Vc) the eyes, or by rotating the visual surround to follow body sway (Vs) (from Woollacott, Shumway-Cook, & Nashner, 1986).

The dramatic declines in postural stability in older adults under conditions in which information from two senses is reduced and/or unreliable (VcSs and VsSs) may be due to several reasons. First, older adults may depend more on a redundancy of sensory information for balance than young adults. If two of the three major balance senses do not provide accurate information, their balance function may thus decline. Second, older adults may rely predominantly upon visual and somatosensory information to maintain balance. Thus, when these two sensory inputs are both reduced, accompanied by the inability to re-weight the input from the vestibular system, older adults are more likely to suffer from postural instability. Whether the inability to re-weight sensory input is due to peripheral vestibular deficits or deficits in higher level sensory organization processes, however, remains controversial.

Woollacott et al. (1986) observed that although most of the older adults investigated lost their balance when they first encountered sensory conflict conditions, they adapted to the condition and could prevent further falls in the same condition after repeated trials. This may suggest that the sensory integration abilities of these older adults remained intact, though adaptation required more trials than the young.

Using a different paradigm, Stelmach, Teasdale, DiFabio, and Phillips (1989) investigated the difference in automatic postural response characteristics when young versus healthy older adults were asked to balance during large and fast versus small and slow platform rotational perturbations. The large platform rotation was expected to evoke reflexive postural responses, initiated primarily by somatosensory inputs, while the small and slow platform rotations were expected to require higher level sensory integrative mechanisms to reorganize the available sensory information. Stelmach and colleagues (1989) found that both the young and older adults adapted to the repeated large and fast platform rotations by decreasing the amount of sway. Yet, they showed different types of changes in sway during the trials using small and slow platform rotations. While the young adults were able to attenuate the sway in response to the slow and small platform rotations after repeated exposures, the older adults were unable to show such a decrease. The results suggest that there may be a problem with sensory integration in these older adults.

Additional evidence for age-related changes in sensory integration abilities is reflected in a study by Teasdale, Stelmach, Breunig, and Meeuwsen (1991). They investigated the way in which older adults adapted to a transitional situation during which visual information suddenly became available or unavailable. It was found that when visual information suddenly became available after conditions in which vision was absent, young adults were able to reorganize the sensory information and reduced sway, compared to the no vision condition, whereas the older adults presented an increased amount of sway.

Decline in motor organization abilities

The temporal and spatial patterns of muscle responses in a balance task are often used to indicate the motor organization abilities of balance control. Using a movable platform paradigm, Nashner (1977) observed that young adults presented relatively fixed muscle activation patterns in response to the instantaneous platform movements.

While encountering forward platform translations which induced backward body sway, young subjects first activated the ankle dorsiflexor, tibialis anterior (TA) muscle, at a latency of 90 to 120 ms (Nashner, 1977). This distal leg muscle activity was followed by the activity of the quadriceps (Q) with a time lag of about 20 ms. Young adults also showed a similar distal-to-proximal muscle activation pattern (gastrocnemius (G) – hamstring (H)) while responding to backward platform translations.

Using a similar experimental paradigm, Woollacott et al. (1986) found that healthy older adults showed three significant differences in motor organization of balance responses to platform perturbations, as compared to the young adults. First, healthy older adults showed a small but significant increase in the onset latency (109 ms) of the distal leg muscle (TA) compared to young adults (102 ms), while responding to forward platform translations. Second, some older adults demonstrated a proximal to distal sequence of muscle activation. Third, whereas the young adults showed a more consistent ratio (0.82 – 1) of electromyogram (EMG) magnitudes between the distal and proximal postural muscle activity, older adults revealed a greater variability of this ratio (0.12 – 0.86). This EMG magnitude was calculated by integrating the muscle activity over a 75 msec interval following the onset of muscle activation.

Altered motor organization in older adults involves not only the temporal but also the spatial organization of balance control. By the term spatial organization, we refer to the organization of the muscles that are activated together in a task. More recently, Manchester and colleagues (1989) found that older adults more often showed co-activation of the agonist and antagonist muscles (for example, increased TA activity, along with an increase in G activity, in response to a backward platform translation) when compared to young adults in compensating for threats of balance. Older adults appeared to use co-activation in an attempt to increase joint stiffness to overcome the postural perturbations.

Balance control during walking

Balance control abilities are the prerequisites for successful locomotion. Postural adjustments during walking serve the following two functions: 1) to maintain equilibrium between the moving body and the base of support; and 2) to quickly adapt ongoing locomotor patterns to environmental changes by using anticipatory strategies when prior information about the changes is available or by using reactive responses when environmental perturbations are unpredictable (Patla, 1991; Shik & Orlovsky, 1976; Winter, 1989; Winter, McFadyen, & Dickey, 1991).

Accordingly, there are two ways to assess balance control abilities during walking. The first one is to evaluate walking performance when the environment is not changing. For example, one could examine certain gait parameters that are highly affected by balance control mechanisms and thus reflect equilibrium control during normal walking.

Using interrupted-light photography, Murray, Kory, and Clarkson (1969) evaluated walking patterns in 62 active and healthy men between the age of 20 and 87. Even though all of these older adults were in good health, they showed characteristic gait changes compared to the young adults. The changes included a slower walking speed, longer stance time, less vertical excursion of the head, greater stride width, decreased elbow flexion, and an extended shoulder posture. These changes, however, did not resemble the pathological gait observed in patients with neurological diseases. Instead, these walking patterns appeared to be caused by an effort to increase stability. However, whether all of these changes in gait patterns of healthy older men were related to gait control itself or to balance control remained unclear.

In another gait study, Winter, Patla, Frank, and Walt (1990) examined the kinematic and kinetic parameters of gait for 15 active and healthy older adults aged from 62 to 78 years. Compared to their previous data on young adults, the results revealed that these older adults had a similar cadence to that of the young adults, but these older adults had a shorter step length, increased double-support time, decreased push-off power, and more flat-footed landing at heel strike. Furthermore, Winter, Ruder, and MacKinnon (1990) recognized that in healthy young adults, the hip moment was highly varied during the stance phase of walking, indicating a compensatory balance mechanism to control the upper body balance during walking. The changes in the hip moment are often compensated by changes in the knee joint moment to maintain a relatively constant support moment during the stance phase. Winter, Patla, Frank, and Walt (1990) thus used the covariance between the hip and knee moments as an index of dynamic stability during walking. The older adults in the study showed decreased covariance between the hip and knee moments, suggesting a decline in balance control abilities.

The second way to assess balance abilities during walking is to examine postural adaptation in response to various environmental changes, which may or may not be predictable. Readers may refer to Patla's chapter in this book for discussion of this issue.

Whereas there is some literature concerning the impact of deterioration in sensori-motor systems on gait patterns of older adults (Barron, 1967; Critchley, 1948; Koller, Glatt, & Fox, 1985), it is difficult to precisely locate the neural areas in which deterioration could cause impaired balance control abilities during walking. In a thorough review, Nutt, Marsden, and Thompson (1993) suggested that within the peripheral nervous system, the vestibular, visual, and somatosensory systems are critical for balance control during walking. Within the central nervous system, the brainstem, basal ganglia, cerebellum, and cortex are primarily responsible for integrating postural responses into ongoing walking activities.

TRAINING EFFECTS USING THE SYSTEMS APPROACH

Several training studies have been performed in the past few years using the systems approach. In particular, those that focus on sensory training for balance will be discussed below.

Sensory training effects on standing and walking balance

Hu and Woollacott (1994a) recently investigated the effects of a sensory balance training program on balance function in the older adult. The training program involved manipulation of sensory inputs from the vestibular, somatosensory, and visual systems. Twenty-four older subjects (65 – 90 years old) were randomly assigned to a training or control group. The training group received a total of 10 hours of training, carried out over two weeks. Each training session lasted about an hour. During the training, the availability of visual inputs was manipulated by having the subjects open or close their eyes; somatosensation from the ankle was manipulated by asking the subjects to stand on a foam cushion versus a firm surface; and manipulation of the vestibular system was performed by having the subjects either extend the head backward or keep the head in the neutral position. These conditions were combined to create eight different sensory conditions of progressive difficulty. In the first four conditions, subjects were asked to stand on a firm surface as quietly as possible with the following sensory manipulation of each condition: 1) eyes open, head neutral; 2) eyes closed, head neutral; 3) eyes open, head extended back; and 4) eyes closed, head extended back. For condition 5 through 8, subjects repeated the same four conditions, except that they were asked to stand on a foam cushion. Thus, the first condition was the easiest one, and condition 8 was the most difficult one.

After the completion of the training, the trained subjects showed significantly improved stability in 5 of the 8 training conditions. Given the high inter-subject variability, postural stability was measured by the changes (in percentage) in the root-mean-square (RMS) values of sway torque in the anteroposterior direction exerted by the subject on the platform. The RMS value in condition 1 during the first training session was considered the baseline and the RMS values of conditions 2 through 8 on the first and last training sessions were compared with this baseline value. The results showed improvement in all of the conditions involving a foam surface (conditions 5-8), and the eyes closed, head extended condition on a firm surface (condition 4). In addition, the training group had less falls and decreased amount of sway in a platform sensory organization test in which somatosensory inputs were reduced by rotating the support surface in direct proportion to body sway, with the eyes either open or closed.

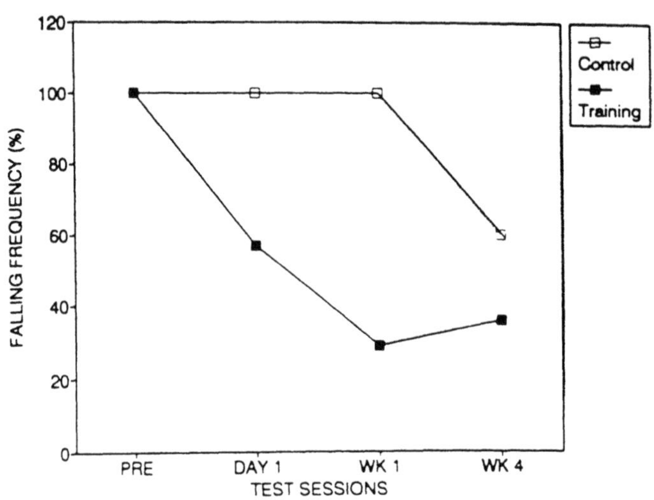

FIGURE 3. Long-term training effect of a multisensory training program. The training group showed a significant decrease in the number of falls in the platform sensory organization test, on the first day (DAY1), one week (WK1), and four weeks (WK4) after the completion of training. The number of falls prior to training (PRE) was set at 100% (from Hu & Woollacott, 1994a).

The training was also found to have a long-term effect. As displayed in Figure 3, the training group showed a significant decrease in the number of falls in the platform sensory organization test, on the first day, one week, and four weeks after the completion of training. In addition, the positive training effects were transferrable to one-legged stance. While the one-legged stance duration in the eyes-open and - closed conditions increased in the training group, this duration in the control group decreased during all post-training tests (Figure 4). However, the average increase in the one-legged stance duration for the training group was only 2 to 3 seconds. Whether this transfer effect has significance for balance control in subjects' daily life or whether it can be generalized to other functional tasks that require balance control (such as walking and bending over) remains unanswered.

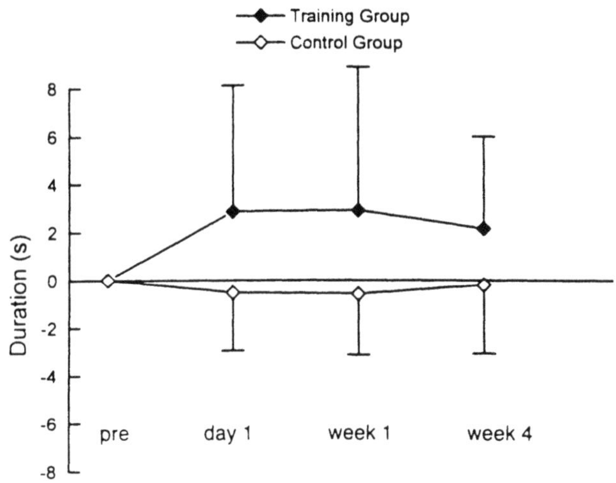

FIGURE 4. Transfer of multisensory training effects on one-legged stance performance. The pre-training score (PRE) was set at zero. The training group showed increased balance time in this test, one day (day 1), one week (week 1), and four weeks (week 4) after the completion of training; while the control group did not reveal differences before and after training (from Hu & Woollacott, 1994a).

Interestingly, by applying a multisensory training protocol to healthy older adults, Hu and Woollacott (1994b) showed that the training not only improved sensory organization, but also sensorimotor integration.

The same groups of subjects mentioned above were also tested for their motor organization of balance control on a movable platform when a sudden platform translation was used to evoke automatic postural responses. After the training period, the training group demonstrated the following changes when compared to the control group: 1) significantly shortened muscle onset latencies of the neck flexors; 2) a decrease in the probability of seeing the antagonist muscle responses (the gastrocnemius, hamstring, and quadriceps when they served as the antagonists); 3) an increase in the probability of response in agonist trunk flexor muscles; and 4) a decrease of the maximal excursion of the ankle joint during the first platform translation.

Given the significant sensory training effects on the balance control of healthy older adults, the next logical question would be whether older people with balance problems can benefit from similar training programs. To answer this question, another balance training study was carried out by Moore and Woollacott (submitted). Thirty-two community-dwelling older adults (65-92 years old) with balance deficits related to impaired sensory processing ability were recruited for this study. However, older adults with diagnosed neurological or musculoskeletal diseases were excluded except for five of the 32 subjects who revealed mild abnormalities in the central or peripheral nervous systems. The sensory processing impairments were confirmed by a balance survey questionnaire and a sensory organization test of balance function (Nashner, 1982). That is, each subject reported insecurity of balance while walking on uneven surfaces, in a crowded environment, or in dimly lit areas, and experienced a minimum of 2 losses of balance on the sensory organization test. The six sensory conditions of the sensory organization test were: 1) normal vision, normal support surface; 2) eyes closed, normal support surface; 3) visual enclosure servoed (rotating to follow the center of gravity sway), normal support surface; 4) normal vision, support surface servoed (rotating to follow the center of gravity sway); 5) eyes closed, support surface servoed; and 6) visual enclosure and support surface both servoed (Figure 5). To minimize pre-training differences between the training and control groups, subjects in the two groups were paired, primarily by matching their age, gender, and number of loss of balance in the sensory organization test during screening.

The focus of Moore and Woollacott's (submitted) balance training was to improve the use of sensory information processing of these older adults, which in turn would improve the speed of adaptation to environmental changes. Although the training protocol was standardized, it progressed in level of difficulty according to the subject's abilities and the rate of improvement during the training period. The progressive protocol assured continuous provision of challenges to the balance systems as the subjects improved throughout the entire training course. Each trained subject attended individual training sessions 3 times a week for 8 weeks, and each session lasted about one hour. During the training, the visual information was manipulated by asking the subject to stand with eyes open, eyes closed, or while wearing goggles which eliminated peripheral vision; vestibular input was controlled by asking the subject to nod or turn the head while maintaining a fix visual target; and the somatosensation at the ankles was altered by using a firm versus foam support surface.

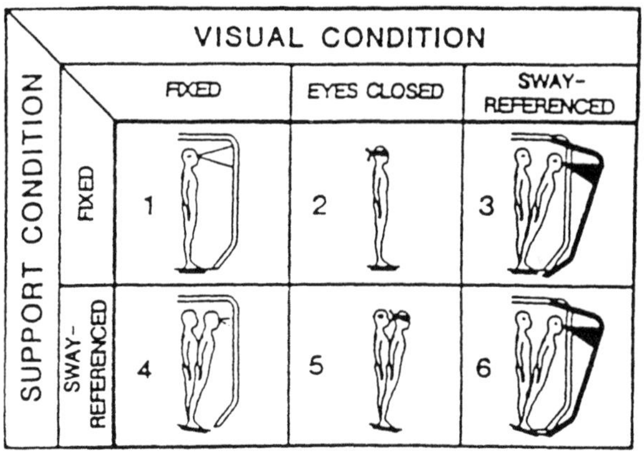

FIGURE 5. Six conditions of the sensory organization test: 1) normal vision, normal support surface; 2) eyes closed, normal support surface; 3) visual enclosure servoed (rotating to follow the center of gravity sway), normal support surface; 4) normal vision, support surface servoed (rotating to follow the center of gravity sway); 5) eyes closed, support surface servoed; and 6) visual enclosure and support surface both servoed (Figure courtesy of Neurocom, International, Inc. Clackamas, Oregon).

Unlike Hu and Woollacott's (1994a & b) training program in which subjects were asked to stand quietly during training, Moore and Woollacott (submitted) incorporated dynamic movement of the whole body into the training program. Thus, while the sensory information was changing during training, the subjects would perform standing quietly, stepping in place, side stepping, reaching up and bending down, turning the body to point to the right or left side, standing with one leg, tandem standing, or stepping forward, backward, or sideways across a foam cushion. These motor tasks, combined with the sensory conditions, formed a standardized training protocol with trials ordered according to increasing difficulty. For example, stepping in place on a firm surface with the eyes open would be one of the easiest training conditions, and stepping sideways with the eyes closed while nodding the head would be one of the most difficult conditions.

Results of training were evaluated at 1, 3, and 5 weeks after training. Several balance tests were conducted to evaluate the pre-training balance abilities and to test the training effects. These included the sensory organization test, functional mobility performance in an obstacle course, and clinical tests. The functional mobility test, modified from Tinetti (1986) and Wolfson, Whipple, Amerman, and Tobin (1990), included daily maneuvers such as rising from a chair, walking on normal and altered surfaces, reaching up, bending down, turning, and sitting down, all performed under different visual conditions. Thus, this test provided insight into how the subjects controlled their balance during voluntary activities, especially during walking. Each maneuver was rated based on a 4-point scale and the intra-rater agreement was 89%. The clinical tests included the range of motion of the ankles and the neck, strength of the ankle dorsiflexors and plantarflexors, and the Romberg and one-legged stance tests in the eyes open and eyes closed conditions.

Among all of these tests, the effectiveness of the training was best reflected in the significant reduction of the number of falls in the sensory organization test at 1, 3, and 5 weeks after training for the training group (Figure 6). The amount of sway in condition 4 (eyes open, surface servoed) of the sensory organization test was significantly reduced (60%) in the training group one week after training, as compared to the control group. There were no significant changes in the amount of sway in sensory conditions 1 through 3 of the sensory organization test for the training group. Both the training and control groups demonstrated a decreased amount of sway in sensory conditions 5 and 6. Clinical musculoskeletal examinations revealed minimal changes in the flexibility of the ankle and neck and in the ankle muscle strength. The improvement in balance control in the sensory conflict condition (condition 4) thus cannot be accounted for by changes in the musculoskeletal subsystem. The finding that the training group did not improve sway more than the control group in conditions 5 and 6 of the sensory organization test may be due to the fact that subjects in the control group had become familiar with the sensory organization test.

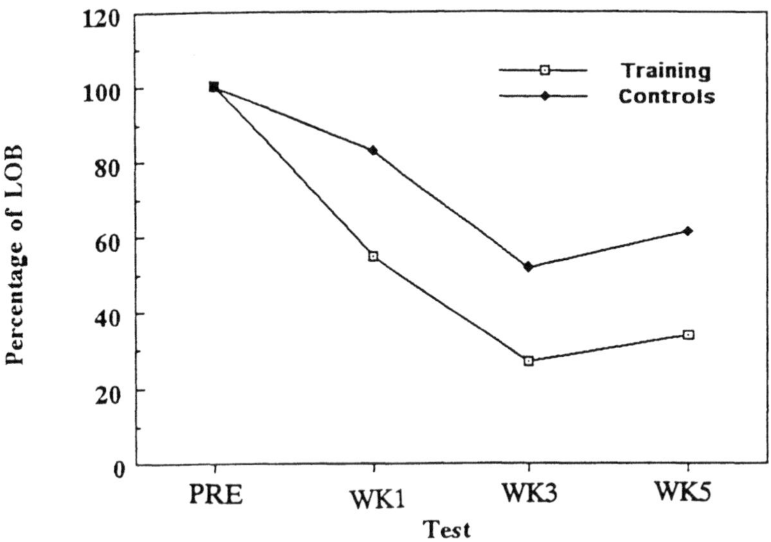

FIGURE 6. Percentage of losses of balance in the training and control groups on the sensory organization test, before (PRE) and after (WK1 = 1 week post-training, WK3 = 3 weeks post-training, WK5 = 5 weeks post-training). The total number of falls in the pre-training test was set at 100%. The training group showed a significant decrease in the number of falls in the sensory organization test, as compared to the control group (adapted from Moore & Woollacott, submitted).

In this study, there were no training effects on clinical balance tests such as the Romberg and one-legged stance. Since almost all of the subjects presented a full score in the Romberg test, it is not surprising that this test was not sensitive to the training effect. The fact that there was no improvement in the one-legged stance could be that ankle muscle strength and ankle flexibility are other important factors for maintaining one-legged stance. These two musculoskeletal factors, however, did not improve in the training group.

Moore and Woollacott (submitted) also did not find training effects on the functional mobility performance, which was particularly designed to test balance control abilities during walking and voluntary movements. They proposed that during training, the subjects were trained on activities that highly challenged their stability limits and thus were forced to present their best balance performance, whereas some of the functional mobility investigated may not be challenging enough for these older subjects to change their strategies for daily activities. Therefore, their best balance performance was not reflected in this test. Another possibility for the lack of transfer of training effects to functional tasks could be due to the fact that most of the balance exercises were practiced while the subjects were standing, whereas the balance performance of the functional mobility test was evaluated while they were walking.

Ledin et al. (1991) in Sweden also conducted a balance training study. Twenty-nine healthy older subjects, aged 70 through 75 years, participated. Fifteen of them were selected for the training group and 14 of them were assigned to the control group by matching their age, sex, smoking history, alcohol and physical activity habits. The training program included various dynamic movements and part of them were aimed at improving the function of the vestibular system. The exercises were: jogging and jumping, walking forward, backward, sideways, on the toes and heels, walking with sudden turns, rising from sitting, standing on one leg with eyes open and closed, visual fixation during neck motion, throwing and catching balls, and bouncing and jumping on a trampoline. The training subjects performed these exercises twice a week for nine weeks. Each exercise session lasted about an hour.

The results showed that after training, the training group improved their balance time in the one-legged stance test with the eyes closed, or while shaking the head, but the control group did not. The training group also significantly reduced sway in condition 4 of the sensory organization test compared to the control group. However, although Ledin et al. (1991) claimed that part of the exercises followed a vestibular rehabilitation program, the training group did not show a reduced amount of sway in conditions 5 and 6 of the sensory organization test. This result was consistent with that of Moore and Woollacott (submitted). The fact that visual and somatosensory information was not reduced or altered during the training and thus subjects were not forced to depend more on the vestibular system, may account for an insufficient challenge or training of the vestibular system. The need for a more challenging training program may be particular true here since the older subjects recruited were relatively healthy. Older adults with severe vertigo were excluded. Ledin et al. (1991) also reported that after training, the training group, but not the control group, could walk 15 meters back and forth (including a 180 degree turn) faster in a 2×15 meter walkway, indicating a better balance control during walking. The fact that Ledin et al. (1991) included walking exercises in the training program may have contributed to the improved integration between balance and locomotion systems.

In sum, sensory balance training programs utilizing the systems approach have demonstrated that balance abilities in the older adult are trainable. In particular, sensory information processing or multisensory integration related to balance control seems to improve by implementing training programs that facilitate the use of balance sensory information. The training effect is most evident in balance tests that are of moderate difficulty (such as condition 4 of the sensory organization test). Balance tests which are too simple (conditions 1 through 3 of the sensory organization test) are often insensitive to the training effects. Further, there is no evidence yet that training effects from standing balance exercises can be carried over to other functional tasks such as walking or can result in a significant reduction of falls in daily life. Apparently, to improve balance control abilities during walking, walking exercises in combination with various balance tasks, may need to be incorporated into the training program. Finally, most of the sensory training programs have only measured postural stability in the anteroposterior direction. Maki, Holliday, and Topper (1994) found that the spontaneous sway in the lateral direction best differentiated older fallers from non-fallers and was also the best single predictor of future risks of falling. Thus, it is suggested that future sensory training studies include measurement of lateral stability in older adults.

Motor training effects on standing and walking balance

As mentioned previously, older adults reveal motor organization problems when they respond to a sudden support surface movement (Woollacott et al., 1986). Although Hu and Woollacott (1994b) reported some positive effects on motor organization after sensory training, no study has yet been done to systematically investigate whether balance exercises that focus on the motor system can improve the temporal and spatial organization of postural muscle activation in older adults (Alexander, 1994).

Fiatarone and colleagues (1990) employed an eight-week high-intensity muscle strengthening exercise program in frail older adults. The results not only revealed a high gain in muscle strength, but also an increase in the speed of tandem gait. Two subjects were able to quit using canes to assist in walking. It appears that muscle strengthening exercise may have some positive impact on balance control in dynamic tasks. More systematic studies, however, are needed.

TRAINING EFFECTS USING THE MULTIFACTORIAL APPROACH

Noticing that multiple subsystems contribute to balance control, some researchers have used a multifactorial approach and incorporated multi-purpose exercise programs in their training program. This type of intervention often not only aims at improving balance abilities but also at reducing the risk factors of falls in the older adults. Tinetti, Williams, and Mayewski (1986) reported that falls in older adults are associated with the number of fall-related risk factors an older person has. This finding implies that the incidence of falls may be reduced by modifying or decreasing the number of risk factors.

Tinetti et al. (1994) conducted a large-scale fall-prevention intervention to test this possibility. Three hundred and one men and women (> 70 years old) living in the community were studied. All of them had at least one of the following fall-related risk factors: postural hypotension, use of sedatives, use of more than 4 prescribed medications, weakness or limited range of motion in the arms and legs, balance impairment, gait impairment, and transfer (to and from bed and chair, or to the bathtub or toilet) difficulty.

The subjects were divided into two groups: intervention (N = 153) and control (N = 148) groups. Both groups had a similar mean number of targeted risk factors. The subjects in the intervention group received standardized intervention protocols targeted at the identified risk factors, whereas those in the control group only received home visits by social-work students. Depending upon the risk factors that the subject had, intervention for the training group included behavioral modification, education, modification of medications, transfer-skill training, strengthening, and balance and gait training. Exercises involving physical training were graded at progressive levels of difficulty, based on the individual subject's improvement. The intervention phase lasted 3 months, during which home visits by a physical therapist and/or a nurse practitioner were provided. In the maintenance phase (from the end of the intervention period to the end of 6 months after enrollment), the researchers kept in touch with the subjects once a month. Dependent measures were the number of falls in the following year, time when the first fall occurred after the intervention, medical care needed after a fall, change in mobility, and change in self-efficacy.

The results showed that the length of time to the first fall in the follow-up year was longer in the intervention group than in the control group (Tinetti et al., 1994). A smaller number of subjects (35%) fell in the intervention group, compared to the control group (47%). Self-efficacy improved significantly in the intervention group compared to the control group. Overall, the total number of risks in the intervention group decreased by 1.1, as compared with a decrease of 0.6 in the control group. Up to about five months after training, a significantly smaller percentage of the intervention group continued to show impaired balance (21%) and gait (45%) as compared to the control group (46% showed impaired balance and 62% showed impaired gait). This study demonstrated that training intervention targeted toward the population that has similar fall-related risk factors has significant training effects on both the reduction of falls and the risk factors per se.

Also using a multifactorial approach, Johansson and Jarnlo (1991) designed dynamic exercises to improve balance abilities of older females, 70 years old, living in the community. The training program consisted of: 1) walking in different directions at various speeds, with a variety of combinations of arms, neck, and trunk movements; 2) dancing steps; 3) weight transfer; and 4) mobility and strengthening exercise for the trunk and lower extremity. The training group received a 10-hour training over 5 weeks (twice a week). Static and dynamic balance tests were used to test the training effects, including one-legged stance with the eyes open and closed, with neck rotation, walking along a beam, walking in a figure-eight, and walking as fast as possible. The results revealed that the training group significantly improved their performance in 6 out of the 9 balance tests, except for the one-legged stance with the eyes closed and the walking along a beam tests. No improvement was observed for the control group.

Similarly, Brown and Holloszy (1991) studied the training effects of a low intensity exercise program applied to healthy sedentary older men and women, aged 60 to 71 years, on the improvement in strength, muscular endurance, range of motion, standing balance, and gait. Sixty-two men and women (30 women, 32 men) were admitted to the training program, while 13 subjects served as the control group. The control and training groups were matched by age, body weight, and body height, but not by their medical or physiological conditions. All subjects were free from orthopedic problems, were non-smokers, and were in good health condition. Measurements of interest included four main categories: strength, and range of motion in the lower and upper extremities and the trunk, muscular endurance, one-legged standing abilities with the eyes open and closed, and gait parameters.

The exercises were multi-purpose and individualized, aimed at improving the strength of all major muscle groups, muscular endurance, range of motion, and balance. The exercises were conducted daily for an hour, five times a week for 3 months. The exercises designed for balance were one-legged stance, walking on heels and toes, and extending the opposite arm and leg while on all fours. The results showed that the score for one-legged stance during eyes open was improved in the older women, but not older men. No change in gait parameters was found. There was improvement in strength, but not in muscle endurance.

Judge, Lindsky, Underwood, and Winsemius (1990) randomly divided 21 older subjects, 62 to 75 years old, into two training groups: a combined training group and a flexibility training group. Subjects in the combined training group received a 6-month exercise program, including resistance, flexibility, balance and walking exercises, while those in the flexibility group only practiced flexibility and balance exercises during the same training period. The results showed that the combined group significantly improve stability of one-legged standing, but the flexibility group did not. A test of stability during two-legged stance, however, did not reveal significant differences between the two exercise groups.

Other global approaches, using nonstrenuous exercise, have also failed to report significant balance training effects in community-dwelling and institutionalized older adults (Basset, McClamrock, & Schmelzer 1982; Crilly, Willems, Trenholm, Hayes, & Delaquerrière-Richardson, 1989; Lichtenstein, Shields, Shiavi, & Burger, 1989). While the training programs may last for 9 to 16 weeks, the exercises that utilized simple total body movement and relaxation and did not challenge the balance ability of older adults often failed to elicit significant training effects.

In sum, the advantage of using the multifactorial approach is that training protocols often include a wide range of balance tasks from static standing to dynamic motor tasks. Further, exercise programs often consist of not only standing or walking tasks, but also strengthening, endurance, and mobility training. The effects of multi-purpose exercise training programs on balance control abilities of older adults, however, are inconsistent. Several possibilities may account for the inconsistent findings. First, the critical component of an exercise form which may be most effective to improve an individual's balance is not identified or targeted. Second, the total duration that a trainee could spend on the balance tasks might be actually reduced because of the practice on other exercises. Future multifactorial approaches may need to place different weighting on different aspects of exercise according to the trainee's needs in an effort to improve the cost-effectiveness of training.

OTHER CONSIDERATIONS RELATED TO TRAINING

According to the systems model, training programs should not only target impaired subsystems, but also be customized to individual needs. Buchner, Cress, Wagner, and de Lateur (1992) introduced the concept of "*a targeted population*" for exercise intervention in the elderly. They recommended that criteria be set for subject enrollment in a particular exercise program, related to improvement within specific targeted systems. The notion is that if the exercise program is targeted toward specific physiological body systems, then those who have impairments in these systems would more likely benefit from the exercise than others. For example, to implement strengthening exercise in older adults in an effort to improve their balance control abilities, the exercise coordinator or researcher should first choose subjects whose balance impairment is mainly due to muscle weakness. Persons with other confounding physiological deficits, such as overweight, alcohol abuse, chronic neurological or muscular diseases, would be excluded from the exercise program.

Along the same line, if group therapy is given to a number of older persons at the same time, as is often seen in many community-based senior exercise programs, it is suggested that the participants first be screened and that the individual subsystems in which they show decreased levels of function be noted. The screening test, for instance, may consist of examinations of all of the subsystems involved in balance control. Participants who present balance impairment primarily due to a deficit in the same subsystem are then grouped together and receive standardized training protocols targeted at the improvement of function in that particular subsystem.

The combination of targeting the population and targeting the subsystems to be trained has several advantages. First, by eliminating the number of unnecessary training programs, it is cost-effective. Second, it prevents overloading or injuring the older adults participating in the exercise. Third, it is likely that the training effect can be optimized. Several previous studies have shown that after balance training, people who have balance impairments are more likely to improve their balance control abilities than those with good balance abilities (Brandt, Krafczyk, & Malsbenden, 1986; Brown & Holloszy, 1991; Johansson & Jarnlo, 1991).

Other issues related to the principles of specificity in training are task- and test-specific issues. To best improve dynamic balance abilities, exercise programs that challenge one's balance control during voluntary movements should be included. Hu and Woollacott (1994a) reported that multisensory training was effective in improving not only balance stability under various sensory conditions, but also one-legged stance abilities that were not included in the training protocol. However, one-legged and two-legged stance can be categorized as static standing balance control, and their balance control mechanisms may be similar. It remains unclear whether the effects of multisensory training can be transferred to other more dynamic tasks.

Finally, a progressive training protocol is preferred over a non-progressive one. Balance control normally operates at a subconscious and habitual level. The potential of balance abilities is thus difficult to reveal unless one is in a challenging environment. Progressive training protocols would facilitate the use and integration of the improved balance abilities into daily activities.

CONCLUSIONS

Balance dysfunction is one of the leading factors contributing to falls in older adults. Falling in older adults often results in serious injuries and subsequent loss of independence in performing functional activities. To reduce the number of falls in older adults, balance training programs may serve as a valid fall-prevention strategy.

The optimization of balance training effects is an important issue in the design of balance rehabilitation programs. The systems model of motor control suggests that training be targeted at the balance-related subsystems that show deterioration in older adults. Also, a training program designed primarily to improve specific subsystems will result in the best training effects if applied to older adults who show deficits in these subsystems. Previous studies using this approach have consistently shown to be effective in increasing balance control abilities in older adults during standing. However, there is not yet evidence that improved standing stability is transferrable to balance control during walking or other dynamic and functional tasks. For the improvement of balance control abilities during walking, exercises that incorporate walking trials are recommended. Future studies are needed to investigate the best strategies to improve the integration of balance control abilities into walking and other more dynamic tasks.

Training programs using a multifactorial approach often include both static and dynamic balance exercises as well as strengthening, mobility, and endurance exercises. This approach, however, has not shown consistent positive training effects. It is also difficult to identify the specific subsystems contributing to the overall improvement in balance control abilities. It is suggested that with this approach, pre-training screening or selection of trainees be performed so that more emphasis can be placed on the impaired systems that are common among the trainees.

ACKNOWLEDGEMENTS

This work was supported by NIH grant AG05317-06 awarded to Dr. Marjorie H. Woollacott. The authors would like to thank Raymond Chong for editing the manuscript.

REFERENCES

Alexander, N. B. (1994). Postural control in older adults. *Journal of American Geriatrics Society, 42,* 93-108.

Baker, S. P., & Harvey, A. H. (1985). Fall injuries in the elderly. *Clinics in Geriatric Medicine, 1* (3), 501-508.

Barron, R. E. (1967). Disorders of gait related to the aging nervous system. *Geriatrics, 22,* 113-120.

Bassett, C., McClamrock, E., & Schmelzer M. (1982). A 10-week exercise program for senior citizens. *Geriatric Nursing, March/April,* 103-105.

Brandt, Th., Krafczyk, S., & Malsbenden, I. (1981). Postural imbalance with head extension: Improvement by training as a model for ataxia therapy. *Annals of New York Academy of Sciences, 374,* 636-649.

Brown, M., & Holloszy, J. O. (1991). Effects of a low intensity exercise program on selected physical performance characteristics of 60- to 71- year olds. *Aging, 3,* 129-139.

Buchner, D. M., Cress, M. E., Wagner, E. H., & de Lateur, B. J. (1992). The role of exercise in fall prevention: Developing targeting criteria for exercise programs. In B. Vellas, M. Toupet, L. Rubenstein, J. L. Albarède, & Y. Christen (Eds.), *Fall, balance, and gait disorders in the elderly* (pp. 55-67). Paris: Elsevier.

Campbell, A. J., Borrie, M. J., & Spears, G. F. (1989). Risk factors for falls in a community-based prospective study of people 70 years and older. *Journal of Gerontology, 44* (4), M112-117.

Campbell, A. J., Reinken, J., Allan, B. C., & Martinez, G. S. (1981). Falls in old age: A study of frequency and related clinical factors. *Age and Ageing, 10,* 264-270.

Crilly, R. G., Willems, D. A., Trenholm, K. J., Hayes, K. C., & Delaquerrière-Richardson, L. F. O. (1989). Effects of exercise on postural sway in the elderly. *Gerontology, 35,* 137-143.

Critchley, M. (1948). On senile disorders of gait, including the so-called "senile paraplegia". *Geriatrics, 3,* 364-370.

Fiatarone, M. A., Marks, E. C., Ryan, N. D., Meredith, C. N., Lipsitz, L. A., & Evans, W. J. (1990). High intensity strength training in nonagenarians: effects on skeletal muscle. *Journal of American Medical Association, 263,* 3029-3034.

Gryfe, C. I., Amies, A., & Ashley, M. J. (1977). A longitudinal study of falls in an elderly population: I. Incidence and morbidity. *Age and Ageing, 6,* 201-210.

Hadley, E., Radebaugh, T. S., & Suzman, R. (1985). Falls and gait disorders among the elderly: A challenge for research. *Clinics in Geriatric Medicine, 1* (3), 497-500.

Horak, F. B. (1991). Assumptions underlying motor control for neurologic rehabilitation. In Foundation of Physical Therapy (Ed.), *Contemporary management of motor control problems: Proceedings of the II Step Conference* (pp. 11-27). Frederiskburg: Bookcrafters, Inc.

Horak, F. B., Shupert, C. L., & Mirka, A. (1989). Components of postural dyscontrol in the elderly: A review. *Neurobiology of Aging, 10,* 727-738.

Hu, M.-H., & Woollacott, M. H. (1994a). Multisensory training of standing balance in older adults. I. Postural stability and one-leg stance balance. *Journal of Gerontology, 49* (2), M52-61.

Hu, M.-H., & Woollacott, M. H. (1994b). Multisensory training of standing balance in older adults. II. Kinematic and electromyographic postural responses. *Journal of Gerontology, 49* (2), M62-71.

Johansson, G., & Jarnlo, G.-B. (1991). Balance training in 70-year-old women. *Physiotherapy Theory and Practice, 7,* 121-125.

Judge, J. O., Lindsey, C., Underwood, M., & Winsemius, D. (1993). Balance improvements in older women: Effects of exercise training. *Physical Therapy, 73,* 254-265.

Koller, W. C., Glatt, S. L., & Fox, J. H. (1985). Senile gait: A distinct neurologic entity. *Clinics in Geriatric Medicine, 1* (3), 661-668.

Ledin, T., Kronhed, A. C., Möller, C., Möller, M., Ödkvist, L. M., & Olsson, B. (1991). Effects of balance training in elderly evaluated by clinical tests and dynamic posturography. *Journal of Vestibular Research, 1,* 129-138.

Lichtenstein, M. J., Shields, S. L., Shiavi, R. G., & Burger, M. C. (1989). Exercise and balance in aged women: A pilot controlled clinical trial. *Archives of Physical Medicine and Rehabilitation, 70,* 138-143.

Maki, B. E., Holliday, P. J., & Topper, A. K. (1994). A prospective study of postural balance and risk of falling in an ambulatory and independent elderly population. *Journal of Gerontology, 49* (2), M72-M84.

Manchester, D., Woollacott, M. H., Zederbauer-Hylton, N., & Marin, O. (1989). Visual, vestibular and somatosensory contributions to balance control in the older adult. *Journal of Gerontology, 44* (4), M118-127.

Moore, S., & Woollacott M. H. (submitted). Sensory organization training for older adults with impaired balance. *Physical Therapy.*

Murray, M. P., Kory, R. C., & Clarkson, B. H. (1969). Walking patterns in healthy old men. *Journal of Gerontology, 24,* 169-178.

Nashner, L. M. (1977). Fixed patterns of rapid postural responses among leg muscles during stance. *Experimental Brain Research, 30,* 13-24.

Nashner, L. M. (1982). Adaptation of human movement to altered environments. *Trends in Neurosciences, October,* 358-361.

Nutt, J. G., Marsden, C. D., & Thompson, P. D. (1993). Human walking and higher-level gait disorders, particularly in the elderly. *Neurology, 43,* 268-279.

Patla, A. E. (1991). Understanding the control of human locomotion: A prologue. In A. E. Patla (Ed.), *Adaptability of human gait: Implications for the control of locomotion* (pp. 3-17). New York: Elsevier Science Publishing Co., Inc.

Patla, A. E., Frank, J., & Winter, D. A. (1990). Assessment of balance control in the elderly: Major issues. *Physiotherapy Canada, 42* (2), 89-97.

Prudham, D. & Evans, J. G. (1981). Factors associated with falls in the elderly: A community study. *Age and Ageing, 10,* 141-146.

Sattin, R. W., Huber, D. A. L., DeVito, C. A., Rodriguez, J. G., Ros, A., Bacchelli, S., Stevens, J. A., & Waxweiler, R. J. (1990). The incidence of fall injury events among the elderly in a defined population. *American Journal of Epidemiology, 131* (6), 1028-1037.

Shik, M. L., & Orlovsky, G. N. (1976). Neurophysiology of locomotor automatism. *Physiological Reviews, 56* (3), 465-501.

Shumway-Cook, A., & Woollacott, M. H. (1995). *Motor control: Theory and practical applications.* Baltimore, MD: Williams & Wilkins.

Stelmach, G. E., Teasdale, N., DiFabio, R. P., & Phillips, J. (1989). Age related decline in postural control mechanisms. International *Journal of Aging and Human Development, 29* (3), 205-223.

Teasdale, N., Stelmach, G. E., Breunig, A., & Meeuwsen, H. J. (1991). Age differences in visual sensory integration. *Experimental Brain Research, 85,* 691-696.

Tinetti, M. E. (1986). Performance-oriented assessment of mobility problems in elderly patients. *Journal of American Geriatrics Society, 34,* 119-126.

Tinetti, M. E., Baker, D. I., McAvay, G., Claus, E. B., Garrett, P., Gottschalk, M., Koch, M. L., Trainor, K., & Horwitz, R. I. (1994). A multifactorial intervention to reduce risk of falling among elderly people living in the community. *New England Journal of Medicine, 331,* 821-827.

Tinetti, M. E., Speechley, M., & Ginter, S. F. (1988). Risk factors for falls among elderly persons living in the community. *New England Journal of Medicine, 319,* 1701-1707.

Tinetti, M. E., Williams, T. F., Mayewski, R. (1986). Fall risk index for elderly patients based on number of chronic disabilities. *American Journal of Medicine, 80,* 429-434.

Winter, D. A. (1989). Biomechanics of normal and pathological gait: Implications for understanding human locomotor control. *Journal of Motor Behavior, 21,* 337-355.

Winter, D. A., McFadyen, B. J., & Dickey, J. P. (1991). Adaptability of the CNS in human walking. In A. E. Patla (Ed.), *Adaptability of human gait: Implications for the control of locomotion* (pp. 127-144). New York: Elsevier Science Publishing Co., Inc.

Winter, D. A., Patla, A. E., & Frank, J. S. (1990). Assessment of balance control in humans. *Medical Progress through Technology, 16,* 31-51.

Winter, D. A., Patla, A. E., Frank, J. S., & Walt, S. E. (1990). Biomechanical walking pattern changes in the fit and healthy elderly. *Physical Therapy, 70* (6), 340-347.

Winter, D. A., Ruder, G. K., & MacKinnon, C. D. (1990). Control of balance of upper body during gait. In J. M. Winters & S. L-Y Woo (Eds.), *Multiple muscle systems: Biomechanics and movement organization* (pp. 534-541). New York: Springer-Verlag.

Wolfson, L., Whipple, R., Amerman, P., & Tobin, J. N. (1990). Gait assessment in the elderly: A gait abnormality rating scale and its relation to falls. *Journal of Gerontology, 45* (1), M12-19.

Wolfson, L., Whipple, R., Derby, C. A., Amerman, P., Tobin, J. N., & Nashner, L. (1992). The effects of age, disease and gender on the balance of healthy elderly. In B. Vellas, M. Toupet, L. Rubenstein, J. L. Albarède, Y. Christen (Eds.), *Falls, balance, and gait disorders in the elderly*. Paris: Elsevier.

Woollacott, M. H. (1989). Aging, posture control, and movement preparation. In M. H. Woollacott & A. Shumway-Cook (Eds.), *Development of posture and gait across the life span* (pp. 155-175). Columbia, SC: University of South Carolina Press.

Woollacott, M. H., & Shumway-Cook, A. (1990). Changes in posture control across the life span – A systems approach. *Physical Therapy, 70* (12), 799-807.

Woollacott, M. H., & Shumway-Cook, A., & Nashner, L. M. (1986). Aging and postural control: Changes in sensory organization and muscular coordination. *International Journal of Aging and Human Development, 23,* 97-114.

Author Index

Subject Index

www.ingramcontent.com/pod-product-compliance
Ingram Content Group UK Ltd.
Pitfield, Milton Keynes, MK11 3LW, UK
UKHW020231290926
15470UKWH00002B/469

9 780444 821010